D0919910

MODELING OF COMPLEX SYSTEMS

An Introduction

OPERATIONS RESEARCH AND INDUSTRIAL ENGINEERING

Consulting Editor: J. *William Schmidt*

CBM, Inc., Cleveland, Ohio

Applied Statistical Methods, *I. W. Burr*

Mathematical Foundations of Management Science and Systems Analysis, *J. William Schmidt*

Urban Systems Models, *Walter Helly*

Introduction to Discrete Linear Controls: Theory and Application, *Albert B. Bishop*

Integer Programming: Theory, Applications, and Computations, *Hamdy A. Taha*

Transform Techniques for Probability Modeling, *Walter C. Giffin*

Analysis of Queueing Systems, *J. A. White, J. W. Schmidt, and G. K. Bennett*

Models for Public Systems Analysis, *Edward J. Beltrami*

Computer Methods in Operations Research, *Arne Thesen*

Cost-Benefit Analysis: A Handbook, *Peter G. Sassone and William A. Schaffer*

Modeling of Complex Systems, *V. Vemuri*

In preparation:

Applied Linear Programming: For The Socioeconomic and Environmental Sciences, *Michael Greenberg*

MODELING OF COMPLEX SYSTEMS

An Introduction

V. Vemuri

DEPARTMENT OF COMPUTER SCIENCE
STATE UNIVERSITY OF NEW YORK AT BINGHAMTON
BINGHAMTON, NEW YORK

ACADEMIC PRESS New York San Francisco London 1978

A Subsidiary of Harcourt Brace Jovanovich, Publishers

ACADEMIC PRESS, INC.
111 Fifth Avenue, New York, New York 10003

United Kingdom Edition published by
ACADEMIC PRESS, INC. (LONDON) LTD.
24/28 Oval Road, London NW1

Library of Congress Cataloging in Publication Data

Vemuri, V.
Modeling of complex systems.

Includes bibliographical references.
1. System theory. I. Title. II. Title: Complex systems.
Q295.V45 003 77-77246
ISBN 0-12-716550-9

PRINTED IN THE UNITED STATES OF AMERICA

To Lord Krishna

Contents

Appendix 2 Elements of Probability

Appendix 3 Elements of Matrix Methods

Appendix 4 Differential Equations

Preface

This book is written primarily for senior undergraduate students and beginning graduate students who are interested in an interdisciplinary or multidisciplinary approach to large-scale or complex problems of contemporary societal interest. Though there is a strong mathematical flavor, this work can also be used by a wide spectrum of students—those who tend to look at human society in terms of manifolds of interconnected problems rather than in terms of specific problems, such as pollution, poverty, or power. The purposes of this book are to help the student become acquainted with the language and framework of modern systems theory, to enable a student to recognize the internal structure of complex systems, and to impart to the student some working knowledge and skill in the use and methods of modeling large-scale systems; it is not meant to be a handbook.

This work was born out of the author's conviction that scientific and technical developments of the past few decades have set the stage for an era characterized by bigness—big explosions of population and pollution and in affluence and effluence. There is no denying that the ecology of the upper atmosphere, problems of land use, control and coordination of surface and air traffic, human communications, and responsiveness of a city to its citizens' complaints represent systems that are more complex than the familiar engineering and physical systems on which so many textbook models are based. These problems are characterized not only by their physical or geometrical largeness but also by a structure in which an essentially technological system is forced to operate in an environment that is constrained by elements that are behavioral or social in their nature. To grow more food for the multiplying myriads of human population, to build cities and make them habitable, to provide health and medical care services, and to perform a thousand other tasks in a complex society, systematic and scientific help is urgently needed. This book is addressed to the new breed of students interested in providing such help.

Because of the nature of large-scale problems, the methods of attacking them are necessarily different. Today, however, there is no generally recognized body of knowledge that can be called large-scale systems theory. As it stands today, the theory of large-scale or complex systems, if any such thing exists, is more a state of mind than any specific amalgam of methods or philosophies. The term large-scale itself is subject to value judgments. This relative state of ignorance (call it "uncertainty" if you wish) within large-scale systems theory has to be accepted at the outset. But since mathematics is the language of science, the author feels that sound mathematical and logical thinking must occupy an important position in any large-scale systems theory. To this end, bits of knowledge have been collected and organized to fill some of the needs outlined above. Since this is meant to be an introductory work, no attempt was made to make it a comprehensive treatise.

Since this work is addressed to a wide spectrum of students of various disciplines, a few relatively simple sections have been included for the sake of systematic argument. Teachers can skip these sections, perhaps giving them as reading assignments. An attempt has been made to explain various concepts by means of illustrative examples gathered from such varied disciplines as anthropology, ecology, economics, engineering, physics, psychology, and sociology. It is important that students, regardless of their primary interest, work through these examples and explore for themselves the similarities, analogies, and differences among systems arising from various disciplines. For the same reason, all the exercises at the end of each chapter should be worked out in full.

Acknowledgments

An author does not write a book alone. His ideas are sharpened and matured by his contact with peers and colleagues; his presentation and his approach are refined through interaction with students. It would be impossible to pinpoint contributions from innumerable sources that have collectively influenced this work. Material presented here has been used several times for classroom instruction at Purdue University and at the State University of New York, and the author is deeply indebted to all those students and colleagues whose comments made the manuscript a better one. Several other individuals deserve special mention here. Included in this group are Mr. Stan Kauffman and Mr. Randy Perlow for their excellent art work, my brother Mr. Narasimha Murthy Vemuri, who caught several substantive errors at the last minute, and finally the staff at Academic Press for their understanding. Of course, these acknowledgments would be incomplete without mentioning my children Sita, Sunil, and Mytilee; it was their curiosity about the manuscript that really gave me the impetus to finish this work.

Suggestions for Term Projects

A term paper is a useful adjunct to the teaching of a course on modeling. Problems of contemporary interest are generally good candidates for investigation by students. It is impossible to list all such problems. However, to set a direction, a small sample is provided here.

1. *Global resource management* It is a well-known fact that the natural resources of this planet are distributed unevenly and are being consumed by the population unevenly and at an alarming rate. One can attempt to build models to predict the impact of various policies regarding the international exchange of resources on the economy and ecology of various nations.

2. *Radioactive dating* Fake paintings have inundated the art market in recent times. This problem therefore has some interest to the connoisseur of art. The authenticity of a painting can be verified by fixing the time at which it was painted using a radioactive dating technique because a radioactive substance, white lead (^{210}Pb), whose half-life is 22 years, is a pigment that was widely used by artists for many centuries. Similarly, carbon-14 dating can be used to fix the dates of artifacts of archeological interest. However, how do we go about detecting fake reproductions of more modern paintings of masters like Picasso?

3. *Spread of technological innovations* Economists, ecologists, demographers, and advertisers have long been interested in the process of how a new technological idea or innovation spreads in a society. For instance, it is useful to know how a new concept of birth control and family planning spreads through a society. This problem is analogous to other related problems such as the spread of a rumor or spread of an epidemic.

4. *Models for the detection of a disease* Compartmental type models can be used for the detection of a diseased condition. Indeed, the compartmental approach is rather widely used to study several problems in pharmacokinetics. For example, diabetes is detected via the glucose tolerance test. In this test, the glucose ingested can be regarded as an impulse (or pulse) applied to the gastrointestinal compartment, and the glucose levels in blood can be viewed as the response of the blood compartment.

5. *Determination of optimal drug regimen* Digitalis glycosides are effective therapeutic agents in the treatment of several cardiac problems. However, there is no well-defined and widely accepted method for the determination of digoxin dosage for individual patients. For example, a normal dosage would be fatal to a cardiac patient who is also suffering from a renal impairment. Similarly, patients with thyroid disease show an altered sensitivity to digoxin. Mathematical models would be extremely useful in such cases.

6. *Population planning* Several aspects of population planning are of interest from the viewpoint of term paper topics: (a) methods and policies for the harvesting of species that are in abundance, such as fish, and of species that are not in abundance such as whales; (b) methods and policies for the preservation of almost extinct species such as the Indian tiger; and (c) methods and policies for weed and pest control in agricultural ecosystems.

7. *Impact of family planning alternatives* There is a widespread belief that there is a lot to be gained via family planning. Are there any long-term effects of family planning that are not widely recognized now? For example, what is the impact of an altered age distribution on the socioeconomic life of a country?

8. *Planning of self-contained buildings and cities* Many innovative architectural ideas pertaining to the construction of "ecologically complete" buildings and/or cities have been sprouting up in recent times. For instance, what is the impact of a large building, which offers all opportunities and services required for normal living, located in the middle of a city? At an isolated spot away from all cities?

9. *Health care services* A frequent problem faced by planners is to decide whether to build a new hospital and, if so, where to build it. In which hospital should one locate expensive diagnostic equipment such as brain scanners and heart–lung machines? This is the so-called facilities location problem.

10. *Relation between complexity and stability of complex systems* It is widely believed that ecological systems with a number of interconnections are more stable. What is the true nature of this relation? Is there similar reason to believe that electrical power systems, for example, with a number of interconnected tielines, tend to be more stable and reliable?

1

An Approach to the Problem

INTRODUCTION

Today, the system is one of the most widely used concepts in scientific investigations. Many different types of systems are familiar to us from everyday experience: mechanical systems such as clocks; electrical systems such as radios; industrial systems such as factories; educational systems such as universities; information processing systems such as computers; medical systems such as hospitals; and many more such as organizational systems, environmental systems, and cybernetic systems. We are thus concerned with a very general concept.

The behavior of systems is not always exemplary. Economic systems are subject to inflation, recession, and depression; biological systems are subject to disease and decay; educational systems are subject to obsolescence; ecological systems are subject to pests and pollution; and hydrosystems are subject to floods and droughts. What can be realized from this recital is that all kinds of systems are subject to external disturbances and do require care and the cost of this supervision is often an important factor in decision making.

There are several ways to improve the quality of performance of systems. One can build a new system and discard the old. This is a common phenomenon in some political systems and can also be done with simple mechanical systems. In several cases replacement is not at all a feasible solution. Examples belonging to this category abound: human bodies, environment, and so forth. An obvious alternative is to attempt to "engineer" the system, that is, steer it into the "proper" direction either by altering its structure or modifying the inputs or both. One way to do this is to observe the output of the system and make the said alterations such that this observed output is as close as possible to the desired output. As these alterations are being made, many of the systems cannot be put out of service. Furthermore, many systems do not operate well in a complete laissez-faire climate. Systematic methods of operating a system economically, efficiently, and in a manner desirable and perhaps acceptable to all concerned parties are needed. This is not always an easy decision-making problem; important organizational, technological, economic, legislative, and legal issues crop up in any discussion of a topic of such a pervasive nature. Some of these issues arising in this context will be discussed here.

1 WHAT ARE LARGE-SCALE SYSTEMS?

The concept of a system is very general. As our facilities do not permit us to investigate all kinds of systems, attention here is confined only to that class of systems that requires specialized approaches because of any one or a combination of the following reasons.

(1) The number of attributes necessary to describe or characterize a system are too many. Not all these attributes are necessarily observable. Very often these problems defy definition as to objective, philosophy, and scope. Stated differently, the structure or configuration of the system is rarely self-evident. In large systems involving, say, people, plants, computers, and communication links, there is scope for many possible configurations and selection of one out of several possibilities has far-reaching repercussions.

(2) The laws relating the properties of the attributes to the behavior of the system are generally statistical in nature. This is particularly true of the disturbances acting on a system. The class of disturbances is very broad, implying that the class of methods or controls used to compensate for the disturbances must be equally broad. For example, classical feedback control methods are quite effective in providing compensation for disturbances in the "technological" variables but they are less so for variations in market conditions, economic conditions, fluctuations due to time delays, and so forth. Restructuring of the system or its operation using mathematical programming techniques (linear programming, for instance) is often used under such circumstances.

(3) Complex systems are not static, they evolve in time. As the environment in which a system operates is not generally under the control of the observer, its influence as the system evolves in time is not apparent at the outset. Any system design must therefore take into consideration the fact that future disturbances may arise which are not present in the existing system, and the control system must itself evolve in order to respond effec-

tively to the future disturbances. Thus, large-scale systems cannot be designed as textbook exercises.

(4) The behavioral (political, social, psychological, aesthetic, etc.) element at the decision making stage contributes in no small measure to the overall quality of performance of the system. Because of this, many large-scale systems problems are characterized by a conflict of interest in the goals to be pursued.

It is important to make a clear-cut distinction between a system's being large and a theory developed for the study of large scale or complex systems. Whether a given system can be considered large or small essentially depends upon value judgment. What is considered complex from one viewpoint could be of simple structure from a different viewpoint.

As a rule, complexity in behavior arises due to a complexity of structure. Thus, a useful indicator revealing the complexity of a system can be found in the complexity of the behavior of a system. Therefore, it is useful to keep in mind that in any procedure aimed at modeling complex or large-scale systems the complexity of the real system should be reflected in the model.

The class of systems thus described is responsible for a new genre of mathematical sciences—the study and control of large-scale or complex systems. A significant practical aspect of the problem is not even a question of control; that is far too ambitious. It is a question of learning enough about a system to permit the development of a meaningful policy for operation. The general problem of operating a large system with limited resources and limited amount of time for observation, data processing, and implementation of control generates new kinds of mathematical questions that have not yet been precisely formulated and certainly not resolved.

Examples of Complex Systems

Although the methods and tools described herein are by no means limited to large-scale systems, the magnitude of time, money, effort, consequences, and significance of large-scale systems frequently require a sound systems engineering approach. For illustrative purposes two examples of large-scale, complex systems are presented here. It is important at the outset to recognize that these examples are not typical and the presentation is only sketchy in detail.

Management of Multipurpose River Valley Systems

In a very simplified sense, the problem faced is the following: if controlled, water in rivers is a valuable resource; if uncontrolled, it causes floods and droughts. Therefore, it is desirable to control the rivers and put the water to beneficial use. A classical method to accomplish this task is to choose a suitable site, build a dam, and use the water impounded for recreation, irrigation, power generation, municipal and industrial use, commercial fishing, and the like. By releasing the water from the reservoir in a controlled fashion it is also possible to control floods in the lower reaches of the river. Thus, in a first analysis it appears that everyone stands to gain by building a dam on a river.

A second look at the problem reveals that any large public project yielding multiple benefits almost always leads to conflicting interests; water resource systems are not an exception to this general rule. In managing water resources, several interests are involved because a natural water system can be altered in several ways. First, the problem of where to construct the dam arises. The task of determining the location of a public facility is not always governed by engineering considerations. It falls in the realm of public policy and the decisions, in general, are essentially political. Political decisions in turn depend upon public reactions, social values, and priorities. Since society is a collection of individuals, methods are required to aggregate individual values and preferences into social values. This is indeed an arduous task, and we do not as yet understand the dynamics of institutional decision making. This is a potentially fruitful area for further research.

Let us consider this problem from another viewpoint. Any large-scale human intervention with nature is likely to produce some side effects or spillover effects. For instance:

(1) A dam causes a decrease in the volume of water in the lower reaches of a river and often disturbs the hydrodynamic balance between the fresh river water and saline seawater. The result is that the lower reaches of the river, close to the sea, become salty. This salt water could irreversibly destroy productive irrigation lands in coastal areas.

(2) Construction of dams and regulation of impounded water is believed to have a variety of biological effects, some good and some bad. In Lake Torron, Sweden, after regulation, the spawning area of graylings was increased because in parts of the submerged area erosion had exposed suitable gravel bottom. Pike, on the other hand, spawn on a bottom covered with vegetation. Because the high-water period in regulated lakes comes later than in unregulated lakes, the water levels in lakes behind dams often fail to reach the vegetated areas in time for pike spawning. Thus, regulation benefits one species and hurts another.

(3) Nondegradable pollutants discharged into the river upstream from the dam tend to accumulate in the lake at higher and higher concentrations, causing the destruction of life therein.

Precisely because there is such a variety of effects from any action taken, the effort involved in the prior assessment of various alternatives is overwhelming. Consequently, computers are indispensable for systems analysis. Besides, there is no clear-cut way to answer some of the questions that arise at the planning stage. How can one compare various planning alternatives when some designs clearly favor one interest group? How can one develop an impartial yardstick for measuring the value or utility of a management policy? How can one assess and measure the actual benefits accrued by a flood protection policy? How can one measure the economic or aesthetic values of a benefit called "recreation"? Can one build a public project of this kind with a profit motive? Whether for profit or not, how can one measure the success or failure of a project of this kind? There are two kinds of problems concealed in these questions. One concerns the judgment regarding the economic soundness of the projects and the second concerns the wise use of a natural resource. For brevity, let us focus our attention on the first problem.

In a multipurpose river valley project water allotted for irrigation often earns minimal revenues. There is no direct revenue from a flood control activity even though the indirect benefits accrued to the society are immense. From experience it was found that revenues earned from selling hydroelectric power are substantial. Therefore, it is of considerable interest to study the mechanics involved in the generation and selling of hydroelectric power that would possibly make the entire river valley project self-supporting.

The gist of the problem can be expressed succintly as a *negotiation problem* between two parties—the manager of the hydroelectric station (MHS) and the manager of a fossil fuel station (MFS), that is, a power-generating facility that uses fossil fuels such as oil or coal. The MHS wishes to maximize the revenue from his reservoir operations over a *planning horizon* by using a strategy of selling energy to MFS. However, as the primary purpose of the project is not power generation, the MHS has to follow certain regulations and guidelines for releasing the stored water:

(1) Releases such as those for power generation and irrigation earn a revenue.

(2) Releases such as those to satisfy the riparian rights of down-stream users, for fish and wild life conservation and salinity control of down-stream aquifers, do not earn a revenue, yet they are mandatory.

(3) Storage in the lake for recreational use may or may not earn any significant revenue but it is unaesthetic to let the shoreline recede during or before a vacation season.

In view of the random nature of stream flows, the amount of energy the MHS can generate is also a random quantity. Therefore, the MHS usually sells his power to MFS on a two-tiered contract agreement. The so called *firm-energy* sales represent a commitment to guarantee the delivery of such energy, while the *dump-energy* sales represent an agreement with no guarantee for its delivery. By failing to meet the *firm-energy* agreement, the MHS not only loses current revenue but also pays a penalty in terms of lost good will and therefore a loss in future earnings. Failure to deliver the contracted *dump-energy* results only in a corresponding loss of revenue and no penalty is suffered. Furthermore, the MHS should always satisfy the release rules engineered to meet emergency situations like flood control activities which usually have a top priority. Thus, the MHS is continuously faced with the problem of making optimal control policies, in near real time, that permits him to earn as much revenue as possible without violating other rules which have a higher priority. How can the MHS do this? One possible way is to base judgments on predictions made by well conceived and realistic models. For instance, it is possible to build a model to predict the precipitation in the catchment of a river by using a weather forecasting model (Fig. 1.1). Then using the so-called rainfall–runoff models one can predict the amount of water available in streams. The flood-routing models then make predictions about the quantity of water in rivers. Armed with this information, it is not difficult to estimate the amount of water that can be impounded by a dam. Some of this water is lost due to evaporation. Some of it is released to generate electricity, irrigate land, and a host of other purposes. The quantity of water available in the reservoir at any time can be estimated by representing the accumulation and discharge of water as a queuing process. Each unit quantity of water entering the reservoir can be imagined to be a person entering a queue. Each unit of water discharged for a given purpose from the reservoir can be imagined as an individual leaving the queue. Of course, most of these processes are random in nature. By appropriately designed models the entire process shown in Fig. 1.1 can be simulated on a computer. The predictions made from this model-building exercise can be subsequently used in decision making.

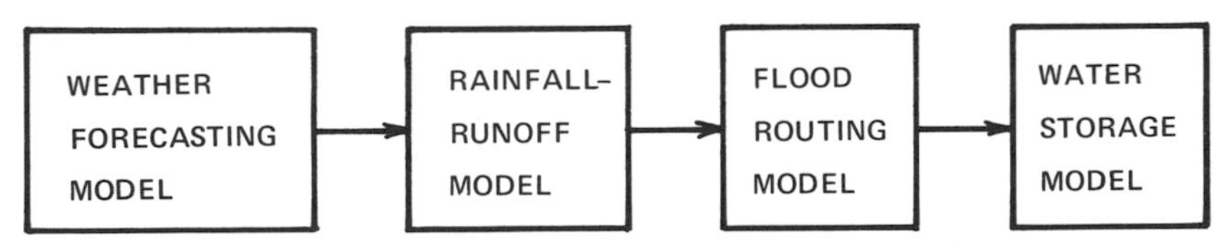

Fig. 1.1 One of many model configurations that would be useful in the management of a river valley system.

This entire process can be rendered more realistic by considering the organizational environment in which the said decisions are to be made. For brevity, only the power-generating aspect of reservoir management is considered here. To put the picture in a proper framework, it is assumed that *all* power-generating divisions in a given region operate semiautonomously under a central power authority (see Fig. 1.2). The managers of all power stations are responsible to their divisional managers. It is also assumed that the output of all stations feed into a region-wide power grid. The task of the central power authority is to coordinate all activities in a way to optimally achieve a stated objective.

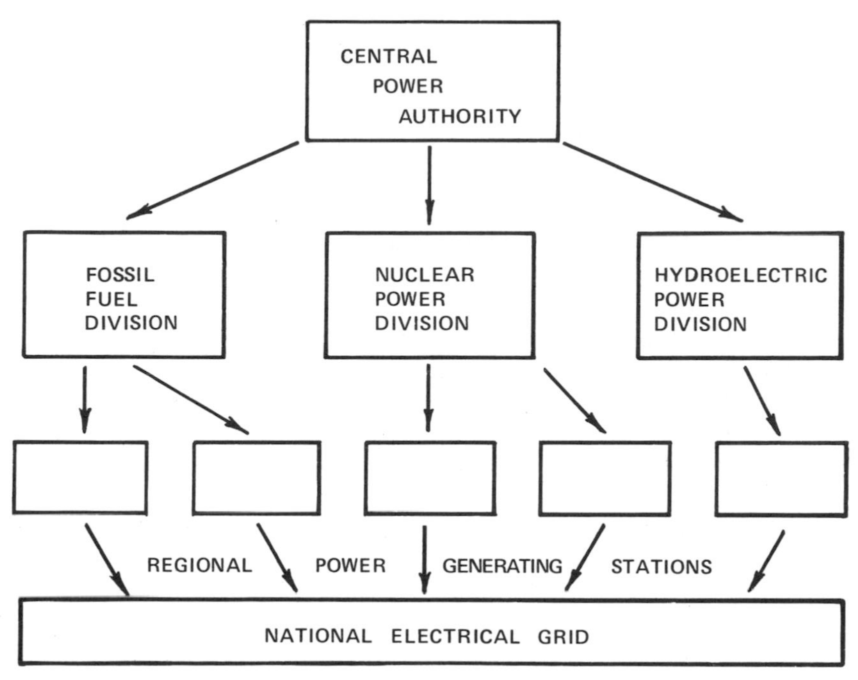

Fig. 1.2 A typical organizational structure.

Environmental Pollution by Trace Metals

The great increase in technological progress has provided many benefits for mankind. This progress, however, set within the framework of an interacting environment, has produced environmental dangers that were not anticipated. One of these dangers is the gradual buildup of toxic trace elements in our environment. Many scientists believe that these toxic metals in the environment pose a more insidious threat to human health than pollution by pesticides, sulfur dioxide, oxides of nitrogen, and other gross contaminants. Typical examples of toxic elements are lead, mercury, and cadmium. Experimental evidence obtained under controlled conditions suggests that a metal like cadmium may function in or may be an etiological factor for various pathological processes including renal dysfunction, hypertension, arteriosclerosis, emphysema, growth inhibition, chronic diseases of old age, and cancer. This brief background description is sufficient to establish a need for a better understanding of the relationship between human and animal exposure to trace contaminants in the environment and the occurrence of disease.

In attacking a problem of this kind, a usual and logical first step would be to systematically collect data about the sources and mechanisms by which cadmium is released into our environment. This should be followed by an understanding of cadmium sinks, routes of transport, rates of flow, and the ecological implications. In other words, the fate and translocation of cadmium in the environment, into plants and animals, into the food chains, and ultimately into man. Having delineated the sources and levels of cadmium in the environment, and the fate and translocation parameters, it is then necessary to identify the legal and economic factors that impede or assist environmental pollution and assess their impact. Connected with this assessment is the identification of public attitudes toward the cost of improving environmental quality through a reduction of cadmium induced hazards versus the cost of not controlling cadmium in the environment.

In the above discussion the problem of environmental degradation has been identified by associating the said problem with several perceivable symptoms. However, to concentrate only on the evident and immediate symptoms is to simplify the predicament by placing the problem in too narrow a perspective. What is important is the long-term cumulative cost to society in terms of health and wealth due to pollution. To apply the systems approach to a problem of this kind really means to follow the interactions among major areas of concern such as trends in industrialization and attitudes toward the depletion of natural resources on the one hand and the effects of pollution on the society on the other hand. Analysis of these trends reveal three basic facts about environmental deterioration. First, long time delays are involved between emission of a pollutant and its ultimate consequences. Second, any imbalance imposed on the environment is likely to have geographically far ranging effects. Finally, it is rather unrealistic to think about the practical possibility of eliminating the environmental problems completely.

Similar preliminary analysis reveals some of the questions to be answered before establishing optimum resource management policies. For instance, if we do not want to squander the options for future generations while maintaining a desirable economic growth rate at the present time, then, how much of the pollutant can the environment assimilate? That is, how much of a pollutant can be safely deposited into the environment? Stated as an alliteration: Is dilution a solution to the pollution problem? How much of a polluting substance can be recycled? How much can be substituted by less toxic or a more abundant material? What are the degradation half-lives, emission rates, differential absorption rates, etc. of the various pollutants? What is the nature and mechanism of synergistic effects?

More specifically, what are the characteristics of the major source–sink paths in the life cycle of cadmium? What are the spatial and temporal distributions of cadmium in a given region? Are there any path segments in which cadmium is retained? If so, how much and at what rate? Are there any closed paths in which cadmium is circulating? What is the nature of these feedback loops? What are the effects of extended exposure to a cadmium polluted environment? What is the effect of specified control measures? What is the effect of legislation that requires recycling of some or all cadmium used by industry? What is the impact of reducing the vaporization of cadmium during smelting?

These questions and others can be answered to some extent by building suitable models to the processes under question. A model-building activity in this problem would generally involve the following steps:

(1) Identification and definition of goals, objectives, and indicators.
(2) Preliminary collection of data and analysis of data.
(3) Identification of the required structure of the model.
(4) Specification of additional data to be collected in order to refine the model.
(5) Development and refinements of the model into mathematical equations and computer programs.
(6) Analysis of the model that leads to suggestions as to the methods of control.

The above presentation is intended only to provide an idea about the nature of large-scale or complex problems. The problems discussed are by no means typical and the description is in no way complete. Relevant aspects of these two problems will be discussed from time to time.

2 HOW TO HANDLE COMPLEX SYSTEMS

There are important conceptual difficulties that one has to surmount in any new endeavor. The study of complex systems is not an exception to this general rule. To visualize the unfamiliar is therefore one important asset for a systems analyst. This requires imagination and intuition. Yet intuition is something difficult to define and even more so to impart by formal instruction, but it usually amounts to an innate ability to make judicious guesses and judgments in the absence of adequate supporting evidence. Years of experience with expensive research clearly shows that a good guess is what all that is needed for a major breakthrough. Einstein's general theory of relativity is a prime example of intuition at its best. While intuition cannot possibly be taught, one can be helped in making inspired guesses by developing a mental framework. Such a framework is an asset to all students interested in large-scale systems. Such a framework need not be precise as long as it is realistic within an order of magnitude.

Consider the task of developing a framework for the basic physical quantity "length." It is easy to visualize the height of a human being, length of an automobile or length of Golden Gate Bridge. This imagination, however, falters if the length involved is too large or too small for human experience. In situations requiring a stretching of the imagination, a picture helps. One can mark the various lengths involved along a line. This, for obvious reasons, is not practical when the range of interest is too large, say, from the diameter of a subatomic particle to the distance of farthest galaxy. A logarithmic scale

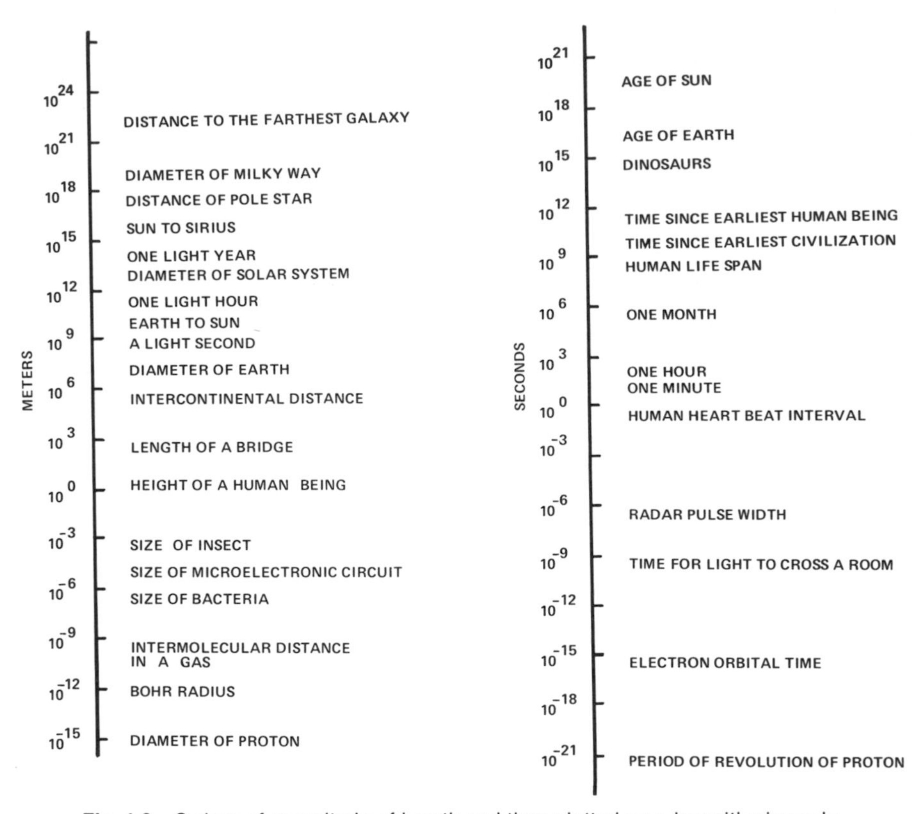

Fig. 1.3 Orders of magnitude of length and time plotted on a logarithmic scale.

obviates this scaling problem. Some of the more familiar "lengths" are shown plotted on a logarithmic scale in Fig 1.3a. Figure 1.3b shows a similar framework to visualize time units.

In the absence of an intuitive grasp of a subject one has to resort to more systematic methods of inquiry. Central to this inquiry is a concern about questions such as the following. What permits us to predict, forecast, or extrapolate the behavior of a system into the future using past or present knowledge about that system? What is the guarantee that the future will behave like the past or present? These questions are ancient and discipline independent. Criminologists, lawyers, engineers, scientists, archeologists, doctors, etc. all have faced this type of questions. Each group sought answers to them in their own characteristic way. It appears that there are currently five basic approaches to the pursuit of scientific truth.

Leibnizian Approach

This approach, usually attributed to the German mathematician Leibniz, is based on the premise that truth is *analytic*. Therefore, a system can be defined completely by a formal or symbolic procedure. In this approach, one attempts to reduce any problem to a formal mathematical or symbolic representation. Laws of physics are examples of models derived from this approach. Mathematical programming models are another example of formal models. While building these models, one almost exclusively directs attention toward exploring the formal structure, and the associated properties of a system.

Among the formal methods of approaching large-scale systems problems, the *decomposition* methods and *aggregation* methods appear to hold good promise. The former attempts to decompose a large system into many smaller subsystems, which are later *composed* or *coordinated* to reconstruct the solution to the original problem. In the latter method, variables of the original system are aggregated or coalesced so as to reduce the dimensionality of the problem at hand. The ultimate goal of these methods is not only to obtain computer solutions in a reasonable time but also to aid in the conceptualization and understanding of large-scale system interactions.

Decomposition is indeed a straightforward operation, and in fact there is no limit to the number of different ways it can be done. However, the problem of coordinating these parts is not always easy and often this difficulty places a practical limitation on the utility of decomposition. For example, in Kron's *method of tearing*, one attempts to "tear" a system along suitably chosen seams; the solution of the problem is then obtained by using a systematic set of rules on the solutions of the subproblems. One can also apply the technique of decomposition to the mathematical model rather than to the system itself. Decomposition of large linear programming problems is an example belonging to this category. Another type of decomposition goes by the name *decoupling*. Separating the normal modes of vibration of a pair of coupled pendulums using some sort of mathematical transformation is an example of decoupling. This decoupling becomes rather tedious if several nonidentical bodies are involved. In such cases one has to resort to approximations. An ancient, yet quite useful, example that reveals the spirit of decoupling via approximation was provided by Newton. The problem faced by Newton was one of establishing relationships (from observational evidence) to describe the motion of planetary bodies. In general, one can possibly have 2^n relations among n bodies. For the solar system, $n = 10$ (including the sun), and therefore one can have $2^{10} = 1024 \approx 1000$ relations. Therefore, one should consider approximately 1000 equations to describe the observed behavior of planets. However, Newton stated his universal law of gravitation in a single equation

$$F = G(m_1 m_2 / r_{12}^2)$$

where F is the force of attraction between a body of mass m_1 and a second body of mass m_2 separated by a distance r_{12}, and G is the universal constant of gravitation. Implicitly this equation states that the force of attraction between two bodies is in no way dependent on the presence of a third body so that pairs of bodies can be considered in turn and their effect summed up. Thus, the ten-body problem is decomposed into a set of 45 two-body problems (45 being the number of ways two objects can be chosen out of ten). By suitably coordinating the solution of these 45 equations, one can approximately describe the motion of any planet. Newton did not get satisfied with this degree of decomposition. He went one step further. As the mass of the sun is far bigger than the mass of any individual planet, Newton reasoned that one can ignore all the equations in which the mass of the sun did not appear explicitly. This leads to a set of ten equations that are completely decoupled. Thus, a 100-time reduction in computation is achieved by a combination of intuition and insight into the peculiar characteristics of a given problem.

What type of problems are well suited for a Leibnizian approach? Problems involving a simple and well-defined structure and problems in which the underlying assumptions are clearly definable lend themselves well for a formal inquiry. For instance superposition of pairwise interactions, as was done in the preceding paragraph, is not always feasible. Consider a sociological problem in which one is interested in studying the behavior of a society. For simplicity consider a family of three. Knowledge about the behavior of the father and child, mother and child, and father and mother is not sufficient to predict the behavior of the whole family treated as a unit. In this case, the Leibnizian approach is not suitable.

Lockean Approach

Central to this approach, due to the English empirical philosopher Locke, is the assumption that truth is *experimental*. That is, a model of a system is necessarily empirical. This implies that the validity of a model does not rest upon any prior assumptions. If the Leibnizian approach is considered a deductive process, the Lockean approach is an inductive process. Statistical techniques are a good example of the Lockean approach. In statistical work, one collects data and, using such techniques as regression analysis, analysis of variance, and correlation analysis, attempts to forecast the future tendencies or trends.

The Delphi Technique

Among the Lockean methods of approaching complex systems problems, special mention must be made about the delphi technique. A basic difficulty in studying complex problems lies in the identification, definition, quantification and measurement of relevant attributes. If some attributes are intangible or shrouded in uncertainty, there is generally no clear-cut way of deciding the relevancy of an attribute, much less measuring it. Decisions tend to be subjective and biased. The delphi technique attempts to bypass this difficulty by fostering a communication process among a large group of individuals. In assessing the potential trends in a novel area of research, a large group (typically in the tens or hundreds) is asked to "vote" on what they think, what the relevant attributes are, and when certain events will occur or some such issue. A fundamental premise underlying this approach is the assumption that opinions from a large group are required to treat any issue adequately.

In general, the delphi procedure has three features: anonymity, controlled feedback, and statistical group response. Anonymity reduces the effect of a dominant individual. Controlled feedback implies conducting the delphi exercise in a sequence of rounds or iterations. First, a set of questionnaire is distributed to the group. The group responds by answering the questions and a summary of the results is made available to the participating members. Use of statistical methods in defining group response is a way of reducing group pressure for conformity. Furthermore, statistical group response is a device to assure that the opinion of every member of the group is represented in the final response. Within these three basic features, of course, there are several variations to the technique.

Kantian Approach

This approach, propounded by the German philosopher Immanuel Kant, is based on the assumption that truth is *synthetic*. That is, experimental data and theoretical base are inseparable. Theories cannot be built without experimental evidence. Data cannot be collected unless a theory tells what data to collect. An important feature of this approach is that it requires one to examine at least two different representations or models and then choose one. This is a noteworthy feature because complex problems cannot be solved with a closed mind and one well-structured approach. In predicting the future, one is dealing not with concrete realities but with hopes, plans, aspirations, and human frailities. Examples of the Kantian approach are abundant in modern technological world. Normative forecasting, program planning and budgeting system (PPBS), cost effectiveness analysis, and benefit–cost analysis are all examples of this approach.

Kantian Delphi

Some of the characteristics of Kant's philosophy can be incorporated into the delphi technique. The original delphi technique was characterized by a strong emphasis on the use of consensus by a group as *the* means of converging on a single model or position on some issue. Instead of merely seeking a consensus, Kantian delphi attempts to elicit alternatives on which to base a comprehensive overview of the subject. Thus, a Kantian delphi procedure explicitly encourages the group not to stop with a mere "yes" or "no" vote; the group is asked to provide alternatives within a given framework. For this reason Kantian delphi is sometimes referred to as "contributory delphi."

Kant's approach is ideally suitable for inherently ill-structured problems and for problems that defy a formal analytical treatment. By the same token, this approach is not suitable for problems that admit a single, clear-cut solution or in situations in which a proliferation of models is too costly.

Hegelian Approach

This approach is derived from the dialectical idealism of the German philosopher Hegel. Central to Hegel's philosophy is the precept that any system can be visualized as a set of logical categories and these logical categories generate their own opposites. That is, truth is *conflictual.* The truth value of any system is the result of a highly complicated process that depends upon the existence of a thesis and an antithesis. The union of these opposites leads to a more adequate grasp of the nature of things until finally all possible points of view with all their seeming conflicts become the constituents of one comprehensive system. One important fact underlying the Hegelian approach is the recognition that *data* is not *information.* Information results from an interpretation of data. Therefore, the Hegelian approach starts with the construction of two opposing Leibnizian models. Then these models will be used to demonstrate that the same data set can be used to support both the opposing models. The hope is that out of a dialectical confrontation between opposing interpretations, the underlying assumptions of the Leibnizian model will be brought to the surface for a conscious appraisal by the decision maker.

An interesting application of the Hegelian approach can be found in the *management games* that some companies are using. These management games are really a popular version of the so-called dialectical inquiring system (DIS). Consider a large company on the verge of making a crucial decision. Then the DIS attempts to locate groups within the company with fundamentally differing plans as to how to cope with the situation. As neither plan can be checked out or verified, DIS attempts to conduct a face-to-face debate on the validity of underlying assumptions. The debate may not result in an optimal solution to the decision-making problem, but any subsequent decision is likely to be better than implementing the viewpoint of a single group.

The basic differences among the four philosophies can be brought out via an amusing example. Consider a group of managers discussing a project. A typical conversation may run like this:

Lockean Manager: Give me the following resources and I will do the job.
Leibnizian: Do the job with whatever resources we have.
Kantian: Why do you want to do the job?
Hegelian: What is the advantage of not doing the job?

3 ISSUES AT LARGE

The one characteristic that makes large-scale systems unique and different is the inclusion of society and its activities as an integral part of the system. The emphasis here is not so much on an individual human being acting as an operator or as a subsystem but on the society acting as a decision making body. Within this context, the terms man, human being, and society are used synonymously in the following paragraphs.

Organizational Issues

It is customary to curse modern science and technology as the responsible agents for all the ills of modern society. Indeed, science has become a most pervasive force in modern society, having a widespread influence over all human activities. It has had an impact on work environment, standard of living, health, social institutions, and recreational activities—in short, on the quality of life. Scientific and technological achievements have had a similar impact upon business organizations. Major advancements occurred, in almost all industries, not only in products but also in the means of production. These achievements have greatly influenced decisions made by governments on such matters as health, education, and welfare.

One of the major factors underlying these achievements has been the ability of humanity to develop social organizations for the achievement of its goals. However, as the problem becomes more complex, organizing and managing these projects become more difficult. Furthermore, there is a growing awareness that the development of very large and highly complex systems are now producing problems of far-reaching consequences for the human race itself! Some of these problems are:

(1) A need to translate scientific achievements into beneficial and usable products or services and a need to achieve ethical and moral controls over the utilization of these achievements.

(2) A need to allocate the available resources in an optimal fashion. The means (resources) available are generally never sufficient to meet the wants (objectives), and hard choices have to be made.

(3) A need to devise methods to resolve conflicts between management and scientists as well as between interest groups.

Realization of the implication of these problems has been responsible for a major shift in the structure of values of people in most of the industrialized world. Improvement of the physical and social environment—often referred to by the ubiquitous phrase *quality of life*—emerged as a highly desirable aim of public policy. Concurrently, growth—industrial or economic—measured in such terms as Gross National Product—was denegraded into the background. Many nowadays question the possibility—let alone the desirability—of unhindered growth.

The quest for environmental quality contributed to a closing of the circle in the evolution of public opinion that welfare should always be the proper goal of public policy. Growth should only be a means to that end. This wisdom in recognizing the semantic and other differences between means and ends should not lead one to accept the thesis that economic growth *conflicts* with social welfare. It would be pessimistic to assume that one is gained by sacrificing the other. Well planned economic and technological growth should always contribute to the quality of life.

It is also well to remember that measurement of the quality of life or welfare is replete with horrendous philosophical problems. The quality of physical environment is, of course, only a part of the concept of "quality of life" that includes social and psychological factors. *The physical environment* includes, among other things, the physical and chemical properties of the land, air, and water that surround us; the distribution of plant and animal life therein; the disposal of wastes; the use of land and spatial relations of structures and of people; noise; and aesthetic qualities of living space. Thus, the physical environment conceivably affects social and psychological attitudes and behaviors in thousands of ways—not all of which are well understood—and play a central role in determining the quality of life.

For an effective management of the quality of life what kind of organizational set up would be desirable? No attempt to answer this question will be made here except to point out the existence of a need to contemplate on this point.

Technological Issues

Technology is knowledge of how to do things. Once technology has taught us how to do something, the decision to apply that technology is made in the matrix of our social institutions. This decision is often made to some extent by consumers of technology and producers of technology, as well as by politicians. In any event, almost all the decisions are made through an institutional structure. Therefore, if one is pleased with the contributions technology has made to the quality of life, the credit should naturally go to the social institutions that made it possible; if one is distressed by the undesirable side effects, the blame also goes to the same place. The role of a systems study is not to assign praise or blame but to understand the fabric of decision making.

It is useful at this stage to examine the nature of complaints leveled against technology and technologists. Consider, for instance, the problem of water pollution. So much has been said and written on this topic implying that technological growth has been the main culprit responsible for the deplorable state of affairs of our water bodies. It is well known that organic substances in water decay by extracting dissolved oxygen from water. As water becomes deficient in oxygen, it cannot support fish and other marine life. This state of affairs not only contributes to an eradication of some animal species but also makes the water stink. Therefore, pollution of water with excessive organic wastes is undesirable. If inorganic compounds containing phosphates (fertilizers and detergents contain phosphates) are dumped in water bodies, periodic undesired blooms of algae occur. This blooming of algae is referred to as *eutrophication* and is not considered good. There are other kinds of water pollution—by trace elements, by heat, and so forth.

Present technological knowledge appears to be sufficient to combat the oxygen depletion problem. Therefore, water pollution due to oxygen depletion has long since passed from the realm of technological to that of economic, and political, problems. The case with eutrophication is somewhat different. The chemical and biological mechanisms associated with eutrophication are not sufficiently well understood to allow for confident prescription of remedies. This example illustrates the predicament faced by mankind when there is too little technology—too little knowledge of the fundamental processes governing the phenomenon of interest. Similar analysis shows that some of the outstanding problems of environmental quality, such as air pollution by automobile exhausts, are as much a result of technological growth as they are due to too little technology to combat the problem. What is the reason for an abundance of technology that pollutes the environment and a lack of it to combat it? It is an elementary principle of economics that the basic research and development of a society cannot be left entirely to private industry, because then there will always be an underinvestment in research and development. Many of the economic benefits of research and development cannot be captured fully by the developer, and he will not take account of these valuable public goods that flow from his efforts in his investment decisions. Therefore, it is good economics (and good politics) to provide some government subsidy to research and development and allow some public dialogue before major technological development decisions are made.

Economic Issues

It should be clear by now that the various issues facing a systems analyst cannot be discussed in isolated categories. It is apparent that organizational and technological issues are closely intertwined. As money is vital and quite often a scarce resource, sooner or later fundamental economic questions crop up and dominate the decision-making scene. The economic issues raised in the context of large-scale systems are an order of magnitude different from the conventional problem of maximizing profit or maximizing sales subject to resource constraints. While solving large-scale systems, the economist is faced with the difficult problem of separating the matter of income distribution from resource allocation. This difficulty arises primarily because the outputs of public investment actions are almost always "public" goods as opposed to "private" goods bought and sold in competitive markets. These public goods are provided to large groups of people simultaneously and in roughly equal physical amount. For example, a program to reduce atmospheric pollutants affects an entire region and one breathing of the cleaner air does not deprive the other from doing it simultaneously. This is different from a private good like an orange—if one eats it, the other does not have it. This phenomenon goes by the name "jointness of supply" and has to be coped with in large-scale systems. Whenever this "jointness of supply" phenomenon exists, there is in general no way to be sure that equating an incremental cost with the sum of incremental "willingness to pay" will lead to maximum welfare. Therefore, whenever the supply of a public good (such as clean air or clean water) changes, both resource allocation and distribution are invariably effected, and the preexisting distribution of income cannot be maintained except through an elaborate system of side payments tailored to each individual. This is a horrendous problem with fundamental legal implications and is almost always a practical impossibility. To overcome this difficulty, applied economists proceed with an implicit assumption that it is erroneous to consider individual situations in isolation in as far as the distribution of public goods is concerned. If the whole complex of public goods is considered together, there will be an averaging effect on the distributional effects. If one accepts this line of argument, it follows that the society that makes its decisions based on efficiency criteria (such as maximum net benefit disregarding distributional problem) will be the one in which a majority of the members will be better off in the long run than the one in which criteria that foreclose efficient solutions are used. This line of argument invariably raises ideological problems.

Even if one tends to ignore the distribution of income problem and the difficult problem of estimating the willingness to pay for public goods, that would still fail to resolve the difficulties. For instance, in an air quality improvement project, ability to estimate the incremental willingness to pay is not sufficient. Since there are physical links among air, water, and solid residuals the willingness to pay scheduling should really be expanded to encompass liquid and solid, as well as gaseous, residuals.

The scope and difficulty of empirical measurements required plus our inability to handle distributional questions makes it almost impossible to achieve "Pareto optimal" allocations of public goods by studying models and simulating markets. Political entreprenuers who are concerned with general taxation and allocation of government budgets must assume general responsibility for the provision of public goods. Therefore, in the final analysis, evaluation of public goods is necessarily a political problem. In a thoughtful essay in the role of modeling in environmental management, Robert Dorfman [8, p. 19] says:

> Now you will charge me with having dragged politics into the problem at the last minute, as if fluid mechanics, biochemistry, meteorology, economics, and computing science were not enough. Politics was there all along, although I was too polite to mention it at the outset. The environment is politics; not only now, when it is in fashion, but always because we all share the same public environment and most

> of us share the costs of controlling it and the benefits of exploiting it. I forbid myself to impose my own hobbies on unprepared, defenseless audiences but I have to say that the clear implication of facing the problem of interpreting environmental models is that they have to include political considerations. I have already said that the weakest part of the environmental model are its behavioral elements, the components that describe how people will react to different environmental stimuli and control measures. Now I tighten that screw one more turn. A proposed control policy that is politically unpalatable is no better than one that violates the law of the conservation of mass. I'll even propose the paradox that economic and political laws are more absolute than physical ones; at least we have displayed a great deal more ability to overcome physical obstacles than social ones.

It appears therefore that the political and organizational structures within which policy decisions are made contributes in no small measure to the *value* problems associated with environmental problems. To see what kinds of new organizational and governmental structures are more suitable, there is a need for a theory of *collective choice* or *social choice.* Kenneth Arrow, who pioneered the modern mathematical study of collective choice, laid down a set of properties one might feel a social choice mechanism could reasonably be expected to have. They are:

(1) *Collective rationality*: In any given set of individual preferences, the social preferences are derivable from the individual preferences.

(2) *Pareto principle*: If alternative A is preferred to alternative B by at least one individual while no one else prefers B to A, then the social ordering ranks A above B.

(3) *Independence of irrelevant alternatives*: The social choices made from any environment depend only on the preferences of individuals with respect to the alternatives in that environment.

(4) *Nondictatorship*: There is no individual whose preferences are automatically society's preferences, independent of the preference of other individuals.

Arrow analyzed voting situations rigorously in view of these conditions and found that no mechanism could be devised that consistently met them all. This is his famous *Impossibility Theorem.* Many people took this lead and attempted to come up with mechanisms by which one can aggregate individual choice functions to social choice functions. But it should be remembered that collective choice decisions must be made by existing legislative bodies.

Legal and Legislative Issues

Problems involving public concern invariably raise conflicting interests. Where there is a conflict of interest, law has a role to play. Unraveling the labyrinth of legal and legislative issues is a delicate and arduous task. To make the context simple, in this section, attention will be focused primarily on the problem of pollution.

First, the discovery of the health effects due to pollution had to precede any concern about law. Until the relationship between the presence of a pollutant within a product and a resultant disease has been established, there is no need for the attention of law. This is the conventional wisdom of the times. From the viewpoint of the legal profession, the problem perhaps can be viewed from two angles: (1) the Workman's Compensation Law and (2) the larger law of general health and ecological effects. As early as the 1880s, insurance to protect workers against the ill effects of their work was established in a few continental countries. The common law in Anglo-American countries gave the worker no protection in the absence of gross negligence by the employer. All of this is history. Today, in many countries, there are laws on industrial hygiene, workers' health and safety, workmen's compensation, and many air and water pollution laws. In spite of these efforts there are many conditions that fall beyond the realm of legislation and litigation.

Consider the problem of using lead in household paints. It was found that the presence of lead imparts a sweet taste to the paint flakes, which induces children to eat them. Large enough quantities of lead in slum children was found to adversely affect their intelligence and possibly even cause death. Legislation banning the use of lead in paints has been suggested and implemented. This legislation did not result in any significant change in the statistics. Further research pointed out that children in inner city slums suffer from a condition called pica that induces them to put all sorts of unsuitable stuff, including paint flakes, into their mouths. Psychiatrists point out that these children suffer from parental neglect, and apparently the only effective way to cure this problem is to induce the parents to show more interest in the child. Requirements like this are hard to legislate. Still the unanswered question is: Why do children from slum areas alone suffer from pica? Investigations revealed that lead from automobile exhaust fumes is sufficient to induce initial symptoms, and the problem is compounded further in slum areas because of poor parent–child relationships. This example illustrates several points. First, legislation banning an obvious source of a pollutant is merely going to provide a panacea, not a cure. Second, the law is not clear as to who should share a major portion of the responsibility for negligence: the child? the parents? the paint manufacturer? the automobile user? the automobile manufacturer? the society at large?

The latter question is further elaborated by the following example. Consider the problem of pollution due to asbestos. In medical parlance, *asbestosis* is the name given to the disease associated with exposure to asbestos. Clearly the ones most endangered are those whose employment regularly causes them to come into contact with it. However, exposure need not be prolonged in order for the consequences to be serious. Indeed, the effect may be noticed so long after the initial exposure that the individual may have forgotten when the exposure occurred. Indeed, the risk of exposure is not confined to a worker. In the construction of high-rise buildings, for example, asbestos is sprayed on the steel structure prior to closing it with concrete. Invariably asbestos dust floats about the neighborhood in spite of precautions and regulations. What is the effect of the public being exposed to such dust over several weeks or months such construction takes? This is not an easy legal question to answer. (For instance, in one case, reportedly a six-week exposure to asbestos dust triggered mesothelioma fatal 23 years later.) The time lag between exposure and onset of the disease makes the burden of proof complicated.

EXERCISES

1. Discuss the advantages and disadvantages of using the solar system as a model for the atom.

2. Discuss the advantages and disadvantages of using

(a) a fluid model for the flow of electricity,

(b) an electrical circuit model for the flow of a fluid.

3. You are asked to write a detailed report on the procedure you would recommend for building a mathematical model for the growth of bacteria in a test tube. Discuss some of the important milestones of this project. What kind of models do you recommend?

4. Devise a simple technique to describe the propagation of a disease through a population. State your assumptions clearly.

5. Models are not always used for predictive purposes. It is often of value to utilize models that are known to be false for the purpose of establishing bounds on the solutions sought. For instance, a rat is placed in a long black box and trained to run the length of the box in a repetitive manner. Each time the rat performs this run, it becomes more skilled and so it can do the run faster. For example, it takes 1 min the first time, $\frac{1}{2}$ min the next time, $\frac{1}{4}$ min the next time, and so forth. How long will it take before the rat can be at both ends of the box at the same time?

6. Conventional wisdom among model builders prompts one to summarize a task of model building as follows: "It is not possible to build a model until data are available, yet it is not feasible to collect data until a model has been exercised." Discuss the truth value of this argument by giving an example to support the statement and a counterexample to oppose the statement.

7. Prepare an order of magnitude chart based on (a) weight, (b) money and expenditure, (c) energy.

In (a) take the average weight of a human being, say, 100 kg, as a starting point and work through many orders of magnitude and go up to the weight of our galaxy and that of the universe on one end and down to electrons and subatomic particles at the other end.

In (b) take your or your parents' weekly income as a starting point and go through many orders of magnitudes, by including such expenses as the operating budget of your school, state, and country. Include such things as waste caused by, say, road accidents, hunger, inefficiency, etc.

In (c) start with the Boltzmann's constant $k = 1.38 \times 10^{-23}$ J/°K and Plank's constant $h = 6.626 \times 10^{-34}$ J sec. Calculate kT at room temperature and compare this with the energy of the power sources around us such as batteries, electrical power stations, lightning discharge, space rockets, power of a human being, atomic explosions, and so forth.

8. In economics one says that a technological *externality* exists if some activity of industry A imposes a cost on industry B for which A is not charged by the price system of a market economy. Give some simple examples of externalities in the context of large-scale systems.

9. The fundamental theorem of welfare economics holds that if N consumers and R firms each act as perfect competitors and maximize their individual utility and profit respectively, the result is optimal in the Pareto sense. That is, no increase in one consumer's utility can be made without reducing some other consumer's utility; no increase in one commodity can be made without reducing some other commodity. Discuss this theorem in the context of environmental pollution. Give examples (or conditions) under which this theorem is not valid.

SUGGESTIONS FOR FURTHER READING

1. For a discussion on the definition of system, see
G. J. Klir, *An Approach to General Systems Theory*. Van Nostrand-Reinhold, Princeton, New Jersey, 1967.
2. For a source book on a variety of systems related problems, see
W. Buckley (Ed.), *Modern Systems Research for the Behavioral Scientist*. Aldine, Chicago, 1968.

3. Some elementary, but typical, large-scale systems are discussed in
   R. Bellman, *Some Vistas in Modern Mathematics*. University of Kentucky Press, Lexington, Kentucky, 1968.
4. For an in-depth discussion of the material in Section 2, see
   C. W. Churchman, *The Design of Inquiring Systems*. Basic Books, New York, 1971.
5. For a comprehensive guide to the delphi technique, see
   H. Linstone and M. Turoff, *The Delphi Method and Its Applications*. American Elsevier, New York, 1973.
6. A commentary on Immanuel Kant's philosophy can be found in
   B. Russell, *History of Western Philosophy* (Chapter 20), Simon & Schuster, New York, 1946.
7. For an elementary discussion of economic concepts relevant to planning, evaluation, and modeling of complex systems, see
   R. de Neufville and J. H. Stafford, *Systems Analysis for Engineers and Managers*. McGraw-Hill, New York, 1971.
8. The quotation of Dorfman on page 19 was taken from
   R. Dorfman, *Problems of Modeling Environmental Control Systems*, Center for Population Studies. Harvard Univ. Press Cambridge, Massachusetts, 1970.
9. An in-depth mathematical discussion on collective choice can be found in
   K. J. Arrow, *Social Choice and Individual Values*, 2nd ed., Wiley, New York, 1964.
10. The two examples presented in this chapter are but a sample from a rich source of problems one can find. For other interesting examples, see
   R. Bellman and C. P. Smith, *Simulation in Human Systems: Decision Making in Psychotherapy*. Wiley, New York, 1973.
   B. C. Patten, ed., *Systems Analysis and Simulation in Ecology*, Vols. I and II. Academic Press, New York, 1971, 1972.
   H. H. Werley, *et al.*, eds., *Health Research: The Systems Approach*. Springer Pub., New York, 1976.
   R. J. Salter, *Highway Traffic Analysis and Design*. Addison Wesley, Reading, Massachusetts, 1974.

2

Language of System Theory

INTRODUCTION

The concept of "system" is perhaps as old as civilization itself. Aristotle expressed this concept by stating that "the whole is more than the sum of its parts." *Webster's Dictionary* defines a system as "a collection of objects united by some form of interaction or interdependence." Thus, a system may be regarded as a set of interrelated elements. The nature of elements involved and the broad kind of relations that might exist among the elements, of course, depends upon the type of system under consideration. The fact that a system can be regarded as a set is indeed very useful because this allows one to draw freely from the edifice of knowledge available in set theory, topology, and function theory. This is exactly the thing that happened to a system in the hands of electrical engineers and mathematicians.

Even though the concept of a system can be traced back to antiquity, the "systems approach" concept appears to be of relatively recent origin. The phrase "systems approach" is usually associated with a cadre of disciplines that have arisen since World War II, the most familiar being operations research, management science, and systems analysis. In the 1960s, the term "system" has attained such widespread usage that it almost became fashionable to throw in this word to impart an air of respectability to any communication. Eventually, "systems" and "systems approach" stood to mean "anything."

In this unfortunate state of affairs, a glimmer of hope remained. Almost all of the disciplines that used a systems approach also used the concept of a model as a basis for developing solutions to problems, the differences among them being more semantic than methodological. Aided by the advent of the computer, the systems approach and modeling of systems rapidly evolved with an impressive record of success in many problem areas. The purpose of this chapter is to present a brief overview of some of the more important ideas and the language in which these ideas are often expressed.

1 GOALS, OBJECTIVES, AND INDICATORS

One of the first steps of the systems approach is that of problem specification. On first meeting a human need, one must define the outlines of the need, the resources available to meet it, and the level of performance necessary to justify the solution. In other words, one must define the problem in terms of the *goals*, *performance criteria*, and *constraints*. Nothing is more important in systems studies than defining the right goals. Working out solutions, however elegant the methods are, with improperly conceived goals is tantamount to answering the wrong questions, seldom a useful exercise. Goal setting is therefore important in order to provide frames of reference for decision making. The next step is to supplement the goals by *objectives* or performance measures because, when a plan is prepared with a particular goal in mind, there must be means of measuring the progress toward that goal.

Consider the goal of "providing the most convenient pattern of shops" in a shopping center. This statement, although aesthetically pleasing (especially the visual aspect) and politically appealing, fails to provide a clear basis for the design of a plan. In order to translate this goal into an objective it is necessary to identify indicators. Typical indicators of shopping convenience are the amount of pedestrian travel and duration of shopping time. Then a typical objective would be to minimize the pedestrian travel, or shopping time, or a combination of both.

Using this simple example as a vehicle, the following observations can be made. First, to be able to measure the degree of success in reaching a goal it is necessary to translate the goals into objectives which can be expressed in terms of some quantifiable attributes. Second, to actually measure the success achieved in reaching a selected objective, it is necessary to have a yardstick with which one can measure the distance or error between the status attained and the objective selected. It is often useful to display goals, indicators, and objectives pictorially. Two commonly used techniques are shown in Figs. 2.1 and 2.2. Identifying appropriate indicators and formulating objectives is not an easy task by itself. The complexity attendant with the selection of goals, objectives, and indicators is illustrated by means of the following example.

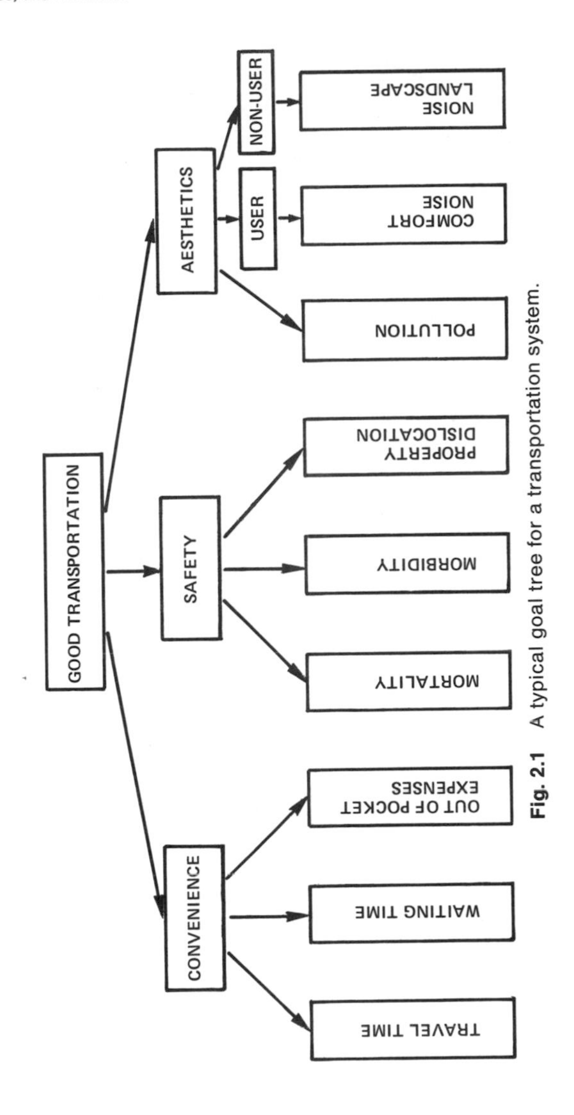

Fig. 2.1 A typical goal tree for a transportation system.

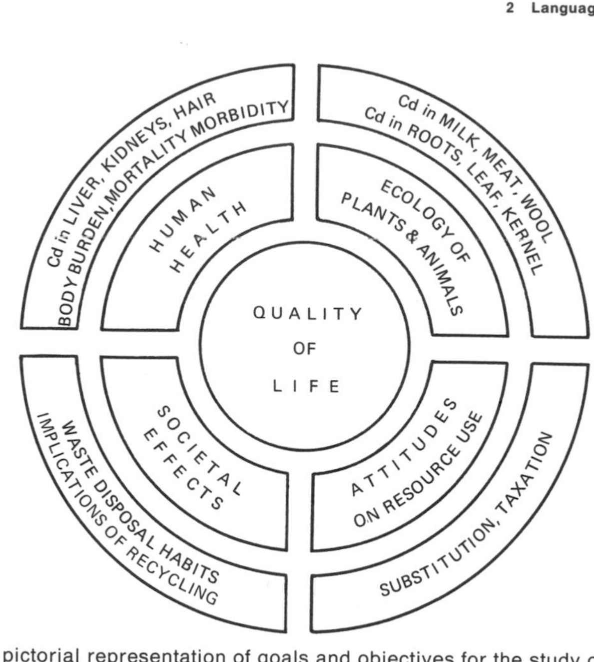

Fig. 2.2 A pictorial representation of goals and objectives for the study of the effects of environmental pollution.

Example 1. Consider the concern about the plight of the poor in many parts of the world. Some of this concern stems from social, rather than egalitarian, reasons. In planning for social change, typical questions that one might require to be answered are: What are the true dimensions of poverty? What are the side effects, if any, of governmental intervention in the lives of the poor? Is there a relation between poverty and social unrest? How can social and psychological consequences of a program be assessed and compared?

A traditional indicator of poverty has been the *standard of living index*. That is, the goal of eradicating poverty has been traditionally reduced to the *economic objective* of maintaining a certain standard of living. It should be kept in mind that this economic definition of poverty should by no means be regarded as universally acceptable. A more recent view is that poverty also includes other dimensions of life. This change is in part due to the growing recognition that (a) quality of life is not solely determined by economic factors, (b) people cannot be motivated to change their behavior solely by economic incentives, and (c) that economic solutions to poverty do not automatically reduce many of the social ills of a society.

To emphasize the importance of noneconomic factors, Galbraith defined the term *insular poverty*. Insular poverty has something to do with the desire of a comparatively large number of people to spend their lives at or near the place of their birth. This homing instinct in an urban slum confines an individual to an area that is intrinsically limited in opportunity. This environment perpetuates its handicaps through poor schools, evil neighborhood influence, and bad preparation for life.

From this viewpoint poverty eradication implies an altering of poverty culture. Establishing social programs that successfully produce cultural change is replete with difficult scientific, moral, psychological, and political problems. For instance, there is no general agreement on techniques to measure cultural deprivation as opposed to economic deprivation. Implementation of any *ad hoc* decisions will undoubtedly result in a redistribution of political and social power within and between various segments of society. This example was meant to illustrate the difficulties in goal setting and objective formulation in a social context using monetary scales. ■

Although a large number of indicators cannot be described in monetary terms, many can be quantified by other means. Typical indicators widely used in social systems are summarized in Table 2.1. Examination of Table 2.1 leads to the conclusion that the conceptual quantification of social and environmental phenomena is not always unfeasible. Also, it is not always easy to determine which parameters or indicators are most appropriate to a given social goal. For instance, consider the goal of improving water quality. Selection of a single parameter to quantify this objective is seldom obvious. Dissolved oxygen concentration, algal growths, alkalinity, and temperature are but a few of the possible water quality indicators.

TABLE 2.1

Some Examples of Indicators

Aspect of life to be measured	Typical indicators
Social mobility	Correlation between the occupational status of parents and children
Health	Life expectancy, mortality, morbidity
Environmental quality	Pollution levels, pollutant diversity
Attitude	Galvanic skin response, pupillary size, record of past behavior
Noise pollution	Intensity, duration, frequency

Several attempts have been made to devise a single indicator to describe a broad range of environmental quality objectives. For example, a composite air quality index (AQI) can be defined as

$$\mathrm{AQI} = (I_c^2 + I_s^2 + I_p^2 + I_n^2 + I_o^2) \tag{1}$$

where I_c, I_s, I_p, I_n, and I_o are indicators, respectively, of the concentrations of carbon monoxide, sulfur dioxide, particulate matter, nitrogen dioxide, and petrochemical oxidants. The indicators of the concentration of individual pollutants can be defined in such a fashion as to exceed unity if a specified air quality standard is exceeded.

As an illustration of a different type of index, consider a water quality index (WQI) defined as

$$\mathrm{WQI} = \frac{w_1 Q_1 + w_2 Q_2 + \cdots + W_n Q_n}{w_1 + w_2 + \cdots + w_n} \tag{2}$$

where Q_i is the ith quality constituent (such as dissolved oxygen, chlorides, etc.) measured on a 0–100 scale and w_i is a weight indicative of the relative importance of the ith quality constituent. Evidently Eq. (1) is essentially an objective criterion, whereas Eq. (2) is somewhat subjective because selection of w_i is often based on expert judgment.

Variety of Performance Measures

It was stated that a measure of performance is usually a function of the indicators. As it is quite possible to define a variety of such functions, it is necessary and convenient to choose one function, called the *objective function* with a view to maximize, minimize, or optimize the value of this function. Evidently, the selection of a properly formulated objective function is of paramount importance in reaching the stated goal. The particular objective (or objective function) used has a direct influence on the model obtained and therefore on the subsequent decisions made.

Usually an objective reflects the overriding concern of the designer or analyst in achieving the stated goal. This task can or cannot be accomplished depending on the nature of the problem and the imposed constraints. A *constraint* is merely a limit on a required characteristic imposed on the system for any reason. Under such constrained conditions, it is quite conceivable that there exists no sufficiently strong means for influencing the system to achieve the goal. In such cases an analyst relaxes the constraints in the hope of finding a solution. A problem is said to be properly formulated if a means exists to reach the stated goal subject to the prescribed constraints.

There are many possible approaches and viewpoints in the selection of a performance measure. Some of these are discussed below.

System Effectiveness

From an operational viewpoint, system *effectiveness* is a measure of the extent to which a system may be expected to achieve a set of specific mission requirements. It is a function of system availability, dependability, and capability. Availability is a condition of the system at the start of a mission. Availability can be defined as

$$\text{availability} = (\text{MTBF})/(\text{MTBF} + \text{MTTR}) \tag{3}$$

where MTBF is the mean time between failures and MTTR is the mean time to restore the system to an operating state. The MTBF can be obtained from

$$\text{MTBF} = (\text{system operating time})/(\text{number of observed failures}) \tag{4}$$

Dependability is a measure of the condition of the system at one or more points during a mission given the condition of the system at the beginning of the mission. *Capability* is a measure of the ability of a system to achieve the objectives of a mission, given the system condition during the mission. Then system effectiveness can be defined as

$$\text{effectiveness} = (\text{availability})(\text{dependability})(\text{capability}) \tag{5}$$

For instance, consider an aircraft maintenance problem. In a calendar year, the craft may be on an alert status for nine months and on a maintenance status for an aggregate period of three months. Then the availability of this craft is $\frac{9}{12}$ or $\frac{3}{4}$. Similarly, dependability and capability can be suitably defined.

Profitability

Another method of judging the performance of a system is from the profit or loss viewpoint. Defining profit as

$$\text{profit} = (\text{value received}) - (\text{cost expended}) \qquad \text{or} \qquad p = v - c \tag{6}$$

an optimum system would then be one which maximizes the profit. Alternatively, the rate of return $R = (v - c)/c$ can be maximized. The optimization of the rate of return R may lead to unsatisfactory results as R has its maximum at the origin, implying that an optimum solution would be to build no system.

Cost Effectiveness

A third method of judging the performance of a system is to use the cost–effectiveness ratio as a criterion. The cost–effectiveness ratio (CER) is defined as

$$\text{CER} = (\text{cost expended})/(\text{value received}) \tag{7}$$

The cost–effectiveness ratio has widespread use in judging military systems. The cost–effectiveness ratio is time dependent in the sense that the short-term ratio is generally different from the long-term ratio.

Benefit–Cost Ratio

In resource management problems, or in problems involving public investment, the so-called benefit–cost ratio is in widespread use. The benefit–cost ratio (BCR) is defined as

$$\text{BCR} = (\text{present value of all benefits})/(\text{present value of all costs}) \qquad (8)$$

One should note that there is no significant conceptual difference between CER and BCR.

All the performance measures described thus far have some drawbacks. The usefulness of these measures are contingent upon reconciling a common unit of measuring items such as dependability, value received, benefits accrued, and so forth. Furthermore, the use of a criterion such as BCR has ethical and social objections. For instance, how would one go about measuring the "value" of a benefit?

Not all benefits can be translated into monetary units. The moral objection to BCR is that it is concerned with the total benefits and costs and is insensitive to their distribution. According to the BCR method, a project is acceptable as long as the ratio is large even if all the benefits have accrued to millionaires and all costs are borne by the poor. Construction of an express highway connecting suburbs to an inner city could be cited as a case in point. The highway renders the inner city easily accessible to the rich suburban residents while it displaces the inner city poor. This is not a new or unique phenomenon. Economists have been facing this kind of problem for a long time. The costs and benefits of a system that accrue to the general public but not directly to the decision maker, or the parties and people the decision maker represents, are called *externalities* or *spillover effects.* Externalities are, by definition, either benefits (such as the increase in land values as a result of the construction of a new highway) or costs (such as pollution) that are not forced back onto the owners or operators of the polluting system.

Social Welfare

Difficulties such as the above led people to consider social welfare as an alternative to benefit–cost ratio. The *social welfare function* is defined as the unique function that expresses the total welfare of the society as a function of all the goods and services available. Evidently, the explicit nature of the social welfare function depends upon the moral and ethical foundations and aesthetic tastes of a society. Therefore, the determination of the exact form of the social welfare function ultimately depends upon the conventional wisdom of the members. Questionnaires may be used to collect information about individual tastes and preferences, but there is no sound way of translating this data into a social welfare function.

From the preceding discussion it should be clear that in large-scale systems, such as those involving public investment and interest, there are two types of objectives: social and technical. Evidently technical objectives are usually subordinated to social objectives. No one would build a dam on a river simply because it is technically feasible. However, once the decision to build a dam has been taken within the fabric of a society, the technologist has the responsibility of implementing the decision in a proper fashion. At this stage one has to look at objectives from a purely technical viewpoint.

In devising an objective function in order to get a well-formulated problem, the following considerations are generally useful.

(1) One must be able to achieve the goal in a reasonable period of time.
(2) The energy or resources necessary to reach a goal must be constrained.
(3) The intensity of effort or power required to effect the required transition must lie within reasonable bounds.

In purely technical problems, such as guiding a spacecraft from the earth to the moon, the above three considerations lead to objective functions that reflect technical details such as minimizing the fuel consumed, minimizing the time of travel, minimizing the distance traveled, or maximizing the safety of the astronauts. In the context of public systems, the above three considerations lead to objective functions with economic overtones such as pure economic efficiency, redistribution of income, or fulfillment of "worthwhile" desires. It is customary to call objective functions that reflect concern about the technical performance of a system the *performance index*, while the term *cost function* is preferred if economic factors dominate the problem. In practice, both these names, and many others, are used interchangeably.

Each objective function has its own merits and must be considered in its own light. First, the choice must be realistic. The second consideration is mathematical and computational. While studying dynamic systems, some of the more commonly used objective functions are:

(1) *The integral square error* (ise) criterion:

$$J_{\text{ise}} = \int_0^T [\hat{x}(t) - x(t, p)]^2 \, dt, \tag{9}$$

(2) *The mean square error* (ms) criterion:

$$J_{\text{ms}} = \lim_{T \to \infty} (1/2T) \int_{-T}^{T} [\hat{x}(t) - x(t, p)]^2 \, dt \tag{10}$$

(3) *The integral of the absolute value of the error* (iae) criterion:

$$J_{\text{iae}} = \int_0^T |\hat{x}(t) - x(t, p)| \, dt \tag{11}$$

where $\hat{x}(t)$ stands for the experimentally observed value of an attribute at time t and $x(t, p)$ stands for the value predicted by the model. The letter p in the argument of x merely states that x depends upon the parameters p used in the model.

The J_{ise} criterion is more often used in parameter identification. The J_{ms} criterion is easy to handle mathematically. This is particularly useful if the inputs of a system are statistical in nature. However, J_{ms} is relatively insensitive to small changes in parameters. Due to this fact, J_{ms} criterion is not generally recommended for parameter identification and modeling problems. However, while designing a system relative to a given performance index, the system should be designed so as to minimize degradation of performance due to parameter variations and under these circumstances J_{ms} is useful. The J_{iae} criterion weighs large errors less heavily and small errors more heavily than J_{ise}. However, J_{iae} is computationally more difficult to implement. Indeed, one can derive an objective function tailor-made to a given situation. The only difficulty with such made-to-order objective functions is that so little is known about their behavior and one has to be careful.

Uncertainties, Value Conflicts, and Multiple Criteria

In the design, planning, and evaluation of many complex systems, it is not only difficult to choose representative goals but it is also difficult to settle on one criterion. This is a particularly vexing methodological difficulty because arbitrary selection of the "best" of several alternative actions in which each action will result in one of several possible outcomes or consequences has a good deal of subjectivity.

For instance, consider the goal to improve the quality of health services in a community. Does it mean decreasing the cost of treatment? Reducing the level of discomfort in bed? Minimizing the recuperation period? Minimizing the probability of relapse or complication? How is this minimization or optimization to be accomplished—with respect to the patient's family as a frame of reference or with respect to the entire society? When all the objectives are of a cooperative nature, it appears at first sight that there is nothing to be worried about as all the objectives drive toward the same goal. However, there is an element of uncertainty and fuzziness in human decision making that one has to take into account.

Consider, as another example, an individual whose goal is to land in a "good job." In the subject's opinion, the attributes of goodness are job satisfaction, job location, job security, starting salary, and opportunities for advancement. In order to make a choice among possible job offers, the candidate should first arrange these attributes in an order of preference. This is indeed a difficult thing to do. Assuming that the individual in question succeeded in arriving at a preference structure such as job security, starting salary, job satisfaction, opportunities for advancement, and job location in that order. Then the candidate is faced with an even tougher problem. How is he going to quantify these attributes when faced with major trade-off questions such as: (a) How much is he willing to lose in starting salary for a job that appears to be more satisfactory? (b) How important is it to be in a higher rank than in a lower rank but more secure position?

As a third example, consider a public project such as the problem of managing a river valley system described in Chapter 1. What, for example, is the most satisfactory way to compare the several dimensions of alternative plans for the management of a river valley system: Cost? Time stream of regional and national economic benefits? Loss of environmental quality? Some of the benefits and costs may be incommensurable entities. The trend in public planning has been to integrate commensurable and noncommensurable entities into a spectrum ranging in scope from political, social, and psychophysical aspects at one end to environmental, economic, and technological aspects at the other end. Of the environmental attributes, for example, pollution costs are most prominent. Additional considerations might be the magnitude of ecological disturbance and the rate of resource exhaustion. Social effects include contributions to dislocation on the one hand and to welfare and longevity of the social environment on the other. Psychological attributes such as satisfaction, attitude, and comfort also play a fundamental role in shaping public opinion, which in turn constitutes a major input in a public planning process.

The decision maker facing a problem of such complexity has to cross several hurdles. For instance, what constitutes an exhaustive set of alternative plans? What attributes are relevant in characterizing each plan? A fundamental question underlying the selection of the attributes is whether the attributes the decision maker thinks as important are in fact the same as those people actually use in responding to the decision maker's actions. This leads to another query as to how and when the public should participate in the planning process. Answers to questions of this kind depend on the ability of the decision maker to gauge public values and attitudes. This is a complex measurement problem.

To appreciate the nature of complexity involved, it is useful to think in terms of specifics. Assume that a finite and exhaustive set of plans

$$P = \{p_1, p_2, \ldots, p_m\} \tag{12}$$

exists from which the decision maker must choose one plan. Each plan will be evaluated with respect to (a) a finite set of attributes

$$A = \{a_1, a_2, \ldots, a_n\} \tag{13}$$

or (b) a finite set of criterion functions

$$F = \{f_1, f_2, \ldots, f_n\} \tag{14}$$

where the f's could be nonlinear functions of the attributes. For concreteness, let $x_{i1}, x_{i2}, \ldots, x_{in}$ denote the "values" the attributes assume for the ith plan. Then the set of values the attributes would attain for all plans can be arranged in an array as shown in Table 2.2.

TABLE 2.2

Raw Attribute Scores for Various Plans

Plans \ Attributes	a_1	a_2	$\cdots$	a_j	$\cdots$	a_n
p_1	x_{11}	x_{12}	$\cdots$	x_{1j}	$\cdots$	x_{1n}
p_2	x_{21}	x_{22}	$\cdots$	x_{2j}	$\cdots$	x_{2n}
$\vdots$	$\vdots$	$\vdots$		$\vdots$		$\vdots$
p_i	x_{i1}	x_{i2}	$\cdots$	x_{ij}	$\cdots$	x_{in}
$\vdots$	$\vdots$	$\vdots$		$\vdots$	$\cdots$	$\vdots$
p_m	x_{m1}	x_{m2}	$\cdots$	x_{mj}	$\cdots$	x_{mn}

In the problem of managing a river system the attributes $a_j, j = 1, 2, \ldots, n$ may stand for drainage area, velocity of stream, width of stream, river fauna, water condition, erosion, sewage, disturbance to benthic layers, visual and aesthetic aspects of pollution, etc.

Some of the attribute properties such as velocity and width, are physical in nature; they can be measured and numerical values assigned. It is not so easy to assign numerical values to such attribute properties as river fauna, erosion, channelization, scour, and disturbance to benthic layers. It is even more difficult to quantify intangibles such as the diversity of environment in terms of color, form, and contrast, which are best described linguistically. The decision-making problem becomes extremely complex while processing nonquantifiable information using conventional physical theories. The difficulty arises not because human beings are incapable of processing non-quantified information but because of the difficulty involved in constructing mathematical and computational procedures using such information.

Subjective Judgment of Performance

In the study of systems, models are often used to determine what *will* happen given a set of conditions. What *should* be done to attain a certain goal is clearly a human decision. Models, however versatile they are, only assist the decision maker to distinguish what is "good" and what is "bad." However, judgments such as "fair" and "unfair," "good" and "bad," and "large" and "small" reflect a relativism that belies their absolute grammatical form. The pervasiveness of this relativism is much greater than is commonly believed. Because people ordinarily try to increase the relative frequency of those events they like best, it appears that, at least in principle, it is possible to establish a scale of performance for the said events.

In physical sciences, the establishment of a hierarchy of precedence is relatively easy, whereas it is extremely difficult in a behavioral environment. Carefully controlled experiments reveal that imbalances in satisfaction apparently depend upon the frequency distribution of outcomes that form the context of judgment. The *range–frequency theory* asserts that a subject's

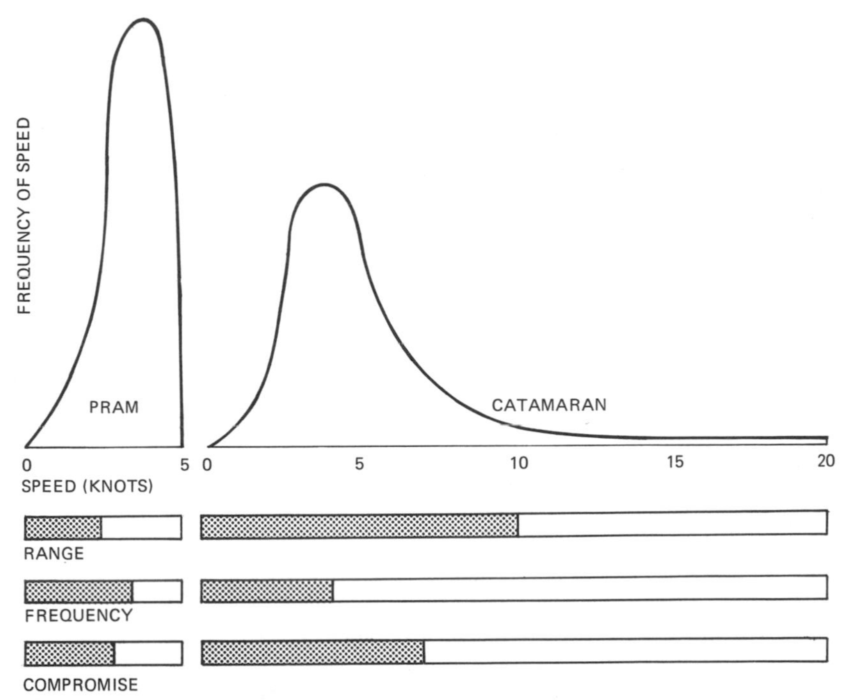

Fig. 2.3 The principle of compromise represented by a judgment of satisfaction for a hypothetical situation in which a devotee of sailing has a choice between a pram (left) and a catamaran (right). The pram has a narrow range of speeds but usually operates at the upper limits of the range; the catamaran has a wide range of speeds but rarely attains the upper region. The bars at the bottom show how a sailor would rate satisfaction if he based his judgments (1) solely on the range of speeds or (2) solely on the frequency with which top speeds are attained and (3) the actual satisfaction that would result from a compromise between range and frequency. (Adapted from A. Parducci, The relativism of absolute judgments, Copyright © 1968, **Scientific American 219** (6), 89, all rights reserved.)

judgment depends not only on the range of values to be judged but also on the frequency with which events in a given range occur.

Consider a specific example of transportation on water for pleasure and let it be assumed that the subject derives more pleasure from the speed of the craft. Would he derive more pleasure from the racing catamaran, which is the fastest sail boat under many conditions, or from the blunt-nosed, slow pram, which is the slowest of the popular boats? The catamaran attains its peak performance on rare days, whereas the pram operates on the upper range of its speed most of the time. Carefully conducted experiments revealed that sailing *experiences* in the pram are more "satisfactory" than in the catamaran. On the *range principle*, the sailor would rate as satisfactory only those occasions on which the speed of the boat was above the midpoint (or median) of the range: 2.5 knots for the pram and 10 for the catamaran (see Fig. 2.3). Under a variety of actual conditions, this performance will be achieved by the pram most of the time; by the catamaran only on rare occasions. On the *frequency principle* the speeds rated as satisfactory would be those above the median, that is, the faster 50% of the sailing occasions, which the catamaran achieves infrequently.

What all this suggests is that happiness or satisfaction has a negatively skewed distribution. That is, if the best can come only rarely, it is better not to include it in the range of expectations at all.

2 ATTRIBUTES AND RESOLUTION LEVELS

After the goals and objectives of a system study are identified, one is ready to plunge into the business of building a model. To be useful, a model must be operational. That is, a model should be flexible so that all relevant sets of conditions can be handled with ease and without ambiguity. Second, a model should exhibit clearly defined ties to reality so that any conclusions drawn from studying a model can be readily transferred to decisions in real world situations. Thus, if a model responds to data from the real world, operates on reasonable assumptions, and produces information that is of interest to the planner or decision maker, and can distinguish between significantly different alternatives, then the model can be called useful. Finally, a model should aid in gaining insight into the quantitative aspects of the structure and behavior of a system.

In building a model, a reasonable starting point would be to pose and try to answer the following questions:

(1) What attributes of the objects of which the system is composed need to be considered?

(2) What is a desirable resolution in measuring the attributes?

(3) What are the mathematical relations among the relevant attributes of each object in the system?

(4) What are the relations representing the couplings of various subsystems?

These are perhaps the four basic questions that arise in the study of a large variety of systems. An attempt will be made in the subsequent sections to present a systematic procedure of learning how to answer these questions.

After goal formulation, one of the earliest decisions to be made in systems modeling is the choice of relevant attributes. For example, one who is conversant with mechanical systems would know that relevant attributes of a particle in motion are its mass, position, velocity, acceleration, and applied force, while the relevant attributes of a spring are its spring constant, force-free length, etc. While studying urban systems, the relevant attributes could be automobiles, households, land use patterns, and attitudes of people. It is not always necessary to observe or measure all the attributes of an object. All one does is to measure the *values* (not necessarily the magnitudes) of certain *quantities.*

When investigating large systems, one generally confines attention to some part of the system that is of interest at a given time. Even though *properties* of a particular part are of interest, it is the *behavior* of a part or *object* in its *environment* that is more significant to a systems analyst. Thus, a spatial and temporal specification of an object with respect to its environment is a relevant attribute. Definition of a *space–time specification* is not sufficient; it is also necessary to decide the level of accuracy and the frequency with which to observe an attribute. The more accurate and/or frequent the observations, the higher is the *space–time resolution* level. This resolution level is sometimes limited by the instruments used in measuring the attributes. At other times, it is deliberately chosen low, which implies that sometimes some of the data are disregarded.

Consider the problem of weather forecasting, or the problem of the transport of a pollutant through the atmosphere. For these purposes some significant questions are: What and how many meteorological attributes are significant and susceptible for measurement? How dense should the network of data collection stations be? How often should the weather data be sampled? What degree and precision in the measurement is necessary and sufficient? The above sequence of questions can be elaborated. Suppose that the temperature T, relative humidity H, and barometric pressure P are considered to be significant attributes and are also susceptible for measurement. The next step, perhaps, is to select locations on the earth (i.e., the latitudes, longitudes, and perhaps elevations) where the values of these attributes are to be measured. A third step, perhaps, is to decide whether it is necessary and/or feasible to obtain continuous records of variations in the selected attributes. If this is feasible, for how long should a record be taken? If not, is it possible to sample the variations at discrete instants of time? If a sampling procedure is adopted, how often is it necessary and sufficient to sample? What kind of instruments are available to perform these operations to the required degree of precision?

The basic physical equations, which are nonlinear, presuppose an extremely detailed knowledge of the state of the atmosphere at the beginning of the forecast. For example, the viscosity term in the basic hydrodynamic equations (the Navier–Stokes equations) is of fundamental importance because it is ultimately responsible for the frictional dissipation of kinetic energy in the atmosphere. However, it can perform a vital function in a numerical calculation only if the latter includes motion on scales as small as a millimeter. Analogous difficulties occur in other equations, especially those describing the condensation of water vapor and precipitation (where the fundamental physical laws apply to individual raindrops) and radiation effects (where the molecular spectra are extremely complicated). The most important weather phenomena, on the other hand, have horizontal scales of 10^5–10^7 meters, and experience has shown that it is necessary to consider conditions over almost an entire hemisphere to predict weather several days in advance. It is obviously impractical to allow for this range of 10^{10} in any conceivable computational scheme. To deal with these difficulties of scale many kinds of approximations in the field of dynamic meteorology have been developed. They are all concerned with including or omitting certain physical quantities from the model. This range–resolution problem can be tackled only by introducing approximations. For instance, the entire range of acoustic waves can be excluded from consideration if the interest is only in large-scale cyclonic motions.

Three typical weather models and the order of resolution level they represent are shown in Table 2.3. Suppose that the interest is to study the pattern of pollutant (such as cadmium particulates from smoke stacks) dispersion and diffusion around a smoke stack. Then a microscale model

TABLE 2.3

Resolution in Meteorological Models

Kind of model	Spatial resolution	Temporal resolution
Microscale	Meter	Minutes
Mesoscale	5–10 km	Hours
Macroscale	100–1000 km	Days

would be useful. Suppose that the problem is to study the environmental effects of a pollutant in a given region. Then it would be appropriate to choose a mesoscale model. If the problem is centered around the global effects of dispersion of pollutants with large "half-lives" (such as the effect of DDT), then a macroscale model would be more appropriate.

Similarly, the resolution level on some of the meteorological variables can be specified. For instance, a popular way of studying air pollution problems is via simulation of weather patterns on a digital computer. For this purpose useful variables of interest are wind speed, wind direction, and stability category of wind masses. As there are an infinite number of possibilities for wind speeds, directions, and stability classes, it is necessary to specify the desired accuracy by specifying a resolution level. A resolution level on wind direction can be specified, for instance, by stating that only 16 possible wind directions (N, NNE, NE, NEE, E, etc.) are significant. Similarly six wind speeds and five stability categories are generally considered significant in mesoscale models. Thus, the resolution levels considered give $16 \times 6 \times 5 = 480$ possible weather conditions (or factors, as statisticians often call them) and meteorologists generally consider that any given weather condition can be reasonably approximated by one of these 480 factors. Selection of the resolution level, as described above, may simplify the mathematical problem by reducing the scope of the problem.

Resolution versus Measurement

Specification of the resolution level is not always a straightforward procedure. A low level of resolution implies a high level of aggregation and data obtained at such coarse aggregations may not shed adequate light on the structure or behavior of a system. On the other hand, any attempt to get finer resolution means monumental measurement problems. The dimension of this problem can be readily seen from an example. In urban planning, one is primarily concerned with land use data, population data, and economic data. Suppose that 20 attributes of the population of a city (number of people, number of males in the age bracket 20–29, number of people in medical and health care services, square footage allotted for residential areas, etc.) are recorded by census tract, and if there are 189 tracts, then a minimum of $189 \times 20 = 3780$ items of information are required. Census tracts represent a fairly high level of aggregation, and 20 attributes do only a sketchy job in describing a city.

This does not mean that a high level of resolution (i.e., decomposition) is always desirable. Decomposition is an important consideration in the use of secondary information such as that obtained from census or survey questionnaires. If data are collected at too fine a resolution level, then one has to go through awkward data processing operations to extract data that has less resolution. For instance, it is awkward for someone who wants the number of elderly people in a city to have to add up five age categories, two sex categories, and three race categories (30 items). Decomposition at the measurement stage has consequences throughout the model building stages. A view widely held by modelers is that decomposition produces stronger relationships, but the reverse is as often true. Too much decomposition produces data that are too thinly spread out. Finely decomposed entities are often so similar that relationships are often diluted making the relation meaningless. For instance, it is easy enough to say that households do not behave identically in their location preferences, but it is quite difficult to define household types and gather data that distinguish between types in a reliable way.

From this discussion it is clear that a complete investigation of all the factors is an arduous task if a fine resolution level is used. If, however, one is willing to settle for a less than complete investigation, techniques such as fractional factorial, Latin square, and Greco-Latin square designs can be used with profit.

3 SYSTEMS MEASUREMENT

Measurement plays an essential role in modeling. In ordinary life the act of measuring seems completely transparent. For instance, when one measures the length of a desk with a ruler, there is no reason to suspect that the result is significantly different from the true length of the desk. In many areas of physical sciences actual measurements are generally reduced via the elaborate superstructure of physical theory to comparatively indirect observations. Consider, for instance, the task of measuring the size of soot particles in a smoke stack, or fog droplets in a cooling tower. Evidently direct use of a ruler is impractical. One must resort to some indirect scheme of measuring, such as the scattering of light when the particles are illuminated. Now the quantity that is actually measured, or read on a dial, might be a voltage in an electrical circuit. Deduction of the relationship between the voltage readings and the particle size may not be a trivial exercise. However, it can be done and validated by suitably conceptualizing a model for the scattering process. The magnificent edifice of modern atomic and nuclear physics, for example, was founded on indirect measurement schemes not too unlike the one described above.

Sciences, especially those having to do with human beings, approach the task of measurement with considerably less confidence. In behavioral and social sciences, it is not entirely clear as to which variables are measurable and what theories to apply to those which are believed to be measurable. As in some areas of physical sciences, one can simplify the measurement task in behavioral sciences by substitution. That is, one measures an easily measurable attribute instead of one that is difficult to measure. For instance, how can one measure attributes such as hunger, anxiety and averseness? Typically one could measure hunger by counting the hours of deprivation of food. Anxiety and averseness can be inferred by measuring the electrical resistance of the skin of the individual in question. This kind of substitution makes sense if there is a superstructure of theory to substantiate the underlying assumptions. For instance, is there a basis to deduce the anxiety of a group of people from data obtained from measurements of the skin resistance of the individuals in the group? Such questions cannot be meaningfully tackled in the absence of a theory to support the measurement process. While modeling complex systems, the modeler should be aware of this fundamental limitation while measuring social and behavioral variables.

Extensive Measurement

When measuring some attribute of an object (or event), one normally associates numbers with the object such that the required property of the attribute in question is faithfully represented by a number. This practice is perhaps partly responsible for some of the notions of measurement that are so well entrenched that their validity is rarely questioned. For example, one would hesitate to ask a question like: Is mass really an additive numerical property?, because numerical representation of mass and the associated properties of numbers are taken for granted. The important question is: How far can one extend the basic ideas of measurement, as they are known in physical sciences, to nonphysical sciences? This question has not yet been solved to any degree of satisfaction. In the physical sciences, a basic procedure of fundamental measurement is the ordinal measurement.

Ordinal Measurement

Central to the techniques of ordinal measurement are the concepts of order and relation. For example, consider two rods designated a and b. One assigns a real number $\phi(a)$ to rod a and $\phi(b)$ to rod b such that $\phi(a) > \phi(b)$ if and only if $a \succ b$ is true, where the relation $\succ$ is interpreted to mean "longer than." Implicit in this type of measurement is one's ability to arrange a set of rods in an order. That is, the set of measuring devices constitute an ordered set. This is the reason for the name *ordinal measurement*.

A difficulty with the above procedure arises if neither $a \succ b$ nor $b \succ a$. That is, $a \sim b$. (read this as "a is equivalent to b"). As $\phi(a)$ and $\phi(b)$ are real numbers, it is possible that $\phi(a)$ and $\phi(b)$ lie arbitrarily close to each other without being equal to each other. However, practical measuring processes are subject to a resolution level and are therefore insensitive to small disparities in length. Therefore, it is theoretically possible that $a \sim b$, $b \sim c$, and still $a \succ c$, which in practice implies that

$$\phi(a) = \phi(b), \qquad \phi(b) = \phi(c), \qquad \text{and } \phi(a) > \phi(c)$$

which is clearly impossible. Therefore, ordinal measurement is suitable only when the sensitivity of the comparison process exceeds the disparities in the length of the rods.

The above procedure of ordinal measurement can be applied to any attribute of an object provided a suitable comparison process leads to relations such as $\succ$ and $\sim$. For instance, consider a relation such as a P b (a is preferred to b). Using relational statements such as a P b one can order a set of choices available. Unlike in physical sciences, this kind of preference ordering is difficult in a social context. Preferences and attitudes may change over time, or with feedback (see the delphi technique). Attributes such as lengths are additive if the objects in question are cascaded or concatenated together. Are preferences of two individuals additive? The reason for this situation is that the concept of concatenation or cascading has not been defined either implicitly, empirically or by convention in the social context. This leads to the concept of extensive measurement. An attribute of an object is said to be *extensively* measured if the concepts a R b (a is related to b), and $a \circ b$ (a is concatenated with b) are defined empirically. Thus, such attributes as length, mass, angle, time, and electrical resistance are extensively measurable.

Extensive Measurement in Behavioral Sciences

Basic to the technique of extensive measurement is the concept of concatenation. The lack of an adequate interpretation for this concatenation operation is perhaps the major stumbling block for the application of the technique of extensive measurement in the behavioral and social sciences. Are nonphysical attributes, such as loudness, risk, utility, and intelligence, extensively measurable? There was considerable success in measuring some psychological attributes such as risk and subjective probability, but a majority of societal attributes are elusive and escape strict definition.

In measuring length, the additive operation is equivalent to the concept of concatenation. Obviously psychological attributes such as aesthetic value, intelligence, or attitude cannot be concatenated in any satisfactory way! The reason for this is a lack of ordering (partial or complete) in the underlying set. Stated differently, the aesthetic values of a *set* of people, say, of different cultural backgrounds, cannot be ordered in any meaningful way. These difficulties, and many other subtle details, are the major drawbacks for the application of extensive measurement techniques in the social and behavioral sciences.

Difference Measurement

One key idea on which extensive measurement is based is the concept of ordering in a set of elements such as ordering a set of measuring rods, or measuring cups. A similar and almost equivalent idea is to consider the set of all differences or intervals between the elements of a set.

The idea and importance of interval or difference measurement is best illustrated by an example. Imagine a special kind of an electrical resistor made up of a material of irregular cross section and manufactured from a material of known or unknown composition. Such resistors can indeed be made by painting a resistor compound on an insulator base. The problem is to mark off points along the resistor such that the resistance between adjacent pairs of points is the same (but left unspecified). To bring in an element of subjectivity, it is also assumed that the only instruments available are one or two light bulbs with wires of negligible resistance attached to them and some batteries.

One method of handling the problem is to connect one bulb and a battery in series with an arbitrarily chosen segment of resistor material (interval *a*, *b* in Fig. 2.4a) and another bulb and battery in series with another segment (*c*, *d*) and vary the position of contact at point *c* until both bulbs are of the same brightness. The points thus marked by the test bulbs define segments of equal resistance. The point is that one can use the difference in the brightness of light to measure an attribute such as resistance. If the brightnesses are judged using standard photometers, the problem really falls in the realm of physics.

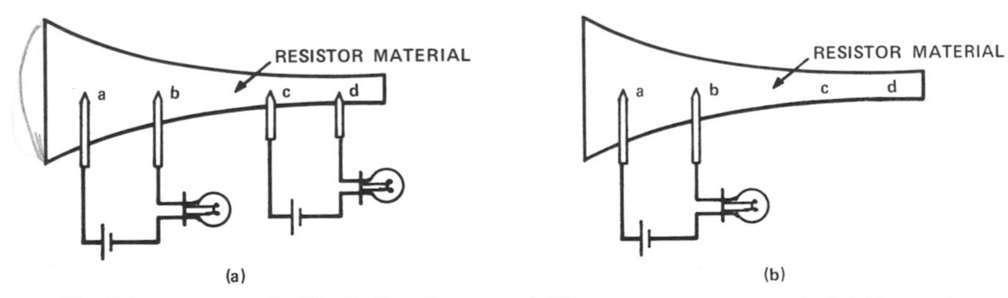

Fig. 2.4 An example illustrating the use of difference measurement: (a) Measuring an attribute such as resistance by simultaneously observing two stimuli, namely, the brightness of two bulbs. (b) Measuring resistance by observing the brightness of the same bulb in a sequential fashion.

A second method of conducting the above experiment is to use only one light bulb, in sequence, on the intervals *a*, *b* and *c*, *d* in Fig. 2.4b, and the subject is asked to match brightnesses. This problem is more difficult because the result now depends not only on the precision of judgment the subject uses but also on the status of the psychomotor and sensory apparatus of the subject because direct comparison of brightness is no longer possible. The subject has to remember what the previous brightness was.

The concept of difference measurement is more widely used in psychophysical experiments than in any other activity of behavioral research. In the *method of bisection*, a subject is asked to choose a stimulus whose quality (brightness, loudness, etc.) is halfway between two given stimuli. In the *method of cross-modality matching*, a subject is asked to match two types of sensations, say, brightness and loudness. If the brightness B_1 of a light was previously matched to the loudness L_1 of a sound, and if the observer is now shown a light of brightness B_2, the underlying assumption is that the subject would choose the sound of loudness L_2 such that the "brightness ratio" B_1 to B_2 somehow matches the "loudness ratio" L_1 to L_2. The quotes used in the preceding sentence imply that the word "ratio" still carries an undefined meaning. If the brightnesses and loudnesses are associated with some numbers (assigned on the basis of subjective perception, or objective measurement), then the "ratios" stand for numerical ratios.

The idea of ordering sensation intervals is not new or exclusive to the field of psychophysics. Utility intervals play an analogous role in economics where one attempts to order the perceived differences in utilities using probability concepts. Indeed the concept of marginal utility is closely related to the concept of ordering differences of utilities. This idea is creeping into social sciences as well. Consider, for example, a set of individuals (or nations) designated by $A = \{a_1, a_2, \ldots, a_N\}$. For various reasons, it is very likely that some pairs of individuals may meet more often than others. This concept can be quantified by considering all pairs of individuals, that is, by considering elements of the Cartesian product $A \times A$, and ordering the pairs according to the frequency of the pair's social or political contact. However, this kind of ordering along a one-dimensional continuum is not always easy. The difficulty can be clearly seen in a political context in which nations professing democracies undertake subversive activities and support military dictatorships abroad while some totalitarian regimes crave to join hands with democracies. A reason for this baffling behavior can be found by looking at what a nation really wants. Professing principles from platforms is one thing and taking a hard look and making an assessment of the true national priorities is another thing. Therefore, it is not always practical to arrange national priorities as a linear array. Stated differently, the objective function of a nation is not scalar valued—it is vector valued. In such situations multidimensional representation is warranted where the coordinates of a point conceivably represent the "profile" of a nation and ordering of pairs (a, b) could conceivably be achieved by some kind of proximity or distance measure.

A more familiar method of measurement based on ordering of differences or intervals is the so-called *rating scale* so often used to gather public opinion through questionnaires. While conducting surveys using questionnaires, it is important to remember that rating scale responses are not just ordinal; more structure is involved in the definition of the scale. For instance, consider the following two questions, one of which is to be included in a questionnaire.

Question 1: Rate your reaction to noise from overflying jet aircraft on a scale from $+2$ (mildly unpleasant) through 0 (indifferent) to -10 (extremely distressing).

Question 2: Rate your reaction to noise from overflying jet aircraft on a scale from $+5$ (mildly unpleasant) through 0 (indifferent) to -5 (extremely distressing).

It is very likely that a majority of people would consider the second question to be unfair because for their perception the effective interval (mildly unpleasant, indifferent) seems to be much smaller than the interval (indifferent, distressing). Similar situations arise in international politics in which a country may profess a neutral stand on an issue by equating the aggressor and aggressee.

More typically, questionnaires contain more than one question, and the reaction of an individual to a particular question is likely to depend upon the conditioning of his attitude by the tone or content of the previous questions. Another difficulty is that all respondents do not respond to the same question in comparable ways. Some people have generalized tendencies (built-in biases) toward extreme ratings; some have numerical preferences. Such issues baffle systematic analysis.

Conjoint Measurement

When no extensive concatenation is apparent, one must look for some other structure in measuring an attribute. One of the most common alternative structures is for the underlying entities to be composed of two or more components, each of which affects the attribute in question. Momentum, which is the product of mass and velocity of an object, is an example from physical sciences. Comfort, which depends on various humidity and temperature combinations, is another example from nonphysical sciences. Since simultaneous measurement of objects and their components is involved, this type of measurement is called *conjoint measurement*. The concept of conjoint measurement is applicable only when the values of the components involved can be selected independently. More formally, the name conjoint measurement applies to techniques of measuring the effects of nominally scaled independent variables from an ordinally scaled dependent variable.

Scales for the Measurement of Attributes

In the past, engineering and economic considerations guided decisions concerning public planning involving such activities as construction of highways, dams, power transmission lines, and locating power stations for generating electrical energy. With the growing recognition of environment-related problems and in response to mandates to identify the environmental impacts of any proposed public policy decision, regulatory agencies are getting interested in planning methodologies that incorporate public attitudes as well as engineering and economic considerations. Decisions, which require a balancing of concern for preservation of aesthetic qualities in the environment with concern for economic constraints, are required to reflect the attitudes and values held by the public.

Therefore, there is a need to conceptualize, define, and measure attributes such as "attitude," which, incidentally, is one of the most ubiquitous of all the terms used in social sciences. There is a good deal of controversy among social psychologists about the definition of the term itself. However, it is almost universally accepted as axiomatic that attitude (or attitudinal behavior) is learned. If attitude is a learned trait, some argued, then it is also capable of modification by further learning. This particular line of reasoning dominated experimental social psychology in the late 1960s and early 1970s.

At this stage it is useful to invoke some of the system theoretic ideas to gain a better understanding of the term "attitude." For instance, what are the various components of attitude? Do these components have a well-defined structure? Does the behavior of this structure become dependent upon its environment? Can one deduce the properties of "attitude" by studying its component parts and structure? Many other questions can be posed. One of the classic methods of studying a system is by observing the system's response or behavior due to an external excitation or stimulus. Similarly, an understanding of attitude can be obtained by studying attitudinal behavior. Indeed, social psychologists use the terms "attitude" and "attitudinal behavior" in a practically synonymous manner.

One of the classic divisions of attitudinal behavior in psychology has been the division into cognitive, affective and connotive elements. The cognitive component refers to the way an individual perceives an object and conceptualizes it and thus represents an individual's beliefs. The affective component is concerned with the emotional underpinning of these beliefs and reflects the positive or negative feeling a person has toward the object. The connotive component refers to an individual's intention to behave in a certain way. Further subdivision of these three components is possible. Some theorists claim that attitudes have motivational properties so that attitudes not only channel and direct people's behavior, but also drive them to respond positively or negatively toward an object. It is evident, therefore, that attitudinal behavior is both complex and diverse. The conceptualization of attitude in a given context therefore influences the form of measurement which is to be used.

Basically, measurement of attitude consists of gathering observations about people's behavior and allocating numbers to these observations according to certain rules. The process of measuring attitude can be visualized in terms of four progressive stages. During the first stage, the investigator forms an initial image of the concept called "attitude." This was done in the preceding paragraphs. During the second stage one tries to specify the relevant dimensions of the concept to serve as a basis for measurement. In the case of attitude measurement, there is, as yet, little consensus as to what these dimensions are. The third stage is marked by a search for a set of indicators for the various dimensions. The final stage involves a combination of these indicators into a single unidimensional index to represent the underlying attitude. To implement this step, one ascribes scores to the various indicators and combines these scores.

The issue of ascribing scores to indicators is not an easy task. In assigning numbers to observations, the investigator is imposing *scale properties* on the data. Such scale properties define the relationship between numbers which are assigned to various observations and the quantity of some attribute (attitude, here) which these numbers are used to represent. In this context four different scaling techniques are traditionally employed: nominal, ordinal, interval, and ratio scales.

Nominal Scale

This involves simply naming of elements, which implies the not necessarily trivial ability to distinguish like from unlike elements. The role of numbers in nominal scaling is purely descriptive. In land use planning, for example, one may wish to label all farm land by the number 5, residential area by 7, an industrial zone by 3, recreational land by 1, and so forth. This does not imply that farm land has more of a particular attribute than recreational land, but merely that farm land is distinguished from the others by a distinct label.

Ordinal Scale

An ordinal scale uses numbers to rank objects from greatest to least with respect to some criterion. Consider the statement, "A strip of interstate highway leads to greater air pollution per passenger mile than a strip of railroad track." Here there is no indication of how much more, but there is an ordering. Similarly, a person who received an attitude score of 2 is assumed to have a more favorable attitude than one who received a score of 1. There are no value implications. To illustrate this point, an ordinal scale has been applied to the physical quantity length in Fig. 2.5. Since the lengths are rank ordered, the only requirement is that the size of numbers increase, say, from left to right. Inspection of this scale shows that when it is applied to attitude measurement there are two things an ordinal scale does *not* imply. First, it does not imply that a person with a score of 4 is twice as favorably disposed toward an object as a person with a score of 2. Second, ordinal scale will only perform rank ordering and the distances between different ranks are not necessarily equal.

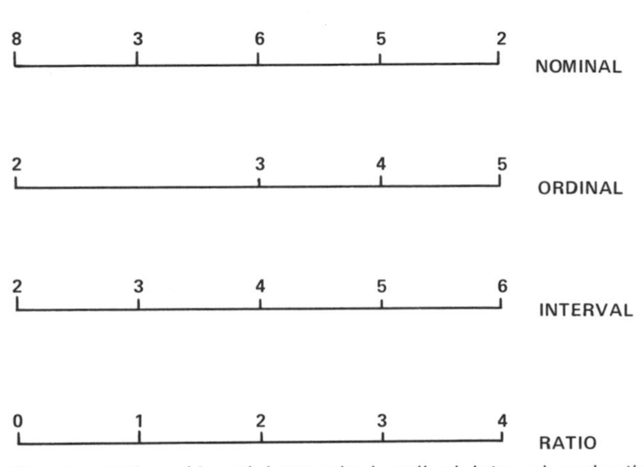

Fig. 2.5 Representation of length by nominal, ordinal, interval, and ratio scales.

Interval Scales

The units of measurement in an interval scale are equal. Using this scale one can state that an interstate highway produces so many units more pollution than a railroad. Here, one can only deal with *differences* between levels, not ratios, since zero level is not defined on an interval scale.

Ratio Scales

In ratio scales, the units of measurement are equal and the scale is calculated from some arbitrary zero. For example, if one plan consumes 10 acres of farm land and another plan consumes 5 acres, than it is relevant to use ratios and say one consumes twice as much as the other.

This classification leads to another question: Which type of scale is appropriate for the measurement of attributes such as attitude? Environmental impact due to visual pollution? There is no obvious and immediate answer to this general question. This is not an unusual situation faced by social scientists and a particular solution is usually found to suit a given set of circumstances.

Scale Conversions

Most analysis techniques assume a homogeneity of scale types, whereas real data sets often appear with mixed scales. One approach in handling such problems is to choose a particular scale type and suitably transform variables to achieve homogeneity. Scales usually are ordered in the sequence nominal, ordinal, interval, and ratio, with the progression reflecting increasing information demands for scale definition. Hence, promoting a variable up the hierarchy of scales implies the utilization of additional information or acceptance of a new assumption. In some cases transition from one scale type to another involves nothing more than a change of viewpoint. Occasionally one can substitute one variable by another which is measured on the desired scale. Some of the relatively simple scale conversion techniques are briefly described here.

Ratio to Ordinal

For example, education is a ratio variable when measured in terms of years of education or number of highest grade completed. This can be converted into an ordinal scale by replacing the ratio variable by ordered categories such as:

(1) No grammar school.
(2) Elementary school completed.
(3) High school completed.
(4) Bachelor's degree holder, etc.

Ratio to Nominal

Weight is a ratio variable. To convert this to a nominal scale, one can measure weight, of children, for example, in terms of how much it exceeds or falls short of an ideal weight. The deviation from the normal (for the children's age and height) can be expressed as

-10 to $+10$:	normal weight
-25 to -11 and $+11$ to $+25$:	some counseling needed
less than -25 and more than $+25$:	remedial action warranted

which possess some ordered property.

Ordinal to Nominal

Suppose five students in a class are ordered as $A\ B\ C\ D\ E$ using the score in an examination as a basis. These ordered categories can be converted to nominal categories by labeling the central student as normal. Thus, we get

C	normal
$B \cup D$	moderately deviant
$A \cup E$	extreme

Another familiar example is the reorganization of an ordered set of numbers as even and odd.

Nominal to Ordinal

In contrast to the preceding three cases, this case involves a promotion of scale type and is therefore more difficult. Unordered categories can be arranged in an order only by using additional information. Typically this new information takes the form of selecting a reference variable. Suppose that the scores achieved by students are initially labeled as "moderate" and "extreme." These nominal categories can be converted into ordered categories by splitting the "extreme" group as "low" and "high" and subsequently ordering the scores as "low," "moderate," and "high." More complicated problems can be handled using techniques such as *ranking*.

Multidimensional Scaling Techniques

One method of overcoming the difficulty alluded to in the preceding paragraph is by using the so-called multidimensional scaling (MDS) techniques. The term MDS is used to refer to a class of techniques in which respondents judge the degree of similarity between pairs of stimuli such as a pair of questions, or a pair of photographs. In MDS, similarity between a pair of stimuli, say, plans, is assumed to relate to some form of "distance" defined on a psychological scale. The above "distance" can be measured on an ordinal scale, interval scale, or even on a ratio scale, and this choice leads to several types of MDS procedures such as metric MDS, nonmetric MDS, etc. In short, MDS can be defined as a data analysis technique that constructs a configuration of points in space from some kind of information about the psychological distances between stimuli. Hence, the fundamental data for MDS are an array of numbers s_{ij} representing similarity (or dissimilarity) between two stimuli i and j. From this data MDS attempts to discover the number of dimensions relevant to perception of the stimuli under consideration.

It is conventional to distinguish two different types of MDS: metric and nonmetric. The more recent nonmetric MDS procedures differ from the classical MDS primarily in that similarity judgments in the former are assumed to be only monotonically related to each other. In nonmetric MDS, no attempt is made to suggest how, or on what basis, a person is to judge the similarity between stimuli: the basis for judgment is respondent generated.

4 THE TAXONOMY OF SYSTEM CONCEPTS

The classification of systems and models is often a useful pedagogical exercise even though no classification is complete or unique. The classification presented in the sequel is quite general and applies to systems, their component elements, and models as well.

Variables

Attributes that vary with time and/or space are of particular significance in the study of systems. In particular, attributes that are *allowed* to vary may be called variable attributes or simply *variables.* In short, a variable is a quantity that varies over a range—spatial or temporal. A number of qualifying adjectives can be prefixed to the term "variable" to render it more descriptive.

When the value of a variable y depends upon the value of another variable x and if a functional relation such as $y = f(x)$ is known to exist between x and y, then y is called the *dependent* variable and x is called *independent* variable.

An *exogenous* variable is one whose value is determined outside the system or its model. Exogenous variables usually characterize some aspect of the substantial environment. Stated differently, exogenous variables are the independent or input variables to a system. An *endogenous* variable, on the other hand, is one whose value is determined within the system. Stated differently, endogenous variables are the dependent or output variables.

Exogenous variables could be of two kinds: *controllable variables* and *uncontrollable variables.* Controllable variables are those on which a model builder could exert some sort of control, whereas uncontrollable variables are those that are being accepted as existing but not necessarily known. *Policy* or *decision variables* are typical examples of controllable variables.

Example 2. Consider an agricultural system. The success or failure of crops is determined by such factors as quality of seed, proper application of fertilizers, and timeliness and amount of rainfall. The status of crops in turn determines the prices of various commodities. In a model of this kind, typical exogenous variables are weather conditions, demand for the commodities, etc. In projecting the future state of the agricultural economy, an economist should realistically take into account any possible fluctuations in rainfall. However, rainfall can be neither predicted nor controlled. Therefore, rainfall is an example of an uncontrollable exogenous variable. Demand preferences, population growth, etc. can also be considered as exogenous in economic models. In this model, endogenous variables might include output, cost, sales, and profit. Decision variables in this example are quality of seeds, fertilizer application, etc. ■

There exist many other kinds of variables. For example, a *real variable* is one whose values are real numbers whereas the values of a *complex variable* are complex numbers. A *Boolean variable* or *logical variable* may be defined as one that takes on only two possible values which are not necessarily numerical in nature. Typically a Boolean variable takes on values like TRUE or FALSE; YES or NO; 1 or 0. Boolean variables of this kind are also referred to as *binary variables.* Relations between Boolean variables are expressed by the so-called *logical functions.*

In many physical sciences, the terms *across variable* and *through variable* are widely in use. An across variable relates the conditions at one point within a system to that at some other point within the system or some universal reference point. Temperature, which is really a measure of temperature difference between an object and absolute zero, is an example of an across variable. A through variable, on the other hand, assumes the same value everywhere within an element of a system. Current flowing through an electrical circuit element, such as a resistor, is an example of a through variable. In socioeconomic sciences an across variable is sometimes referred to as *propensity* and a through variable as *flow rate.*

As systems are sometimes characterized by a set of interconnections between entities called objects, often it is convenient to describe an individual object A by what are called the *terminal variables.* If the terminal variables of an object A can be divided into two categories—*input* and *output* variables—then A will be called an *oriented object.*

By varying the input variables (or causes) x, an experimenter can observe the output variables (or effects) y, and can derive, at least in principle, a set of *input–output relations,* which can be written abstractly as $x \mathrm{R} y$. This relation in an oriented object can also be represented pictorially, as shown in Fig. 2.6.

Suppose that an experimenter performed an experiment on an electrical resistor by passing a current i through the resistor and measuring the voltage v across it. The result of this experiment can be summarized by the relation

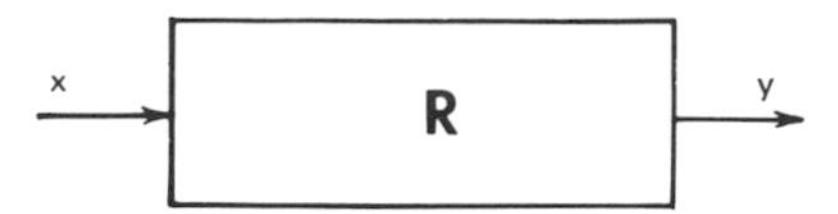

Fig. 2.6 Block diagram representation of an input–output relation R.

$v = \mathrm{R}i$. It should be emphasized that this relation remains valid, regardless of which of the variables, v or i, acts as input and which acts as output. That is the roles of v and i as input and output can be interchanged. Thus, depending on the context, the above relation can be regarded as an input–output relation of an oriented object or as a terminal relation of any object.

From the preceding discussion it is not very difficult to observe that a distinction can be made between *physical* and *abstract* objects. A *physical object* is an object perceived by the senses and its abstraction is the mathematical relationship that gives some expression for its behavior. In making an abstraction it is possible to lose some of the relationships that make the abstraction behave similarly to the physical object. Every physical object can be represented abstractly by an oriented object. However, the converse is not necessarily true. There exist oriented and nonoriented objects for which no known physical counterparts exist. If an abstract object A can be represented by a physical object P, then A is said to be *physically realizable.*

Example 3 Suppose that the terminal relation of an object A is given by $F = ma$, and it is desired to find out whether this terminal relation corresponds to any physical object P. That is, can the above relation be physically realized? A student of mechanics readily recognizes that if F is interpreted as the force applied on a particle of mass m, then a is the acceleration attained by the particle. An electrical engineering student sees this from a different angle. If m is regarded as a resistance of a resistor, then a can be quickly identified as the current flowing through the resistor and F as the voltage drop across the resistor. Thus far, two physical realizations are obtained for the above relation. Such realizations are generally referred to as analogs of each other. Still another realization is an amplifier with input voltage a and output voltage F and amplification factor m. This third realization clearly illustrates that a terminal variable, say, F, can correspond to input in one realization (e.g., the force applied to a particle) and output in another realization (e.g., the output voltage of an amplifier). ■

To be able to build a model to a complex system, it is often necessary to measure the value attained by a variable and represent this value by a number. A *measurable variable* is one that is susceptible to physical measurement; otherwise, a variable is said to be nonmeasurable. Attitudes of people at a bargaining table is a typical example of a nonmeasurable variable. If a variable is nonmeasurable, it can still be taken into account somehow or it may be omitted completely from further consideration.

As complex systems are usually characterized by a large number of variables, it is useful to represent sets of numbers, characterizing the values of variables, in a compact fashion. The vector notation serves this purpose in a remarkable way. A *vector* is merely a set of numbers arranged in a definite order. The order is important and the position of any particular number in the array has significance. The set of numbers so arranged are called the components of the vector. Normally, the components of a vector $\mathbf{x}$ are arranged in a *column*:

$$\mathbf{x} = \begin{pmatrix} x_1 \\ x_2 \\ \vdots \\ x_n \end{pmatrix}$$

On occasion, it is convenient to display the components of $\mathbf{x}$ as a *row*. Then, the above vector is written as $\mathbf{x}^\mathrm{T}$ (read this as "x transpose") where

$$\mathbf{x}^\mathrm{T} = (x_1, x_2, \ldots, x_n)$$

This definition of a vector says nothing about the physical significance of a vector. The numbers may represent any physical quantity in any unit that is convenient. Thus, if a physical event can be described by a set of attributes, then the values of these attributes can be used to define a vector. These values merely describe an abstract vector. A physical representation of the vector can be obtained only after a *basis*, or a *coordinate system*, is established; that is, after one decides, say, that the first component will be temperature in degrees Celsius, the second component will be blood pressure in millimeters of mercury, and so forth. It is only after these decisions have been made that one can actually write down a vector. This distinction between an abstract vector and its physical representation is fundamental. It is important because a physical event can be represented many ways.

Example 4. For example, the condition of a patient may be represented by the variables: oral temperature, heartbeat, and blood pressure. If the oral temperature is 101°F, the heartbeat is 80 per minute, and blood pressure is 150, then the condition (the abstract vector) can be represented by the vector $\mathbf{x}^\mathrm{T} = (101, 80, 150)$. It may be possible to represent the condition of the same patient by the variables: rectal temperature, height of the *PQR* complex in his electrocardiogram, and pulse rate. If the rectal temperature is 102°F, the height of *PQR* complex is 0.9 units (where unit height corresponds to a normal condition), and pulse is 85 per minute, then the condition can be represented by the vector $\mathbf{y}^\mathrm{T} = (102, 0.9, 85)$. Thus the physical condition of the patient is the same; the representation simply took different shapes. ■

This observation can be stated more broadly. A system is described by a certain number N of variables. Of these, only a certain number n can be independently specified, the remaining $(N - n)$ then being determined by the dynamics of the system. The entire set of variables can now be separated in a convenient fashion into two sets of sizes or dimensions, n and $(N - n)$. Now define a *driving vector* or *input vector* whose components come from the subset of independent variables and *response vector* or output vector whose components come from the subset of dependent variables.

The vectors described here do not necessarily carry a geometric meaning. They are merely representations of real physical processes. In many books, a vector is often defined as an entity that has both magnitude and direction. What, for example, is the magnitude of a vector that is 99°C along one axis and 115 mm Hg along the other axis? We do not know until we define what we mean by magnitude and direction. This does not mean that a vector cannot be represented by a magnitude and direction. If it serves some useful purpose, one can certainly do so.

Many physical systems are designed and built to perform certain well-defined functions. Submarines and spacecraft must navigate in their respective environments to accomplish their objective. An electrical power system must meet its load demands. In order to determine whether a system is functioning properly, and ultimately control the system performance, one must know what the system is "doing" at any instant of time. In other words, one must know the *state* of the system. In navigation, the state consists of the position and velocity of the craft in question. In an electrical power system, the state may consist of voltages and their phase angles at various places.

The concept of state generally relates to those physical objects whose behavior can change with respect to time and to which a stimulus can be applied and a response observed. The state of a physical object is any property of the object which relates input to output such that knowledge of this input time function for $t \geq t_0$ and state at time $t = t_0$ completely determines a unique output for $t \geq t_0$. The *state variable*, often denoted by the vector $\mathbf{x}(t)$, is the time function whose value at any specified time is the state of an object at that time. The set of all $\mathbf{x}(t)$ is called the *state space*.

Suppose that a spring is connected to a mass and the spring–mass system is set in motion. If the position coordinate x and velocity y of the mass are given at the present time, the position and velocity of the mass at any future time can be calculated using the dynamic equation obtained by applying Newton's second law. The set (x, y) can be called the state of the spring–mass system and the variables x and y are state variables.

Constants and Parameters

Every quantity that enters a relation is either an absolute or universal constant, an arbitrary constant or a variable. An *absolute* or *universal constant* is a quantity whose magnitude cannot be changed under any circumstances. It may be a pure number such as π, a dimensional number such as the speed of light in vacuum, or a dimensionless constant such as Plank's constant. An *arbitrary constant* is a quantity whose magnitude can change, *but is not allowed to change* in the particular problem under consideration. Thus, whether a quantity is a variable or an arbitrary constant really depends on how an analyst treats it in a given set of circumstances. The term *parameter* is commonly applied to as an arbitrary constant in an equation which expresses a particular form of relationship between two or more variables. For example, in

$$y = a + bx$$

the arbitrary constants a and b are called the parameters of the equation. In other words, a parameter is a number that characterizes one aspect of a collection of objects. Thus, b characterizes a family of straight lines whose slope is the number b. Alternatively, one can define a constant as a measurable variable which is unchanged over time or space. A parameter, then, is a quantity that is not measurable but can only be estimated.

The System

An element (or system) is said to be *linear* if the excitations and responses are directly proportional over the entire range of measurement. For example, an electrical resistor is linear if the current through it doubles when the voltage applied across its terminals is doubled. Thus, an element (or system) is said to be linear if the variables used to describe its conduct can be related via a linear equation, that is, an equation of the first degree. This equation could be algebraic, differential, or integral in nature. Very powerful analytical techniques exist to study linear systems. For instance, the linear functions are the simplest of the algebraic functions. A sum of two linear functions is also a linear function. Linear equations are the easiest of the algebraic equations to solve. The elegant and powerful determinant and matrix methods apply only for treating linear systems. Finally, if a system is linear, it is possible to obtain the response of the system to a collection of excitations by adding the response of the system to each individual excitation acting alone. This principle of adding solutions is called the *principle of superposition.* Nonlinear systems, on the other hand, are much harder to treat mathematically. The principle of superposition does not apply. Without this additive property, the use of series expansions leads to complications. Systems of equations can no longer be treated using matrix methods.

An element is said to be *lumped* if its behavior is specified completely in terms of input–output relationships at its external terminals. In lumped system theory, the system is regarded as being composed of an array of distinct elements interconnected in some specific manner. The lines interconnecting these elements, perhaps linkages in mechanical systems, wires in electrical systems, or communications in social systems, have no special significance other than to indicate the topological relationship of the elements. The physical dimensions and positions of the elements are not of primary significance and therefore are of no direct consequence in the analysis of system behavior. From a mathematical viewpoint, a lumped system (or lumped parameter system) is one whose behavior can be described adequately by a set of *ordinary* differential equations. If the dynamic behavior of a lumped system is of interest, in general, time is the only independent variable. The above kind of analysis is completely inadequate if one is also interested in the internal behavior of the system components. Then it is no longer adequate to characterize a component by its terminal behavior alone. In this case the problem becomes a *distributed parameter system* or *distributed system* or a *field problem*. Thus, in a distributed system, the spatial dimension becomes an integral part of the problem and usually enters the problem as another independent variable. Mathematically, a distributed system is characterized by *partial* differential equations. The problem of determining the conditions under which a system can be represented adequately by a lumped model or a distributed model is often difficult and subjective. For example, electrical engineers are accustomed to approximate physical systems by lumped models when the actual physical dimensions of the system are "much smaller than the significant wavelength of the exciting signal." As wavelengths are inversely proportional to frequencies, it is said that lumped approximations are valid for low frequencies. Many signals of relevance in large-scale systems contain a wide range of frequency components, and the problem reduces to one of deciding which frequency components are relevant. This decision can only be made after a good deal of experience about the system and the techniques is gained.

A *time-invariant* system is one whose input–output relations do not change with time. In other words, the nature of the response of the system depends only on the nature of the excitation, and not on the time of its application. Otherwise, the system is *time-varying*

If an element of a system acts only to dissipate or store energy, it is termed a *passive* element. Elements through which energy enters a system are called *active* elements.

If the inputs and outputs are capable of changing at any instant of time, the system is called a *continuous-time* system. It is important to remember that "continuous time" does not imply that all inputs and outputs are continuous functions in the mathematical sense; they are functions of variables that vary continuously. The so-called *discrete-time systems* are those whose attributes exhibit change only at discrete instants of time, say, every second, or year, or perhaps irregularly. Between these instants of time, the magnitudes of various attributes are either constant or not defined. In either case, the behavior of the system during successive time intervals is of no interest. That is, discrete time systems are event oriented.

It is worth noting that continuous-time systems can be treated as discrete time systems by observing the former's behavior only at discrete instants of time.

Example 5. A person's body temperature is a continuous time signal; however, a patient's temperature on a doctor's chart is a discrete-time signal because the temperature is usually measured or *sampled* only a few times a day.

Stock prices appearing in a newspaper are discrete-time signals because they assume new values only when a new edition goes to press. Usually prices are also adjusted and quoted in levels differing by one-eighth of a dollar. Such adjusted signals are called *quantized* signals.

A good example of an engineering large-scale discrete system is the digital computer. The arithmetic unit consists usually of a basic element (an adder, a multiplier, etc.), a set of auxilary elements (registers, gates, etc.), and some couplings. All the basic elements usually operate only at those instants of time specified by a central clock and are practically inoperative at other times. To an external observer, the operation of a digital computer becomes meaningless if it is observed through a "time window" that opens asynchronously with the clock in the digital computer. ■

In *experimental branches*, such as physics, biology, economics, or astronomy, a system represents an abstraction that must be used when investigating natural phenomena. These phenomena (or systems, if that choice is preferable) can be classified into two categories: *controlled* and *uncontrolled*. Those in the first category include experiments that can be performed in the laboratory, where it is often possible to vary different parameters and thereby test the predictions of the theory extensively. The phenomena in the second category do not permit such freedom of maneuver. Here man can only observe events as they march on, without being able either to control them or reproduce them in the laboratory. To obtain the maximum possible information from such phenomena often requires a great deal of patience and ingenuity on the part of the observer, and conclusions drawn from these observations do not have the same degree of accuracy that one expects from a laboratory experiment.

A theory of electrical conductivity can be tested in a laboratory by subjecting the material under test to a wide range of voltages and currents.

However, many astronomical theories are based on observations of phenomena which belong to the second category.

From another viewpoint, systems can be primarily *physical* such as communication systems, control systems, chemical processes, and air defense systems, or they can be primarily *operational* such as inventory control systems, logistical systems, dynamic scheduling systems, accounting systems, automated teaching systems, and computation systems. Obviously many systems of practical interest have both physical and operational aspects. These two aspects of systems are responsible for two divergent branches of system theory. The *analytical system theory* is one offshoot of classical analysis. Its main tools are differential equations and calculus of variations. The optimization problem falls in this class. The second branch, the *combinational system theory*, deals with logical networks, reliability problems, and the like. The emphasis here is on synthesis—putting together complicated systems out of relatively simple components.

Thus far, no attention has been given to disturbances and measurement errors. These are generally referred to by the word *noise.* If noise effects are ignored, the problem falls in the category of *deterministic systems.* As noise is a random or stochastic phenomenon, systems subjected to noise fall in the category of *stochastic systems.* Since most of the real systems are constantly subjected to external disturbances and observation errors, the theory of stochastic systems plays an important role in the study of complex systems.

The Modeling Problem

One can also look at a system as a *direct* or *inverse* problem. A block diagram representation of a system, as those shown in Fig. 2.7 would be a useful vehicle in presenting the ideas related to direct and inverse problems.

Direct problems are characterized by a complete specification of the contents of the boxes in Fig. 2.7, and one is required to study or predict the response of the box to any specified input or disturbance. This problem is also called the analysis problem. From a mathematical viewpoint, an analysis problem, in general, constitutes the solution of an algebraic or transcendental equation, a differential equation, integral, integrodifferential, or some such equation. In system analysis, the system characterization and the excitation are specified and the system response is to be found. The precise nature of the specification, of course, depends on the particular kind of the system. For example, with a given set of equations describing the dynamics of an aircraft, we may be required to find vehicle response to a given wind gust.

The *inverse problem* is much more complex. Here the response to a particular input or set of inputs is known, but either the equations describing the process are unknown or incompletely defined, or the inputs themselves are

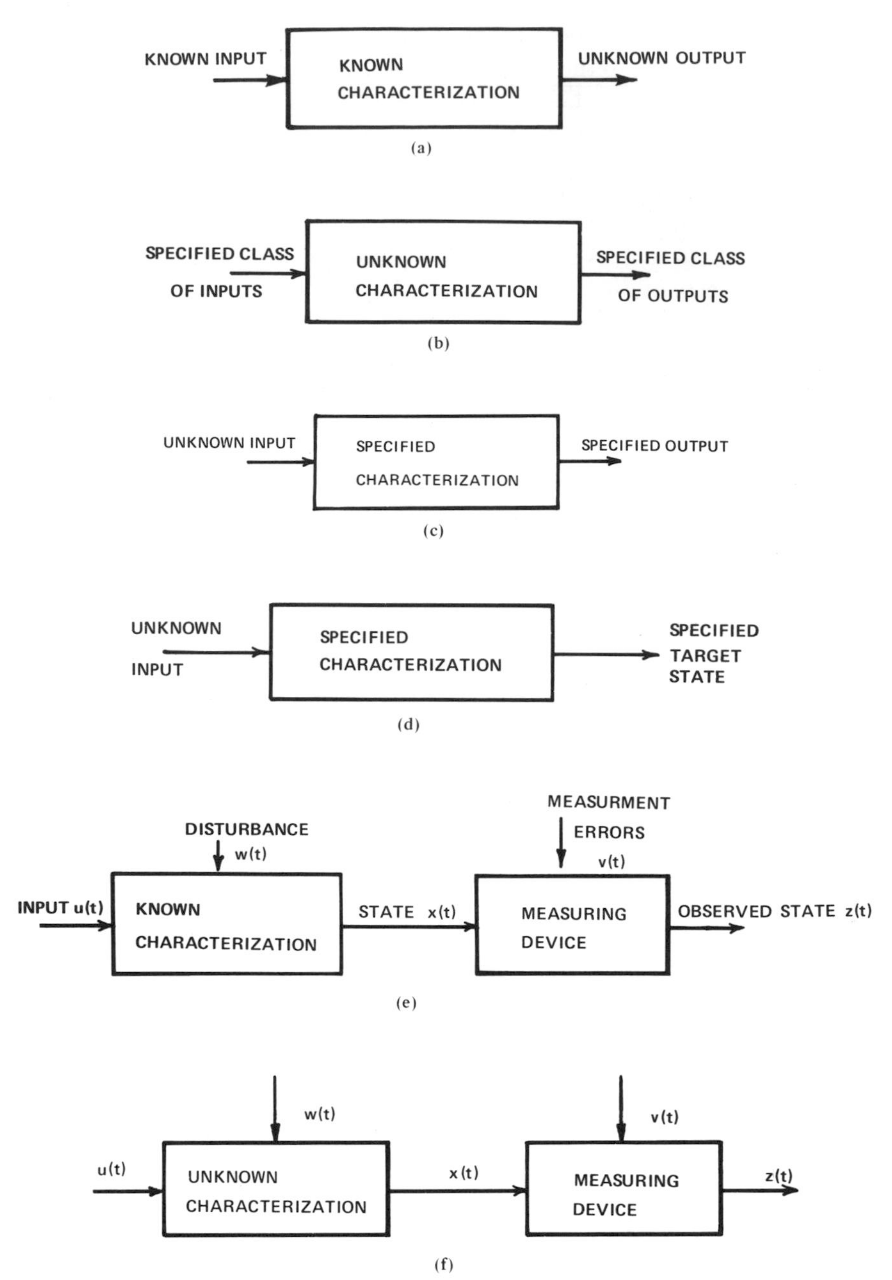

Fig. 2.7 Types of direct and inverse problems. (a) An analysis problem. (b) A design or synthesis problem. (c) An instrumentation problem. (d) A control problem. (e) A state estimation problem. (f) A parameter estimation problem.

unknown. Depending on the specific situation, an inverse problem may present itself as a (a) design or synthesis problem, (b) instrumentation problem, (c) control problem, or (d) modeling or identification problem. While a well-defined and properly posed direct problem generally has a unique solution, if it has any solution at all, the same is not true with an inverse problem. An inverse problem often leads to a multitude of mathematically acceptable solutions out of which one physically acceptable solution, if any such solution exists, has to be selected. Criteria for this selection often depend upon the nature of the inverse problem.

In a *design* or synthesis problem, the nature of the expected excitation and the required response are specified and a physically realizable oriented object having this excitation–response relationship has to be built or designed. For example, it is possible to have many alternative designs to a waste collection and treatment system—each satisfying some stipulated performance specification. To choose one out of many possibilities, it is necessary to establish design criteria. Perhaps it is necessary to minimize cost and maximize efficiency and reliability. Perhaps it is necessary to meet some peak load demands. In each case a different design may represent an optimum solution. Sometimes more than one design may satisfy a given criterion or no design may exist that satisfies all the specifications. In eventualities of this kind decision-making procedures enter the picture and a choice has to be made or specifications have to be altered.

In an *instrumentation* problem, a system characterization and an observed response are given and one is required to find the excitation signal which was responsible for the observed response. For example, if the equations describing the dynamics of an aircraft are known and the telemetry records of response to wind excitation are available, one may be interested in finding the magnitude and direction of the wind gust responsible for the observed behavior. Phrased differently, an instrumentation problem becomes synonymous with the more familiar *control* problem. If the system characterization is known and a desired response is specified, the problem of finding the inputs necessary to produce such a response is the classical control problem.

Finally, the *identification* or *modeling* problem is perhaps the most difficult of all inverse problems. Given a set of inputs and corresponding outputs from a system, find a mathematical description (or a model) of the system. This problem is referred to as the *modeling problem.*

The modeling problem itself may be separated into two categories, depending upon the degree of *a priori* knowledge one has about a system. If the nature of the process is totally unknown then the problem is called a "black box" or *system identification* problem. Strictly speaking, a system identification problem is therefore one in which not even the nature of the equations describing the process are known. In many practical cases, a considerable knowledge of the nature of the underlying physical processes may be available, but the specific value of the state or the parameters or both may be unknown. This is a *system estimation* problem.

In a deterministic system, knowledge about the system equations and parameters can be used to determine the state of a system. Therefore, in a deterministic context, the *system estimation* problem is synonymous with the *parameter estimation problem* and *state estimation problem*. The problem is slightly different in a stochastic environment. In a stochastic context the state of a system not only depends upon the parameters but also on the statistical properties of the noise. Therefore, a state estimation problem is one in which we are given a known system with a random input and measurement noise and we are asked to determine a "best" estimate $\hat{x}(t)$ of the state $x(t)$ using a noise corrupted measurement $z(t)$ of the state $x(t)$. In a parameter estimation problem we are given the statistical characteristics of the disturbances and measurement noise and we are asked to determine the best estimate of certain parameters based upon a knowledge of the deterministic input $u(t)$, the measured output $z(t)$.

An identification or modeling problem generally arises while attempting to control a physically existing system. Therefore, unlike the design problem, the criterion for selecting one out of many possible models is different here. The criterion should be physical *plausibility* rather than physical *realizability*. In general, an identification or modeling problem is characterized by a specification of a finite set of input–output observations. While it is possible to check the validity of a particular design, it is quite difficult to verify the validity of an identified model. The reason for this can be attributed to many possible causes. In many modeling problems the input is generally not under the control of the observer. Furthermore, the period of observation is always limited and therefore the validity of the model is correspondingly limited to the range of variation of input signals which appeared within the observation interval. In other words, a model cannot possibly contain more information than is available in the data from which it is derived.

EXERCISES

1. Develop a goal tree for an educational system using the school you attend and the problems faced by it as a frame of reference.

2. Develop a goal tree for the problem of population planning. Use the United States and India and the problems faced by them as examples in developing your thoughts.

3. The problem of purchasing an automobile can be thought of as a multi-attribute decision making problem. Using actual information available in the market, construct a table such as Table 2.2. Discuss the difficulties encountered by you in filling the entries of this table. If you are actually purchasing an automobile, describe the decision-making process you would use.

4. Arrange the four scales, namely, nominal, interval, ratio, and ordinal, in an hierarchical order such that each scale embodies all the properties of all the scales below it in the hierarchy.

5. Classify the following variables in terms of the scales of measurement (i.e., in terms of nominal, ordinal, interval, and ratio) that are most appropriate to measure them: temperature in degrees Kelvin; temperature in degrees Celsius; weight; height; age; counts such as the number of cars, children, and hospitals; specific gravity; human judgment of texture, brightness, sound intensity, etc.; military rank; eye color; place of birth; favorite actor; and level of education.

SUGGESTIONS FOR FURTHER READING

1. For a discussion of some of the general ideas presented in this chapter, from various viewpoints, consult
   A. D. Hall, *A Methodology for Systems Engineering*. Van Nostrand-Reinhold, Princeton, New Jersey, 1962.
   R. L. Ackoff, S. K. Gupta, and J. S. Minas, *Scientific Method*. Wiley, New York, 1962.
   D. N. Streeter, *The Scientific Process and the Computer*. Wiley, New York, 1972.
   The last reference contains a number of interesting case studies.
2. For a basic presentation of theoretical economic principles that underlie benefit–cost analysis, refer to
   E. J. Mishan, *Economics for Social Decisions: Elements of Cost–Benefit Analysis*. Praeger, New York, 1973.
3. A refreshing discussion on the concept of performance index from a nonmathematical viewpoint can be found in
   J. K. Galbraith, *The Affluent Society*. Houghton-Mifflin, Boston, 1958.
   V. Pareto, *Sociological Writings*. Pall Mall, London, 1966.
4. For an interesting presentation on the subject of subjective judgment of performance, see
   A. Parducci, "The relativism of absolute judgments," *Scientific American* **219**, No. 6, 84–90 (1968).
5. For an elegant mathematical discussion on the mathematical theory of measurement, with emphasis on the measurement of nonphysical quantities, see
   D. H. Krantz, R. D. Luce, P. Suppes, and A. Tversky, *Foundations of Measurement*, Vol. I. Academic Press, New York, 1971.
6. For a descriptive presentation of several aspects of attitude measurement, from the viewpoint of a sociologist, see
   N. Lemon, *Attitudes and Their Measurement*. Wiley (Halstead), New York, 1973.
7. For an introductory book on preference measurement, using psychometric techniques, consult
   R. D. Bock and L. V. Jones, *The Measurement and Prediction of Judgment and Choice*, Holden-Day, San Francisco, 1968.
8. An easy to read presentation on scale conversion techniques can be found in
   M. R. Anderberg, *Cluster Analysis for Applications*. Academic Press, New York, 1973.

9. Material presented in this chapter represents one point of view. For other viewpoints and other definitions not covered here, consult

   A. W. Wymore, *Systems Engineering Methodology for Interdisciplinary Teams*. Wiley, New York, 1976.

   T. G. Windeknecht, *General Dynamical Processes: A Mathematical Introduction*. Academic Press, New York, 1971.

   M. D. Mesarovic and Y. Takahara, *General Systems Theory: Mathematical Foundations*. Academic Press, New York, 1975

   M. F. Rubinstein, *Patterns of Problem Solving*, Prentice-Hall, Englewood Cliffs, New Jersey, 1975.

3

The Modeling Process

INTRODUCTION

Science without an empirical base is impossible. The strength of science is largely measured by its capability in making intelligent observations and predictions. A fruitful method of making these predictions is by building theories using Leibnizian, Lockean, Kantian, or Hegelian approaches. A theory usually deals with aspects of reality that are not immediately evident to the senses. In the physical sciences, when a theory is tested and found to describe adequately the aspect of reality with which it deals, it is called a *law*. In the behavioral sciences it is difficult to make positive statements and behavioral laws seek shelter in probabilities. A theory may be expressed verbally. For instance, a basic theorem in human behavior holds that "men will act on the basis of what they perceive." A theory may be expressed very economically using the language of mathematics, as in $E = mc^2$. Another way of expressing theory is by means of a model. One example of this sort is the physical model of an organic molecule like a methane molecule built by using balls and rods. Another example is the *conceptual model* of the planetary structure of an atom.

There are great and viable differences between theories and models. A theory could state that the subject matter has a structure, but it is a well-conceived model that reveals the structure. For instance, a statement such as "a methane molecule has one carbon atom and four hydrogen atoms," though profound in itself, is far less revealing than a ball and rod model that reveals the three-dimensional structure and the precise nature of the chemical bonding. Thus, a model can be constructed as a specific form of a theory.

Further, if the model in question is a physical model, then it allows one to play around with a theory in a rather concrete physical way. Thus, a model is a representation of a system in a convenient fashion so that conjectures made about the performance of the system can be readily tested. It is important to remember that the word "model" is meant to imply a manifestation of the *interpretation* that a scientist gives to observed facts. Facts remain unchanged, but models change. Thus, the term "model" covers a vast variety of configurations. It can be a symbolic model such as a set of mathematical or logical equations, an iconic model such as a map or a scale model, an analog model such as an electrical or mechanical device, or a logical model such as a flow chart or a computer program.

Recent years have seen intensive attempts to extend the art of modeling in general and mathematical modeling in particular to an ever-expanding range of application areas. This accent on modeling has been stimulated by the increased availability of interactive digital computers and special-purpose simulation languages. A particularly significant impetus toward the development of mathematical models, however, has been the tendency of virtually all physical, life, and social science disciplines to become more quantitative in their methodology. It has been pointed out that all decision making involves an implicit (if not explicit) use of models, since the decision maker invariably has a causal relationship in mind when he makes a decision. Mathematical modeling can therefore be regarded as a formalization of decision-making processes. As a result, models and simulations are now used in many fields that were formerly considered too "unscientific" or vague to lend themselves to such an approach.

The evolution of the modeling art has not been without difficulties and controversies. For example, specialists in the modeling of such "hard" systems as electromechanical control systems have challenged the validity of models in such "soft" areas as economics and sociology; similarly, the relevance and utility of well-established mathematical techniques useful in modeling "hard" systems has been questioned by some of those engaged in modeling "soft" systems. In fact, doubts have been raised as to whether the term "model" really means at all the same thing in diverse areas of application. It is the purpose of this discussion to clarify these problems by providing a unified perspective of the modeling process and the variety of applications of mathematical models.

1 TAXONOMY OF MODEL TYPES

Models can be categorized into a variety of classes and subclasses, and this classification can be done in many different ways. For example, a distinction can be made between *descriptive* models and *prescriptive* models.

Descriptive models, as their name implies, are those which attempt to describe an observed regularity without necessarily seeking recourse to an explanation for the observations made. Thus, descriptive models represent the first stage of rationalization, generalization, and theory building. Descriptive models are generally expressed in a native natural language. The major disadvantage of descriptive models is that the method of prediction is usually internal and thus cannot be communicated easily. The major advantage of descriptive models is that the cost of making predictions is extremely low. Prescriptive models, on the other hand, are normative. Normative science does not stop at describing and generalizing observations. Since the term "normative" implies the establishment of standards of correctness, a normative model is more suitable for predictive purposes. Other types of classification are also in vogue. A sample of a different type of classification is shown in Table 3.1. Examples of physical models referred to in this table can range in scope and complexity from simple floor plans to complicated aircraft models for wind tunnel studies. The procedural models referred to in the last row of this table are nothing but simulators, as they are described subsequently in this chapter.

Mathematical Models

A mathematical model is a set of equations that characterizes a real-life system, the prototype system, in the sense that some of the excitation–response relationships of the prototype system are correctly represented. A subset of all of the prototype system inputs is expressed mathematically and serves as the excitation of the mathematical model; the solutions of the model equations then constitute mathematical representations of the corresponding subset of system responses. In mathematical modeling we are given the excitation and the response, either as mathematical functions or as a set of experimental observations, and we are asked to characterize the system by mathematical expressions. Mathematical modeling is therefore a synthesis problem, and it is necessary to consider what constraints should be specified in order to select the most suitable model from all possible models characterizing the specified excitation–response relationship. And herein lies the major difficulty in devising a modeling methodology applicable to the wide variety of application areas.

In general, the optimum mathematical model is considered to be that one that bears the closest correspondence to the prototype system, that is, the model which can be considered to be the most reliable tool for predicting responses to excitations other than those utilized in constructing the model. At the same time, the realization that modeling is an inverse problem makes it clear that there can never be a unique solution. That is, no matter how detailed and reliable the excitation–response observations, there can be no

TABLE 3.1

Taxonomy of Model Types[a]

Model (form of expression)	Method of prediction	Method of optimizing	Cost	Ease of communication: Technical	Ease of communication: Nontechnical	Limitations
Descriptive (native language)	Judgment	?	Low	Poor	Poor (appears good but often misunderstood)	Cannot repeat the prediction process
Physical	Physical manipulation	Search	High	Good	Good	Cannot represent information processes
Symbolic	Mathematical Numerical approximation	Mathematical Mathematical	Low Medium	Good	Poor	Needs previously developed mathematical structure
Procedural	Simulation	Search	High	Fair	Good	General properties not easily deduced from the model

[a] Reproduced with permission from *Design and Use of Computer Simulated Models*, by J. R. Emshoff and R. L. Sisson, Macmillan, New York, 1970.

assurance that the mathematical model constructed with the aid of these observations actually will be a correct characterization of the system being modeled. The validity of a model constructed using excitation–response observations is always open to question.

Figure 3.1 shows the spectrum of mathematical modeling problems as they arise in a variety of physical, life, and social science disciplines. Near one end of the spectrum, the "white box" end, we find the mathematical models arising, for instance, in electric circuit theory. Here one usually knows the structure of the circuit and most, if not all, of the element values. Using Kirchhoff's laws and similar network theorems one can construct the mathematical model virtually without recourse to experimental data. Occasionally one or more parameter values remain to be identified, but this is a relatively simple and straightforward problem.

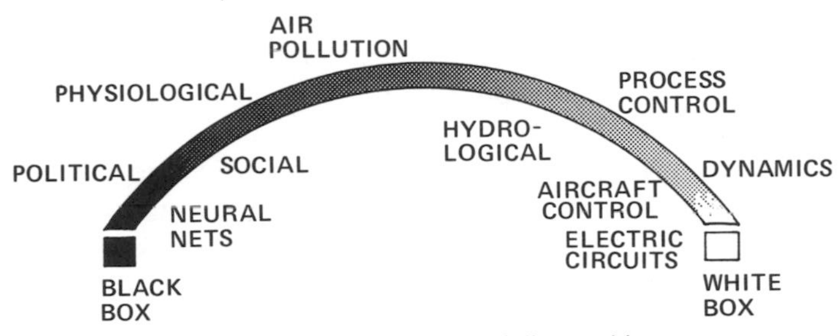

Fig. 3.1 The spectrum of modeling problems.

Proceeding along the spectrum, we encounter the problems in mechanics such as aerospace vehicle control. Here most of the model is well known from basic mechanical principles and knowledge of the dimension and characteristics of the system. However, some parameters, for example, certain aerodynamic functions, must be identified from actual flight experiments. This identification, using excitation–response relationships is complicated by the presence of noise and by lack of control over certain experimental conditions. Proceeding farther away from the white end of the spectrum, we encounter the mathematical modeling problem in chemical process control. Here, basic chemical reaction equations and reaction rates are provided. However, a considerable number of variables and parameters are not capable of being measured or controlled.

Moving farther into the dark area of the spectrum, we encounter the models of so-called environmental systems. Here there is a general understanding of the physical and chemical processes involved (e.g., movement of water in underground reservoirs, diffusion of air pollutants). But the field within which these processes occur is not readily accessible to measurements.

In essence, the phenomena being modeled occur in a medium whose distributed properties are only very imprecisely known. Continuing further into the direction of darkness, a variety of life science models are encountered. Here there is only an approximate understanding of the physical and chemical laws that underlie physiological phenomena, and furthermore the characteristics of the system being modeled are apt to change in time in an unpredictable and uncontrollable manner.

Economic, social, and political system models fall in the very dark region of the spectrum. Here even the basic laws governing dynamic processes, not to mention the relevant constituents of the system, are open to question and controversy. Neural network problems, and specifically the modeling of the human brain using electroencephalogram (EEG) data, are placed near the very black end of the spectrum because here there is considerable controversy even as to the basic cause–effect mechanism by which EEGs are produced.

It is recognized, of course, that there are many types of mathematical models in use in any specific application area, and that there may well be overlaps in the shades of gray applicable to different fields. The primary purpose of Fig. 3.1 is to highlight the existence of a wide range of "gray box" problems, all of which are mathematical modeling problems. Further insight into the distinctions between the modeling methodologies and the models as they appear in different disciplines can be gained by considering the modeling process in more detail.

Deductive versus Inductive Modeling

Formulation of mathematical models is a field of endeavor in which science and mathematics meet. In this activity, there is more science than mathematics; mathematical ideas are involved, but not mathematical methods. It is only through continued study of the various branches of science that the student gradually acquires the knowledge and experience which enable him to express a given set of conditions in the appropriate mathematical form. With increasing experience one attains a better understanding of the ideas and methods, and this in turn leads to a deeper insight into a natural phenomenon.

The construction of a mathematical model entails the utilization of two types of information: experimental data constituting observed excitations and responses, and knowledge and insight about the system being modeled. As indicated above, inverse problems such as the modeling problem do not have unique solutions. A valid model therefore cannot be constructed purely inductively from excitation–response observations, since an infinite number of models satisfy these relations. For this reason, the validity of a model varies directly with the amount of additional insights (constraints) that can be brought to bear upon the modeling process. The more plentiful these insights, the lighter is the shade of gray of the "gray box." The more deductive and the less inductive the modeling process, the more valid is the mathematical model.

Deduction is reasoning from known principles to deduce an unknown; it is reasoning from the general to the specific. In modeling deductively, one derives the mathematical model analytically and uses experimental observations only to fill in certain gaps. This analytical process makes use of a series of progressively more specific concepts that may be broadly categorized as laws, structure, and parameters.

Laws

It is almost impossible to present here even a brief outline of the various physical laws occurring in all branches of knowledge. For instance, Newton's laws of motion are fundamental to mechanics, Avogadro's hypothesis plays a basic role in physical chemistry and Ohm's law in electrical circuit theory (the second law is basic to thermodynamics), and so on. No modeling activity can ignore or bypass such fundamental laws of nature. However, there is still room for generalization. As a system has been visualized as a set of interconnected elements, certain *conservation* and *continuity laws* governing the interconnections can be recognized.

The conservation principle states that if some physical quantity is applied to a system, the total amount of this quantity existing within the system at any time thereafter must be equal to the net amount thus added (or subtracted) plus the amount initially present within the system. Some insight must be used to specify which physically quantity is conserved. In a thermal system, heat energy is conserved. In a system, such as an electrical circuit, electrical energy, however, is not conserved (since some of it is converted into heat); rather electrical charge is conserved. The *continuity principle* states that the through variables must emanate from a *source* and return to a *sink*.

Considering the similarity of the basic laws in various disciplines and the fact that all the major physical areas can be classified under the same general headings, it is not surprising that the mathematical expressions characterizing the behavior of these areas are very similar in form. Once techniques are developed for thoroughly handling the various problems that arise in one area, these same techniques can be applied to virtually all other areas.

The application of a law to a system usually involves the focusing of attention upon a single physical area. For example, when utilizing Kirchhoff's laws in analyzing an electrical circuit, one generally ignores chemical and thermal processes, which may be going on simultaneously within the system. However, modeling of a more realistic situation is more complex. It is common experience that the mechanical working of a solid body generates heat and, conversely, heating a body produces mechanical work. It is also common experience that the properties of many materials may vary with time, temperature, and the rate at which they are deformed. Plastics, polymers, biological tissue, and many metals may continue to deform under constant applied load, and the rate of continued deformation may be highly sensitive to temperature variations. Such thermomechanical phenomena are typically nonlinear in nature. The greater the applicability of well-established laws in such a context, the smaller is the error incurred in the mathematical model, and the lighter the shade of the "gray box."

Structure

The system being modeled is usually regarded as consisting of a large number of interconnected components or elements. This view of a system frequently involves the making of simplifying assumptions or approximations. However, because this approach lends itself so well to eventual computer implementations, even systems that are continuous in space (for example, a water reservoir or the atmosphere) are often viewed as being composed of an array of closely spaced elements. In physical systems these elements can be broadly classified as being either dissipators of energy, reservoirs of potential, or reservoirs of flux. In nonphysical systems there is a wider assortment of possible elements. In any event, the construction of a valid mathematical model demands a knowledge of the types of elements present in the system and how these elements are interconnected. The interconnections specify the paths over which matter or energy flows within the system; the types of elements determine what happens to this matter or energy at different locations within the system. Mathematically, this in turn determines the number of simultaneous equations in the mathematical model, as well as the types of terms in each equation (first derivatives, second derivatives, etc.). Whereas the deductive knowledge of the laws governing a system are obtained from the study of a scientific discipline, knowledge about the structure of a system can only come from insight into the specific system being modeled. The more that is known about the structure, the lighter is the "gray box."

Parameters

The parameters in a mathematical model are the numerical values assigned to the various coefficients appearing in the equations. These are related to the specific magnitudes of all of the elements comprising the system as well as to the boundary and initial conditions, which together with the governing equation constitute a completely specified model. Some of these must be determined inductively even in modeling problems near the white end of the spectrum, although in that case, one often knows that the parameters fall within a certain narrow range, that they are linear, that they are constant in time, etc. As one proceeds to the darker regions of the spectrum, the number of unknown parameters increases progressively.

The shade of gray of the model is a measure not only of the relative amount of deductive and inductive information that is brought to bear on the modeling process; it is also often related to the quality and usefulness of the excitation–response observations. Systems observations may be obtained either actively or passively. In active experiments, the modeler specifies interesting excitations, applies these to the system under study, and observes the response. In contrast, in passive observations one is unable to specify excitations but must accept whatever input–output data is furnished. In some modeling applications, for example, an electrical or mechanical system, active experimentation is widely used and leads to relatively valid models. In other areas, for example in underground water pollution or in economic systems, the actual system is too large or too remote to permit the application of specified excitations. Active experimentation is a particularly powerful tool near the light end of the spectrum because key experiments can be constructed to answer certain specific questions about the model.

Analog Models

An analogy between two things is said to exist whenever there is a resemblance not necessarily between the two things themselves but of two or more attributes, circumstances, or effects. For example, concepts of continuity and conservation of energy are relevant to several fields of study and leads one to suspect the existence of some kind of analogy. Schroedinger's work on quantum mechanics, for example, rested on formal similarity between the theory of light waves and the theory of particle orbits that eventually led to his theory of particle waves. The similarity between the force of attraction between two physical bodies and the force of attraction between a metropolis and people seeking opportunities led to the gravity type models to transportation systems. Specifically, analogies allow the possibility of drawing conclusions about a less understood phenomenon in terms of a well-known phenomenon. It should be remembered that analogies are useful in merely imitating a given system—not in duplicating it. This imitation is necessarily restricted to a subset of the structural features, attributes and behavioral traits of the system.

Specifically there are two types of analogies: direct and indirect. Simply stated, a direct analog is something like a scale model in which the structural, geometrical, and visual characteristics of a system are preserved as much as possible. An indirect analog, on the other hand, is a model in which there is only a mathematical similarity between the system and the model.

2 STEPS IN MODEL BUILDING

Figure 3.2 illustrates the various steps involved in a model-building process. The system under study is first observed and is therefore characterized initially by a set of observed data. In order to make predictions about the behavior of those regions of the system that are unmeasured or not accessible for measurement, a model that is consistent with the observed facts is constructed. Thus, a degree of generalization is involved in this step. This step may be called the *semantic* link (i.e., the study of the relations of signs to the objects represented). To understand the consequences of the generalization,

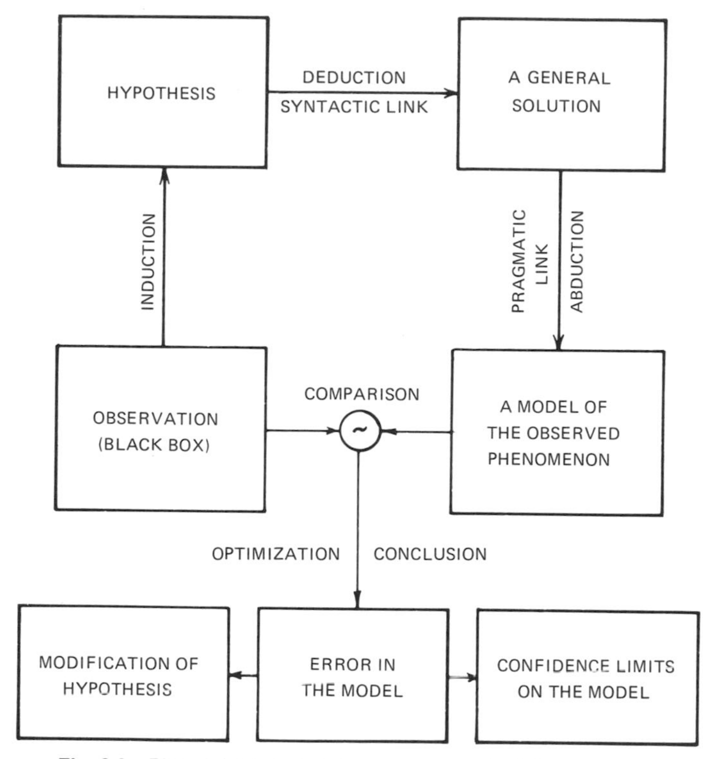

Fig. 3.2 Pictorial representation of a model-building process.

the model is manipulated to get a set of consequences. This step may be called the syntactic link (i.e., the study of relations between signs and other signs). This step usually widens the scope of the problem (due to the generalization), and it is therefore necessary to focus or specialize by introducing additional pertinent conditions called *constraints* or *boundary conditions.* This step may be called the *pragmatic* link (i.e., the study of the relations between signs and the use of signs). This step enables one to draw inferences from the model. The last, but not the least, vital step in modeling is to investigate the validity of models and the inferences drawn from them.

It is important to distinguish between an observable phenomenon and a model of the phenomenon. One, generally, has no influence on what is being observed. However, one can use informed judgment in selecting a model. In building a model, of necessity, one must strive to simplify models either by ignoring details or by focusing on the relevant and important. The success of a model, therefore, depends on whether the details ignored are really unimportant in the development of the phenomenon being studied. It is quite difficult to state with certainty whether or not a given model is adequate before some observational data are obtained. To check the adequacy or validity of a model, one has to compare the consequences predicted by a model with actual observations.

As a first step in deciding the appropriateness of a model to describe a given phenomenon, let us consider what might be suitably called a *deterministic model.* By this is meant a model that stipulates that the conditions under which an experiment is performed determine the outcome of the experiment. If a resistor of resistance R is connected to a battery of voltage E, a mathematical model that describes the flow of current I would be (according to Ohm's law)

$$I = E/R \tag{1}$$

The model defined in Eq. (1) predicts the value of I if E and R are given. If the above experiment were repeated a large number of times, each time using circuit elements of the same quality, one would expect to observe the same value of I. In other words, if the experimental conditions are unaltered, using the same set of instruments under essentially the same environmental conditions, the same reading would be obtained. There are many examples of natural and man-made phenomena for which deterministic models are appropriate. Gravitational laws describing, say, falling bodies, and Kepler's laws describing the movement of heavenly bodies are but two examples of deterministic models.

There are many phenomena for which a deterministic description is simply not adequate. *Nondeterministic* models are generally required for an adequate description of such phenomena. These models generally go by the names *probabilistic models* or *stochastic models*. Suppose that one wishes to determine the amount of precipitation a certain watershed receives as a result of a storm front that passes through a neighboring region. Meteorological data such as barometric pressure, pressure gradients, wind velocities, and stability of air currents do shed some light on the approaching storm. This information, though it may be valuable for making gross predictions (such as "showers" or "heavy rain"), simply does not make it possible to state with any accuracy how much rain will fall. This is an example of a phenomenon that does not lend itself easily to deterministic description.

Nature of Solutions Sought

At this point it is instructive to make a distinction between scientific and engineering approaches to model building. A mathematician, when confronted with a problem, worries about the existence and uniqueness of solutions, the validity of certain commutative multiplication, or some such delicate question without being concerned with the physical meaning of his operations. Mathematics is an edifice built on a set of axioms, and as long as one conducts oneself without violating the axioms, lemmas, and theorems a mathematician would be satisfied. The interests of a scientist, of course, lie in the physical world; however, his attitude is more or less philosophical. The goal of a scientist is to construct theories to explain the observed behavior of the physical world around him and for this purpose he has to take into account all possible alternatives to explain certain phenomena. The goal of engineering analysis is to obtain specific answers to specific questions with a specified accuracy at a minimum cost in time, labor, and equipment. Moreover, while an engineer would certainly like to obtain general solutions, he is generally willing to settle for numerical solutions.

Abstractions in Time and Space

All systems exist in the time and space continuum. Mathematically this means that time and space are continuous independent variables. In physical field problems, as, for example, heat transfer or meteorology or water pollution, this continuous nature is fully retained in the mathematical model. As a result, the model consists of a set of partial differential equations. In many other modeling situations, however, it is expedient to simplify the model by focusing attention on dynamic processes only at discretely spaced points in space and/or in time. In fact, it is possible to classify mathematical models as:

(1) Continuous space–continuous time (CSCT), where all independent variables are maintained in continuous form.

TABLE 3.2 Major Types of Space–Time Abstractions

Type of abstraction / Problem attributes	Continuous space–continuous time (CSCT)	Discrete space–continuous time (DSCT)	Discrete space–discrete time (DSDT)
Typical areas of application	Heat transfer Air pollution Meteorology Hydrology	Electric circuits Mechanical systems Control systems	Traffic flow systems Management systems Economic systems Social systems
Pictorial representation of system	Drawing of field boundaries	Circuit diagram	Block diagram Flow chart
General type of model	Partial differential equations	Ordinary differential equations	Algebraic equations Heuristic description
Examples of typical model equations	$\frac{\partial^2 \phi}{\partial x^2} + \frac{\partial^2 \phi}{\partial y^2} + \frac{\partial^2 \phi}{\partial z^2} = k \frac{\partial \phi}{\partial t}$	$a \frac{d^2 y}{dt^2} + b \frac{dy}{dt} + cy = f(t)$ $y' = f(y, t)$	—
Examples of names of characterizing equations	Navier–Stokes equation Maxwell's equations Diffusion equation Laplace's equation	Kirchhoff's laws Newton's laws	Queuing systems Fuzzy sets
System constituents	Continuous medium	Interconnected circuit elements	Interconnected entities
Dependent variables	Potential flux	Across variables Through variables	Activities
System parameters	Local field characteristics	Element values	Attributes
Digital simulation languages	PDEL LEANS	CSMP CSSL MIMIC	GPSS SIMSCRIPT

(2) Discrete space–continuous time (DSCT), where discrete points in the space domain are used but the time variable appears in continuous form.
(3) Discrete space–discrete time (DSDT), where both the space and the time variables are discretized.

The CSCT method is used primarily in the modeling of field problems such as those arising in environmental problems and in various engineering disciplines; the DSCT method is used very widely in circuit problems as well as in control system design; the DSDT approach has found considerable application in recent years in the modeling of organizations, manufacturing plants, transportation systems, urban systems, and other social science areas. The terminology that is used to describe models differs for each of these three abstractions. Similarly, the nature of the mathematical model differs for each approach. This in turn has led to the development of distinct classes of problem-oriented simulation languages to be used in implementing these models on the computer. These differences in the three major types of abstraction are summarized in Table 3.2.

With respect to the spectrum of mathematical models, the DSCT models are most often used near the "white box" end of the spectrum. They are used to simulate circuits and control systems that are modeled with only a very small reliance upon inductive observations. The DSDT models are used primarily near the dark end of the spectrum, and have been used particularly extensively in social science areas. Finally, the CSCT models usually fall near the center of the spectrum, where they are used to characterize mass transfer in distributed physical systems, such as those occurring in the wide variety of environmental problems. This is shown in Fig. 3.3

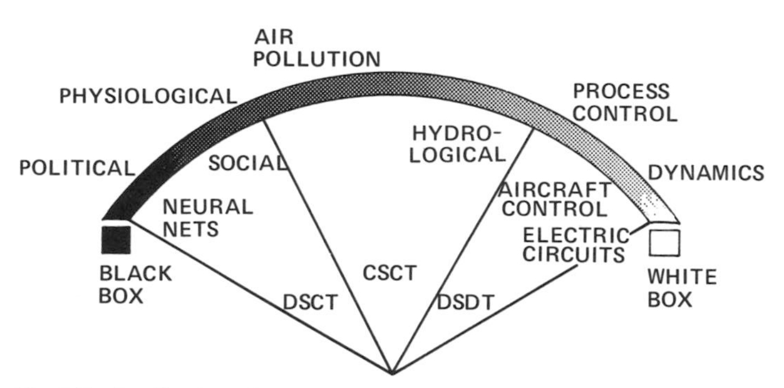

Fig. 3.3 Application of time–space discretization to modeling problems.

Over the years some confusion as to the meaning of the term "simulation" has arisen. In most application areas using DSCT and CSCT models, simulation means experimenting with mathematical models. A model is formulated and implemented on a computer. A variety of excitations is then applied to the model, often interactively, and the resulting responses recorded. Often various parameters and sometimes the structure of the model are varied so as to determine the effect of these variations upon the excitation–response relationship. In contrast, in application areas using the DSDT abstraction, simulation is regarded as an alternative to mathematical modeling. Here a simulation model is constructed by various intuitive or heuristic techniques without attempting at any time to express the system in mathematical form.

Motivations for Modeling

The reasons for constructing a model and the ultimate use of the model differ markedly for different shades of gray. As one proceeds from the light end of the spectrum to the dark end, there is a gradual but steady shift from the quantitative to the qualitative, as shown in Fig. 3.4.

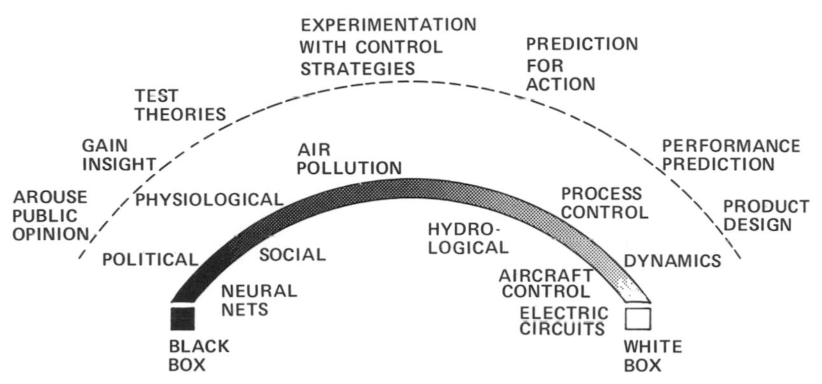

Fig. 3.4 Objectives of modeling.

Near the "white box" end of the spectrum, models are an important tool for design. For example, in electrical circuit design, models permit experimentation with various combinations of circuit elements to obtain an optimum filter characteristic. Here the validity of the model is such that the errors inherent in modeling can be made small compared with the component tolerances normally associated with electrical circuit elements. Similarly, in the area of dynamics, models can be employed to predict to virtually any desired degree of accuracy the response of the system to various excitations. Such quantitatively oriented models can be used with great assurance for the prediction of system behavior.

Closer to the "black box" side of the spectrum, models play an entirely different role. Frequently they are used to provide a general insight into system behavior—behavior that is often "counterintuitive." Thus, systems containing many complex feedback loops may actually respond to an excitation or control signal in a manner that is diametrically opposite that which was expected. Occasionally, the primary objective of the model is to arouse public opinion and promote political action by suggesting that current trends lead to disaster in the not too distant future.

Ranged between these two extremes in the motivation for mathematical modeling lies a plethora of part-qualitative and part-quantitative positions. It is very important to recognize, in evaluating and in using mathematical models, that each shade of gray in the spectrum carries with it a built-in "validity factor." The ultimate use of the model must conform to the expected validity of the model. Likewise, the analytical tools used in modeling and in simulation should be of sufficient elegance to do justice to the validity of the model; excess elegance will usually lead to excessively expensive and meaningless computations.

Computers as Modeling Tools

A mathematical model of a system is only useful in so far as such a model allows solutions. Modern computers and computational methods are concerned with strategies to obtain these solutions in an efficient manner.

It is intuitively clear that computers are useful in the development and testing of models, but it is not always clear about the manner in which they should be used. The majority of individual operations performed on a computer are in no sense different from arithmetic computations performed by hand on a desk calculator. In a computer, one finds a qualitatively new computational medium simply because its speed, accuracy, and reliability are orders of magnitude better than the precomputer state of the art. Thus, the impact of modern digital computers on the field of numerical methods has consisted of far more than mere application of existing numerical techniques to problems of ever increasing complexity. It is useful, therefore, to question the role that this new computational medium can play in the development of models.

Mathematical theory, applied to physical problems, has been extremely successful when the theory (or model) is *linear*, when *symmetry* can be invoked, and when only a few *variables* need be employed. These restrictions are frequently severe and limit the development of modeling considerably. On the other hand, in a computational phase, mathematical properties such as linearity, symmetry, and a small number of variables are not critical in obtaining solutions. What essentially a digital computer demands is that a problem be *finite* and *discrete*. This forces models simulated on computers to be discrete and finite dimensional.

As an illustration of this dichotomy, consider the problem of building a model of a galaxy. This problem can be approached in two basic ways: by the methods of *particles* and *fluids*. In the first method, one visualizes a galaxy as a finite set of interacting stars or "particles," in which each star moves in a gravitational field arising from all other stars. However, the number of stars in a typical galaxy is of the order of 10^{10}. Even if one represents the motion of each star by only one variable, the number of variables involved is generally beyond the processing capabilities of modern computers. The other alternative is to visualize a galaxy as composed of so large a number of stars that, effectively, it may be described as a continuous fluid. Now, using methods of statistical mechanics, one can derive equations describing the evolution, density, and distribution of the galactic fluid. However, to satisfy the finite requirements of a computer, the fluid is divided into a set of elementary cells.

3 SIMULATION

If an entity being studied is conceived of as a system, then the *structure* of a system is the physical arrangement of its subsystems and components at any given time. All the changes that occur in these subsystems, components, and the relationships among them are called a *process*. Just as the structural relationships of a system are revealed by a model, the process characteristics and functional relationships can also be revealed by using a dynamic or operating model. Such a model is called a *simulator*, and the art of playing around with a simulator is termed *simulation*. Thus, computers are tools, models are projections of facts on a conceptual axis, and simulators are physical realizations of these conceptual projections. Stated differently, a simulation is a special kind of model, a model is a special way of expressing a theory, and finally a theory is an abstraction of some aspect of reality.

Why is it useful to simulate systems? It is frequently economical to study a given phenomenon in a model or a simulator rather than in its natural setting. Though simulation is frequently an expensive proposition, it is usually less so than the alternative of attempting to gain the desired information from the actual system. For example, to build a scale model of an airplane, test it in a wind tunnel, and make observations on the flow of air over the foils is expensive, but far less so than to build an actual plane for testing purposes. Costly mistakes can be avoided by simulation.

Using a simulation increases the *visibility* of the phenomena under study. Observing the flow past an airfoil in a wind tunnel is far easier than observing the flow of air in an actual situation. In some instances, a simulation provides the only way to observe a phenomenon. A case in point is simulation of nuclear warfare. Another way by which a simulation could increase the visibility is to bring to the foreground a particular effect. For instance, it is difficult to study the circulation of blood in a living organism, for the system is so complex that any interference with the flow of blood may affect the existence of the system itself. In a simulation, it is possible to block the flow of blood or inject some drug into it and study the consequences.

Simulation allows *reproducibility*. There are two reasons for wanting to reproduce events. The first is to cope with the element of *chance*. Second, simulation gives the ability to reproduce a situation that might never occur again in the forseeable future. Thus, one can design a dam by subjecting a model to a variety of simulated floods and study the consequences under the worst flooding and drought conditions.

Finally, simulations are safe. Not only do simulations allow one to avoid putting humans in dangerous situations, but they also allow a study of dangerous situations themselves without actually creating them.

While studying *behavioral systems*, the need for simulation becomes even more acute. In those areas where it is very difficult to get objective data, most of the predictive capability has to be derived from vague, tentative, and intuitive conjectures. In these circumstances, simulation means squeezing consequences out of a crude structure put together from suppositions. Precisely because the structure is crude and analysis is laborious, computers become indispensable tools. It is important to remember that a simulator can neither generate data nor uncover physical laws. A simulator merely explores the consequences of a structure. Simulation is only a beginning of theorizing, not an end.

Despite the extreme complexity of human behavior and societal structure, one can make inroads by identifying patterns, regularities, and laws even on a microscale. A *piecemeal simulation* is one in which one attempts to simulate a small segment of human or societal behavior and study it intensively. The choice of segments, of course, is based on the existence of fairly well established microtheories in a segment. For instance, a good deal of work was already done on the nature of human behavior, either as an individual or as a member of a committee. Suppose that one wants to understand the conditions under which a persuasive communication is or is not effective in modifying beliefs. One can simulate the situation, systematically vary parameters such as personality characteristics, group structure, and communications flow and study the impact of various configurations.

A second approach to studying complex social systems is by *skeletal simulation*. Here, instead of trying to focus on available microtheories, one attempts to look at the macropicture as a whole. A first step of this procedure is to select those visible linkages or relations about which the available information is greatest. Using that one builds a skeleton of the relations. As more and more data become available, the skeleton can be improved by adding flesh and muscle to it.

Both the piecemeal and skeletal approaches to simulation have their attendant dangers and pitfalls. These dangers are primarily due to human frailty rather than any inherent weakness in the scientific process. Consider the problem of simulating *international relations.* As is usually the case, some detailed information about the comparative military capabilities of two principal antogonists is available. However, nothing is known about the personalities of the chiefs of state, and so the temptation to ignore the variables associated with the personality traits of the decision makers is overwhelming—simply because that information is not available. This could be quickly revealed by a well-designed simulation, as the outcome of any simulation that ignores a very important variable is apt to be absurd.

It is true that building models to simulate human behavior is more complex than simulating systems conceived by humans. Central to any discussion of human behavior however are such concepts as goal, reinforcement, adaptation, strategy, constraint, decision, and environment. These very concepts are axiomatic for the modern edifice of abstract system theory. Therefore, attempts were made along parallel lines to describe human behavior using numerical or algebraic variables as quantifiers of behavior. This approach met little success presumably because human behavior, in general, is much too complex to admit description by numerical variables. A possible way of dealing with the problem is to employ fuzzy models described by fuzzy variables in place of numerical and well-defined variables.

Validation of Computer Simulated Models

Experimental verification of physical models is done by subjecting them to the characteristic signals that the physical system is likely to encounter and examining the response signals. Constructing scale models with precision and observing the results with precision is possible in the case of systems occurring in the physical sciences. In the case of behavioral systems it is not easy to construct true models. If, for instance, the management of a ground water basin is influenced by factors such as socioeconomic or political considerations, it is almost impossible to simulate the total environmental system by using scale models. Mathematical models with a greater degree of abstraction are used invariably to study the behavior of such systems. It is obvious therefore that the "law of similitude" in this respect would mean not the proportionality of parts but similarity of logical processes in the behavior of the system and its model.

In problems of inference from a model to the real system, the question arises whether the model under examination is a valid one, or to what extent it can be considered a valid model. In the modeling of complex systems, most of the efforts thus far seem to have been directed in building complex models on computers, and the problem of validation does not seem to have received much attention. Basically, the validation of a simulated model does not pose a problem different in principle from the validation of any other scientific hypothesis, but the complexity that is typically built into such models is so great that the process of validation is different.

Important features of a scientific method may be briefly summarized as follows:

(1) Careful and accurate classification of facts and observations of their sequence and correlation.
(2) Discovery of scientific laws by the aid of creative imagination.
(3) Equal validity for all normally constituted minds.
(4) Self-criticism.

In order to test the validity of a theory or a model one may use these criteria by putting the following questions and trying to elicit positive answers. Does the theory permit careful and accurate classification of facts and observance of their sequence and correlation? Does the theory provide scope for discovery of scientific laws by creative imagination? Is the theory equally valid for all normally constituted minds? (That is, are we sure that we will not encounter a "Maxwell's demon" as we extend the theory and try to generalize it?) Is the theory or the model capable of withstanding criticism? If one gets positive answers to these questions, it means that the theory has withstood the fourfold test and is scientifically valid. Insofar as computer simulation is concerned, one may say that it satisfies this fourfold test if:

(1) Preparing the model for simulation means writing a suitable program, which in turn involves processing the data and classification and correlation of the data.
(2) Interpreting the computer results and predicting the behavior of the real system involves logical inference and creative imagination.
(3) The logical mechanism of the computer ensures uniform results to anyone who feeds the same data, thus obviating the differences between normal and abnormal minds.

The final and the most important touchstone of validity of a scientific hypothesis is its ability to sustain criticism. A scientific hypothesis must be capable of being disproved. Theories or models should be subjected to tests capable of showing them to be false. As Karl Popper has said [4, p. 86]: so long as a theory withstands detailed and severe tests and is not superseded by another theory . . . we may say that it has proved its mettle or that it is corroborated.

Having come to the conclusion that a scientific hypothesis or model should be subjected to the most severe tests aimed at disproving it, one should discover the relevant tests that are most appropriate to the problem at hand. A physical model is sought to be validated by ensuring similitude in construction and verifying whether the changes in state produced in it are truly representative of the theoretically expected values of the real system. Validation of behavioral models is not that easy, however, because of the unpredictable behavior of the variables involved. The need to include behavioral components in models occurs because of the inherent interaction of scientific and technical variables with such factors as social, political, legal, and aesthetic considerations. When such factors enter into the equations of optimization, then the validity of the overall models can be judged from two points of view, namely, determining the truth value of the results obtained and from a utilitarian viewpoint.

Determination of the truth value involves verification at three stages, to be sure, namely, verification of the validity of the concept, validity of inference, and verification of empirical concordance. Several philosophical theories have been in existence that discuss this problem of conceptual validity. Important among them are *rationalism*, *empiricism*, and *positivism*. Rationalism holds that a model or theory is simply a system of logical deductions from a series of synthetic premises of unquestionable truth values, not themselves open to verification. The phrase *synthetic a priori* has been coined by Immanuel Kant to describe the premises of this type. Thus, for the rationalists, the problem of verification of a theory reduces to the problem of searching for a set of basic assumptions underlying the behavior of the system under study. At the other end of the spectrum is empiricism. Empiricism refuses to admit any postulates or assumptions that cannot be independently verified. Empiricists ask that we begin with facts and not with assumptions.

Validity of Inference

Inference is scientifically valid if the inference can be drawn by every logically trained mind and if the inference is drawn from known things to unknown things. Establishing the validity of inference in simulation problems is of paramount importance.

Experimental Verification

The final stage of verification is the experimental verification. In social sciences, one cannot always conduct experiments under controlled conditions and at a particular instant of time. Industrial processes, ecological systems, or economic activities are always spread out in time, and one has to wait long periods of time to observe the facts and verify the predicted results.

Experimental verification at this stage means testing how far the results predicted by simulation agree with actually observed values. Here one can use the nonparametric methods of statistical inference to test the correspondence between fact and theory and establish confidence intervals on the performance of computer simulated models. For instance, one can use the simple *chi-square test* and decide whether there is any correspondence between observations and predictions at a particular probability level.

Perhaps the best method of verification may be subjecting the model to multistage verification: first, searching for Kant's *synthetic a priori*, second, attempting to verify the postulates subject to the limitations of statistical tests, and finally, testing the truth value and ability of the model to forecast the behavior of the system under study.

4 ALGORITHMS AND HEURISTICS

Setting goals, identifying objectives and indicators, selecting variables and their resolution levels, establishing relations—this sequence of operations can usually be done with the mental facilities, that is, without recourse to mechanical aids. The next step is to build prescriptive models. This involves solving equations, simulating systems, making comparisons, evaluating alternatives, and finally making decisions. As a systems approach implies a comprehensive attack on a given problem, the systematic and comprehensive treatment of complex systems usually involves consideration of myriad alternatives. In the precomputer era such a consideration would have been an impossibility. Indeed the availability of a computer permits consideration of more alternatives and more possibilities than ever before. Precisely due to this reason computers and computational processes become important and indispensable parts in the study of complex systems.

In what way can a computer accomplish this job? A computer is a fast-acting, tireless machine that can perform repetitive operations at immense speed. While studying complex systems where informed knowledge is lacking, one can at least use, as a last resort, brute force or trial and error methods in order to gain some understanding about the behavior of systems. Therefore, it appears that computer aided methods are best suited to problems requiring:

(a) Planning where one must consider a large number of interdependent projects.

(b) Explicit considerations of randomness and fuzziness.

(c) Knowledge of different organizational structures and management systems as they relate to a given problem.

(d) Knowledge of how preferences develop among individuals and how they reflect preferences of a society.

(e) Knowledge of regions of a problem where conventional methods succeed or fail.

(f) Knowledge helpful in the specification of criteria for operation and evaluation.

As of now, no analytical techniques based on a sound theoretical foundation are available to tackle all these problems in a coherent and systematic fashion. Experimentation on a computer, or computer simulation as it is often called, is the only recourse available.

Computer simulation is as much of an art as it is a science. At the heart of computer simulation lies computation or calculation. To perform this computation, a strategy is required. Basically there are two approaches to develop this strategy: the heuristic approach and the algorithmic approach. Stated simply, a heuristic is a common sense method, whereas an algorithm is common sense with hindsight. Faced with a novel problem for which no known solution technique is available, the common human instinct is to try a heuristic method. Both algorithmic and heuristic methods are explained in the sequel.

Characteristics of an Algorithm

The notion of an algorithm is basic in computation. An *algorithm* may be defined as a finite set of rules which defines a sequence of operations for solving a specific type of problem. That is, the computer must be furnished with a specific set of rules that specifies a unique course of action for every possible circumstance that could arise in the course of a computation. An algorithm can be characterized by two basic properties: finiteness and definiteness. Finiteness implies that an algorithm must always terminate after a finite number of steps. (A computational method has all the characteristics of an algorithm except it possibly lacks finiteness.) Definiteness implies that each step of the algorithm must be precisely defined and the domain of applicability of the algorithm must be stated.

In constructing an algorithm, the following technical goals should occupy a central position:

reliability
efficiency
domain of applicability
flexibility
simplicity of use

The most obvious and crucial difficulty in reaching these goals is the selection of an appropriate mathematical procedure upon which to base the algorithm. The question of reliability touches closely upon the choice of the mathematical procedure, but there are some aspects that are somewhat independent. There are many instances in which minor variations in the implemented version of an algorithm result in dramatically different error characteristics. Thus, once a basic approach is chosen, it is often necessary to consider the exact organization and sequencing of the computation to achieve acceptable error characteristics.

A small but crucial step in the construction of algorithms is the specification of the interface between the algorithm and the user. There are several facilities available and the primary criterion in selecting facilities is convenience and flexibility for the user. An algorithm can be made significantly faster sometimes by using significantly larger amounts of main storage. In the past, when only one program at a time was in the main memory, this trade-off was frequently easily resolved because there was an excess of memory available. In multiprogramming operating systems, this trade-off is very real even for small programs.

Finally, the domain of applicability of an algorithm should be as wide as practical. However, there is real danger that an algorithm will be developed which is especially effective for a large class of problems no one wants to solve. A systems approach to this problem would be of some interest. One possible way is to classify the problems and develop an algorithm for each class. The following example illustrates some of the ideas presented above.

Example 1. Consider the problem of constructing an algorithm to determine whether or not a given number is a prime. For definiteness, let the given number n be greater than 2. An algorithm is required to produce a 1 if n is a prime number and a 0 if n is not a prime number.

One method of accomplishing this objective is to first divide n by 2, then by 3, 4, and so on until a number is found that divides n exactly, or until n itself is reached as a divisor. This is admittedly an inefficient procedure, but it works. A flow chart of this procedure is shown in Fig. 3.5.

This algorithm has some of the desirable properties of a good algorithm and lacks some of them. The domain of this algorithm is the set (3, 4, 5, 6, . . .) while the range is the set (0, 1). It is important to note that the domain is *not* the set of all positive integers as the procedure is not good for $n = 1$ and $n = 2$. If $n = 1$, the algorithm never stops. If $n = 2$, the algorithm stops but produces a wrong result by assigning the value 0 to k, saying in effect that 2 is not a prime. Similarly if n is assigned a value such as 14.3, the algorithm never stops. Thus, the algorithm is correct but not general. Careful inspection reveals that its efficiency can also be improved by making several modifications.

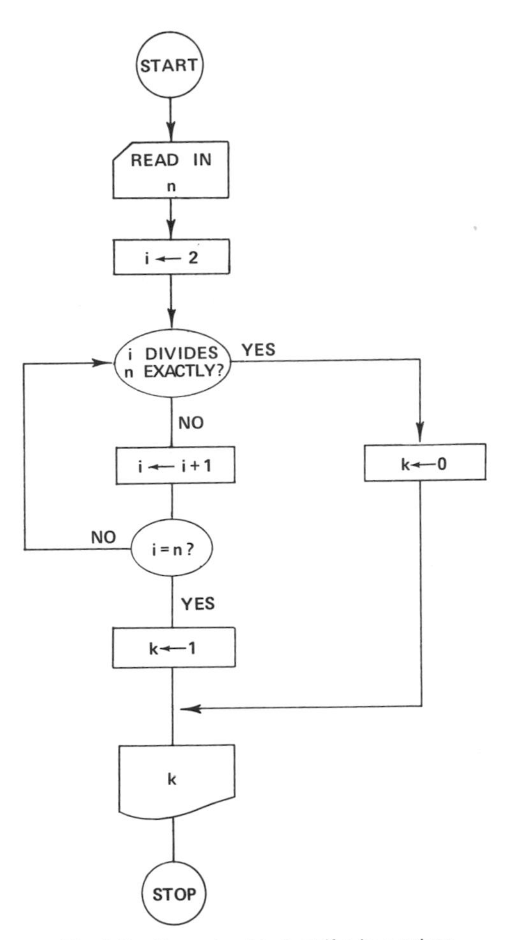

Fig. 3.5 Flow chart to test if n is a prime.

The algorithm in Fig. 3.6 is better than the preceding one in several respects. First, its domain has been enlarged to include the integers 1 and 2. A second and significant improvement is achieved by testing for divisibility by 2 separately, near the beginning, and testing only for divisibility by odd integers 3, 5, 7, and so on. This change in strategy reduces the amount of work by half. The third improvement is achieved by using a little bit of common sense and avoiding brute force. It is a well-known property that there is no divisor of n that is greater than $\sqrt{n}$ if none has been found which is less than $\sqrt{n}$. Therefore, it is useless to test once a value x^2 larger than n has been reached. For large values of n, the amount of work saved is significant. These improvements, though significant, worsened the algorithm by making it more complicated. ■

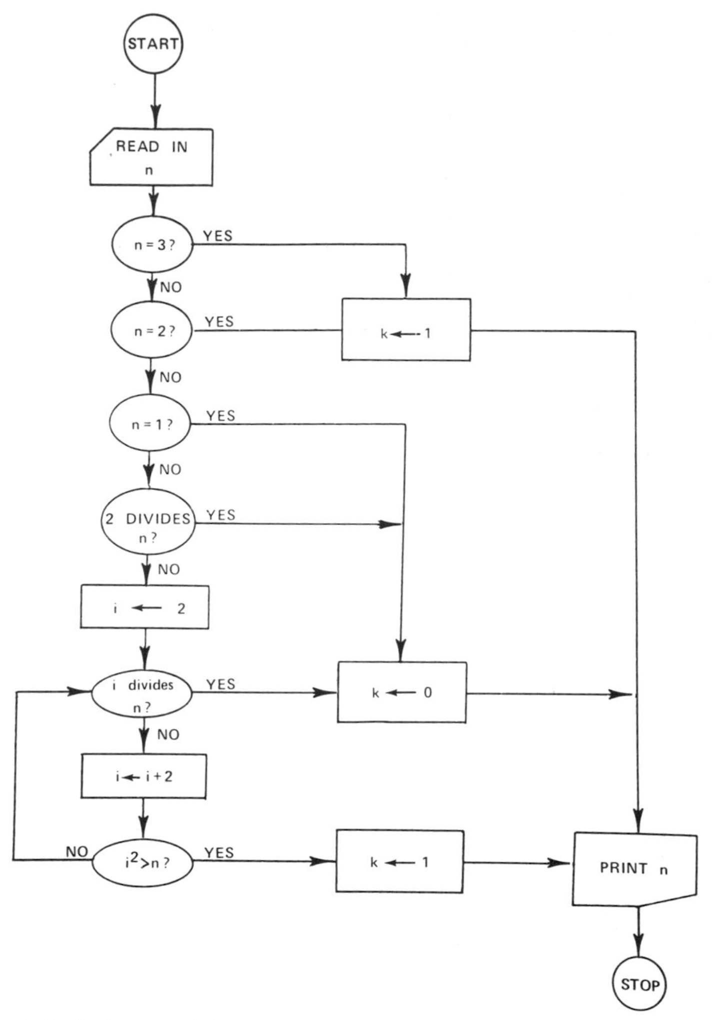

Fig. 3.6 An improved algorithm to determine whether a number n is a prime.

Successive Approximations and Stopping Rules

The finiteness requirement of an algorithm is not always satisfied in a natural way. In such cases it becomes necessary to terminate the computational process by what are called *stopping rules*. For instance, consider the problem of constructing an algorithm to solve equations such as

$$x = f(x)$$

or more specifically, consider the problem of solving the quadratic equation

$$x^2 = x + 1 \tag{1}$$

From elementary algebra, it is well known that the positive root of the above equation is

$$x = (1 + \sqrt{5})/2 = 1.61803 \cdots$$

However, such elementary methods fail if the left side of Eq. (1) is replaced by, say, x^6.

A systematic procedure to solve the general problem using an algorithmic procedure can be developed using the method of *successive approximations.*

As $1^2 < 1 + 1$ and $2^2 > 2 + 1$, it is evident that the positive root of Eq. (1) lies between 1 and 2. Now let us pretend that $x = 1$ is indeed a solution to Eq. (1) and call this *initial guess.* A second guess can now be obtained by evaluating the right side of Eq. (1) using the initial guess. That is,

$$x_2{}^2 = x_1 + 1 = 1 + 1 = 2$$

or

$$x_2 = \sqrt{2.} = 1.414$$

A third guess x_3 can now be obtained by using x_2 on the right side of Eq. (1):

$$x_3{}^2 = 1.414 + 1 = 2.414$$

$$x_3 = \sqrt{2.414} = 1.553 \cdots$$

This procedure can be repeated indefinitely using the general *recurrence formula*

$$x_n{}^2 = x_{n-1} + 1 \tag{2}$$

It can be easily seen that the successive guesses satisfy

$$x_1 < x_2 < x_3 < \cdots < x_{n-1} < x_n < \cdots < x \tag{3}$$

From Eqs. (3) and (1) it can be seen that

$$x_n = \sqrt{x_{n-1} + 1} < \sqrt{x + 1} = x$$

Therefore, there is reason to believe that the successive approximates move closer and closer to the true value of the root x. This phenomenon is called *convergence.* In particular, the convergence here is called monotone. (Why?) Whenever this property of convergence can be established, it is possible to terminate an otherwise indefinite successive approximation procedure using some criterion for stopping. For example, the procedure can be terminated after a specified number of steps, or when x_n differs from x_{n-1} by less than, say, 10^{-3}, or using some other stopping rule of this kind.

Note that this method is the same for all kinds of equations and the procedure takes no advantage of the nature of the equation being solved.

The above procedure is not infallible. If one rewrites Eq. (1) as

$$x = x^2 - 1$$

and attempts to solve this using a similar successive approximation procedure, the process diverges. In many practical problems it is not always possible to tell whether an algorithm converges or diverges until after it is actually tried out. One of the advantages of a computer lies in exhibiting the utility or fallibility of a particular algorithm.

The Newton–Raphson Algorithm

The Newton–Raphson method provides a powerful algorithm for solving (or finding the roots of) linear or nonlinear algebraic equations of the form

$$g(x) = 0 \tag{4}$$

The procedure is very similar to the successive approximation method except now improved guesses are made by utilizing some of the knowledge about the function whose roots are to be determined. For instance, if one is interested in solving

$$x^2 = 5 \tag{5}$$

it is first rewritten as

$$y = x^2 - 5 = 0 \tag{6}$$

A graph of $y = x^2 - 5$ is shown in Fig. 3.7. The goal of the procedure is to locate the position of the point P [the positive root of Eq. (5)] on the x axis using a numerical procedure.

Suppose one makes an initial guess $x_1 = 3$. Instead of making an improved guess blindly, one can utilize the information already contained in Fig. 3.7 or equivalently in the given equation, namely Eq. (5). This is done as follows.

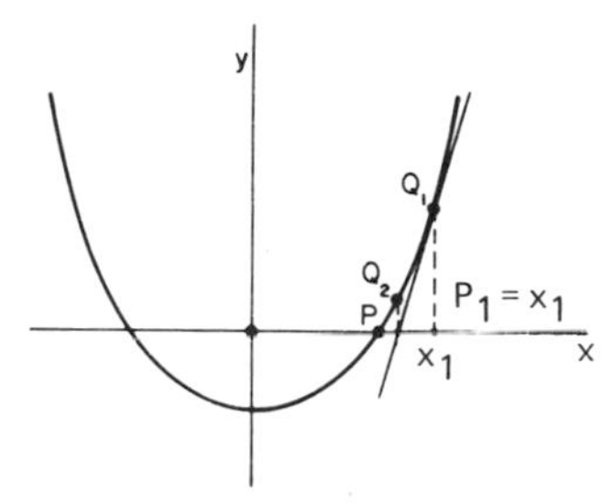

Fig. 3.7 Graphical interpretation of the Newton–Raphson method.

The parabola in the vicinity of the point Q_1 is approximated by the tangent drawn to the curve at Q_1. Intersection of this tangent with the x axis is taken now as an improved guess x_2. This procedure can be expressed mathematically by writing an equation of the tangent at Q_1 as follows.

Let

$$P_1Q_1 = y_1$$

then

$$y_1' \triangleq \frac{dy_1}{dx} = \begin{array}{l}\text{the tangent of the angle the line}\\ P_2Q_1 \text{ makes with the } x \text{ axis.}\end{array}$$

Therefore,

$$y_1' = P_1Q_1/P_1P_2 = y_1/(x_1 - x_2)$$

Solving for x_2

$$x_2 = x_1 - y_1/y_1' \tag{7}$$

Using Eq. (6), the above equation can be rewritten as

$$x_2 = x_1 - (x_1{}^2 - 5)/2x_1 = (x_1{}^2 + 5)/2x_1 \tag{8}$$

The approximation can be further improved by making another guess x_3, using the formula

$$x_3 = (x_2{}^2 + 5)/2x_1 \tag{9}$$

or in general

$$x_n = (x_{n-1}^2 + 5)/2x_{n-1} \tag{10}$$

The Newton–Raphson method does not always converge. Indeed, one can easily visualize a situation where the process of finding a root oscillates between two points (see Fig. 3.8). If x_1 is the initial guess, the tangent to the curve at (x_1, y_1) intersects the x axis at x_2 and the tangent at (x_2, y_2) intersects once again at x_1. Any detailed discussion of convergence is beyond the scope of this work but one can say that the Newton–Raphson method converges if y' and y'' do not change sign in the interval (x_1, x) and if y_1' and y_1'' have the same sign.

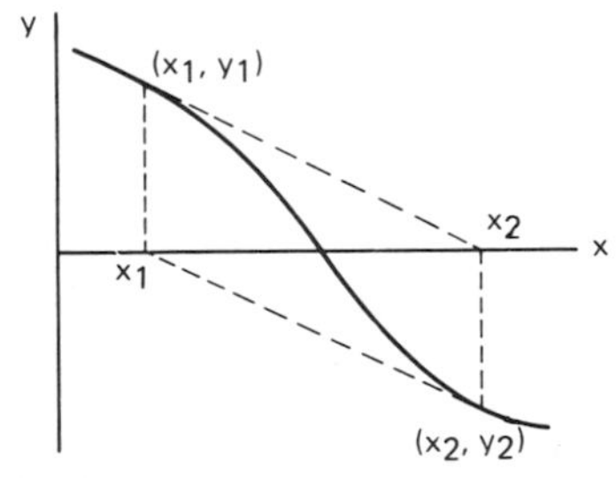

Fig. 3.8 Example of a situation where the Newton–Raphson method fails.

Example 2. To find the root of

$$y = 2\exp(-2x/10) - x$$

one first computes

$$y' = -0.4\exp(-0.2x) - 1$$

Using a general recurrence relation as in Eq. (7), namely,

$$x_n = x_{n-1} - y_n/y_n'$$

one gets

$$x_n = x_{n-1} - \frac{2\exp(-0.2x_{n-1}) - x_{n-1}}{-0.4\exp(-0.2x_n) - 1}$$

Assuming an initial guess of $x_1 = 1$

$$x_2 = 1 - \frac{2(0.8187) - 1}{(-0.4)(0.8187) - 1} = 1 + 0.4802 = 1.4802$$

Similarly

$$x_3 = 1.4802 + \frac{2(0.7438) - 1.4802}{1 + 0.4(0.7438)} = 1.4859$$

which is regarded as sufficiently accurate for the present purpose. ■

Algorithms for Sorting

In using computers to solve large-scale systems problems, data need to be *sorted* frequently. Sorting is nothing but arranging data into a specified sequence. Sorting is necessary not only to arrange raw data into a meaningful sequence but also to rearrange a specific sequence into a different sequence or into several different sequences.

For instance, information concerning reservations on an airline may be fed into a computer as they come in, while the office of the vice-president of operations may want this information sorted by *destination.* The advertising office may want a daily breakdown of these transactions sorted by the age group of the passengers while the sales office may want a breakdown of the passenger miles traveled, sorted by class of travel. The personnel office may want a report of these transactions for the year sorted by the salesman.

There are two kinds of sorting: *internal* and *external* sorting. If the set of data to be sorted is small enough to fit into the main store of the computer, an internal sort is desirable; otherwise, an external sort is required. In large-scale systems studies, a more frequently occurring situation is to sort relatively small sets of data stored on an external storage device, such as a disk, which cannot all be contained in core storage all at one time. This requires a combination of internal and external sorts. Only internal sorting will be described in the sequel. The following definitions are useful to describe the sorting procedure. The attribute (or *field*) to be sorted is called the *key*. The order or rank of numbers, or characters, is called the *collating sequence*. A key may contain any number of characters and fields. A *string* is a group of consecutive records, the keys of which are in collating sequence. A *merge* is simply the process of putting together two or more files that are in the same sort sequence based on the same key.

Internal Sorting

There are several ways to conduct an internal sort. Only the more important things are presented here. In the so-called *scan and move* technique the key of each entry is scanned. The highest ranking key is located and moved to the "top of the list," while the rest are moved down to fill the gap. Then the second highest is moved to the second position from the top and the process is repeated until the entire list is exhausted. As any one can see, for each entry moved up, a large body of entries must be moved down. This is time consuming. If the ordered entries can be stacked in a second area, then the time spent in moving data can be eliminated at the expense of requiring additional storage space.

The second method, called *bubble sort*, is a compare-and-switch method. This method always guarantees a final sorted array after $(N - 1)$ *passes*, where N is the number of entries in the list. Each successive pass examines one less entry than the preceding pass. Therefore, the last pass is a single comparison. Thus, the work involved is progressively reduced.

Example 3. Sort the following four letters into their alphabetic order using bubble sort. Display the state of the list as the sorting progresses:

Start	First pass		
D	*D*	*D*	*A*
C	*C*	*A*	*D*
B	*A*	*C*	*C*
A	*B*	*B*	*B*
Given list	*A* and *B* are switched	*A* and *C* are switched	*A* and *D* are switched

Second pass		Third and final pass
A	*A*	*A*
D	*B*	*B*
B	*D*	*C*
C	*C*	*D*
B and *C* are switched	*B* and *D* are switched	*C* and *D* are switched

■

The third method of sorting is called *radix sorting*. Indeed, a card sorter uses this method. In essence, each key is examined in its own right regardless of its relationship to other entries in the list. This method is best explained by an illustrative example.

Example 4. Sort the following numbers in descending order of magnitude:

24, 37, 02, 32, 10, 26, 18, 15, 29, 07, 05, 16, 21, 23, 20, 36, 22, 23, 08, 12

Because there are ten possible digits, ten hoppers are provided as shown below. The first pass conducts a sort with the low-order digit as its key. The result is

0	1	2	3	4	5	6	7	8	9
		12							
		22				36			
20		32	23		05	16	07	08	
10	21	02	33	24	15	26	37	18	29

The second pass picks the entries out of the hoppers from the highest to the lowest key (i.e., from 9 to 0, in this case) and bottom to top and separates on the next digit position, yielding

0	1	2	3
		20	
		21	
	10	22	
02	12	23	32
05	15	24	33
07	16	26	36
08	18	29	37

The entries are now taken out of their hoppers highest to lowest, bottom to top to yield

37, 36, 33, 32, 29, 26, 24, 23, 22, 21, 20, 18, 16, 15, 12, 10, 08, 07, 05, 02 ■

Algorithms for Merging

Sorting of records to create a string or file is only a beginning in data manipulation operations. As new data become available, it often becomes necessary to update the main files by inserting the data at their proper location. This process involves three steps:

(1) Select the key to be used in sorting.
(2) Sort the new records according to the key selected.
(3) Merge this new file with the main file.

The merging operation calls for comparing pairs of records and inserting of a new record in the main file whenever a match is found. The job of *comparing* pairs of records to establish a linkage between old and new files is a tedious one for a human operator but well suited for computer operation. However, the job of *inserting* a new record in a given place in a file is ill suited for computer operation. A human operator can insert a record in a file by pushing the records on either side and creating space. But with a computer file, either on a magnetic tape or in core memory, insertion can only be done by copying all subsequent records into new locations to create the necessary gap. This single point turns out to dominate the speed and efficiency of many file maintenance operations on a computer.

One of the main reasons for introducing the so-called *list structures* is to overcome the rigidity encountered in allocating storage in a computer memory whenever files are to be merged. Central to the idea of *list structure* is the concept of a *list*. This deceptively simple concept can be explained by way of an illustrative example.

Example 5. Consider a magazine such as a ladies' magazine. If the magazine is opened at random, it is very likely that the top of the page says "continued from page 16" and the bottom of the page says "continued on page 75." Before reading is completed, a reader may go through this hop, skip, and jump process several times. This is a simple example of list structure. ■

Thus, a *list* is a set of entities (pages, in the case of the magazine example) together with indicators that specify the logical sequence of the members of the set. Though a list structure is undesirable (from the reader's viewpoint) in a magazine because of the nuisance of shuffling pages back and forth, this nuisance is absent in a computer. When computers are used to manipulate lists, the magazine analogy can be stretched somewhat. If the principle of hopping around is established and accepted, the interleaving of various articles and stories can be made more flexible than is common in magazines by relaxing the customary rule that a reader turn the pages *backward* only to locate the starting of a story.

The actual implementation of algorithms for merging is a little bit involved and will not be considered here.

Fuzzy Algorithms

The algorithmic approach provides an impressive tool for the analysis and design of a variety of systems. What is still lacking are methods for dealing with systems that are too complex or too ill defined for conducting precise analysis. Such systems not only pervade life sciences, social sciences, psychology, and many other "soft" fields but also appear as subsystems in large-scale traffic control systems, power distribution networks, neural networks, and so on. Perhaps a major reason for the ineffectiveness of classical mathematical techniques in dealing with systems of a higher order of complexity lies in their failure to come to grips with the issue of fuzziness.

The conceptual structure of formal mathematics is too precise for a realistic description of real world situations. For example, the collection of animals clearly includes dogs, cats, horses, etc. as its members and clearly excludes such objects as rocks, fluids, and plants. However, such objects as bacteria, and starfish have ambiguous status with respect to the set of animals. The same kind of ambiguity arises in the daily discourse between human beings in which conversational language is replete with such adjectives as large, small, substantial, significant, and simple. Even phrases with a strong mathematical flavor, such as accurate, approximate, vicinity of, and neighborhood of, are fuzzy in their meaning.

The notion of a fuzzy set is often useful in providing a means of mathematically describing situations which give rise to ill-defined collections of objects for which there is no precise criterion for membership. Sets of this kind have vague or fuzzy boundaries. The sets

$$A = \{x : x \text{ is } \textit{much larger} \text{ than } 5\}$$
$$G = \{g : g \text{ is a } \textit{beautiful} \text{ girl}\}$$
$$L = \{l : l \text{ is a line that } \textit{best fits} \text{ a set of observed data}\}$$

are typical examples of fuzzy sets where the fuzziness is conveyed by the meaning of the italicized words. In each of the above, the criterion for membership in a set is imprecise. Despite the lack of precision in the sense described, fuzzy statements such as "the stock market has suffered a *sharp* decline" and "Corporation XYZ has a *bright* future" do convey meaningful information.

More formally, if $X = \{x\}$, denotes a collection of objects, a fuzzy set $A \subset X$ is a set of ordered pairs

$$A = \{(x, \quad \mu_A(x))\}; \qquad x \in X$$

where $\mu_A(x)$ is termed the grade of membership of x in A. Indeed $\mu_A(x)$ is a function that defines a mapping from X to M, called the *membership space.*

Example 6. Let $X = \{0, 1, 2, \ldots\}$ be the set of all nonnegative integers. A fuzzy set A may now be defined (rather subjectively) as the collection of ordered pairs.

$$A = \{(3, 0.6), (4, 0.8), (5, 1.0), (6, 1.0), (8, 0.6)\}$$

where the first entry of the ordered pair (3, 0.6) stands for the integer 3 from set X and the second entry 0.6 says that the grade of membership of the integer 3 in the set A is 0.6. ■

Here fuzziness should not be construed to mean randomness; it indeed reflects a lack of smooth transition from membership in a class to nonmembership in it. Essentially randomness has to do with uncertainty concerning membership or nonmembership of an object in a set. Fuzziness, on the other hand, has to do with a kind of gradation in membership intermediate between full membership and nonmembership. For instance, a statement such as "Corporation XYZ has a modern outlook" is imprecise by virtue of the fuzziness of the phrase "modern outlook." On the other hand, a statement such as "the probability that corporation XYZ is operating at a loss is 0.3" is not a fuzzy statement.

Thus, fuzziness has some relevance to complex systems studies because when the complexity exceeds a certain threshold, it becomes unfeasible to make precise statements about its behavior even in a probabilistic sense. With this rationale, it is now possible to introduce the concept of a fuzzy algorithm. Fuzzy algorithms underlie much of human thinking. Indeed, we consciously or subconsciously use fuzzy algorithms when we walk, eat, cook food, park a car, and make decisions—just to mention a few. In a fuzzy algorithm the following four concepts occupy a central position and therefore need further consideration. They are:

(1) What is the precise meaning of a fuzzy instruction?
(2) How can one execute a fuzzy instruction?
(3) How can one observe the result of executing a fuzzy instruction?
(4) How can one define convergence in a fuzzy context?

A fuzzy instruction is merely a fuzzy statement presented in the form of a command to a computer. Upon command, the instruction is to be followed if certain conditions are satisfied. Thus, an instruction is a fuzzy instruction if (a) the particular statement is nonfuzzy but the condition to be met is fuzzy or (b) the particular statement is fuzzy but the condition to be met is nonfuzzy or (c) both the statement and the condition to be satisfied are fuzzy.

Example 7. Consider instructions such as

If $X \gg 10$, take two steps.

If you see an elderly lady, yield your seat.

In both these cases, the conditions to be met are fuzzy, but the procedure to be followed if the condition is met is nonfuzzy:

If $X > 10$, take several steps.

In this case, the condition is nonfuzzy but the procedure to be followed is fuzzy:

If you see an elderly lady, be nice to her.

This is a case in which both the condition to be met and the procedure to be followed are fuzzy. ■

Example 8. A blindfolded subject A is required to be transferred from a point X to a point Y in a room. The repertoire of fuzzy instructions is limited to the following set:

(1) Turn counterclockwise by approximately α degrees, α being a multiple of, say, 10.
(2) Take a step.
(3) Take a small step.
(4) Take a very small step.

Construct a fuzzy algorithm.

To construct an algorithm, it is necessary to assume that after the subject executes an instruction, the new position and orientation of the subject are observed fuzzily by the experimenter (this is called *fuzzy feedback*), who then chooses that instruction from the repertoire which fits most closely to the latest observation. With this assumption, the following eight instructions will constitute a fuzzy algorithm:

(1) If A is facing y, then go to instruction 2 or else go to 5.
(2) If A is close to y, go to 3 or else go to 7.
(3) If A is very close to y, go to 4 or else go to 8.
(4) If A is very, very close to y, then stop.
(5) Ask A to turn counterclockwise by an amount needed to make him face y. Go to 1.
(6) Ask A to take a step. Go to 1.
(7) Ask A to take a small step. Go to 1.
(8) Ask A to take a very small step. Go to 1. ■

This example is presented only to give a flavor of the concept central to fuzzy algorithms. Whether such algorithms work is a question that cannot be answered precisely. The algorithm can be expected to work reasonably well as long as the degree of fuzziness in observations is relatively small and the subject executes the instructions in a way that is consistent with the expectations of the experimenter. Such vague statements would be unacceptable to those who expect to prove the convergence properties rigorously. Unfortunately, there may be no alternative if one wishes to get approximate solutions to many complex ill-defined problems.

Characteristics of a Heuristic Procedure

The main difficulty with algorithmic procedures in problem solving is that they may take a vast amount of time to yield a solution. There are many situations of practical importance in which the use of an algorithm, even if it exists in a very sophisticated form, proves to be quite uneconomical. There are other cases in which algorithms are simply not available. Situations involving planning, bargaining, and others in which human decision making is involved are a case in point. In order to use a computer to simulate these situations it is necessary to use some kind of *ad hoc* rules in a computer program. The term "heuristic" is coined to describe such a procedure. A *heuristic* is a rough guide or principle, rule of thumb, hypothesis, or approximation. Playing chess is necessarily a heuristic procedure and so are the majority of decisions taken by management in business and industry. All future based activities are necessarily heuristic; acquisition analysis, investment analysis, diversification analysis, and expansion analysis, as well as all types of sales forecasting in business circles, are inevitably heuristic in the sense that they are approximate and liable to error.

Heuristics could be numerical and nonnumerical. A sales forecasting model is a numerical heuristic model. Artificial intelligence places a lot more emphasis on linguistic and nonnumerical aspects.

In order to understand the difference between heuristic and algorithmic programming, it is useful to invoke the concept of a hierarchical model. For instance, it is easy to see that every heuristic program (or model) must have an underlying computable model. That is to say, at a very elementary machine code level all programs are necessarily algorithmic, but they may subserve higher level models that are heuristic. For example, consider heuristic verbal models such as

(a) "Black clouds usually mean rain."
(b) "Red sky at night means a bright day tomorrow."

These are fuzzy statements. However, underlying such propositions is a counting procedure that states, "Black clouds occurred 100 times and of these 100 times, 90 times it rained." Thus, the counting is indeed algorithmic, but the probability based on the counting is heuristic. This particular point is perhaps the basic difference between fuzzy algorithms and heuristics. Stated differently, a heuristic program is a nonfuzzy approximation, expressed in a computer language, to a fuzzy algorithm. Heuristic programs are necessary to implement fuzzy algorithms on a computer because present day computers cannot operate on fuzzy sets. At this stage, before departing from the topic, it would be useful to illustrate the meaning of a heuristic by way of an example. As an example of a heuristic program, consider the problem of "balancing" an assembly line. The problem may be stated as follows. Given a production rate, what is the minimum number of work stations consistent with the time and ordering constraints of the product?

The procedure can be divided into two parts. First one constructs a hierarchy of increasingly simple line balancing problems by aggregating groups of elements into a single compound element. Then one proceeds to solve a simple line-balancing problem by assigning groups of available workmen to elements and then taking as subproblems those compound elements that have been assigned to more than one man. Thus, a need to use heuristics arises to:

(1) Aggregate groups of simple elements into compound elements.
(2) Solve the simplified problems.
(3) Disaggregate (when the simplified problems fail to work) and bring back the detail of the original problem.
(4) Coordinate the solutions of the individual problems.

The obvious similarities among human problem-solving procedures, hierarchical modeling philosophy, aggregation methods, fuzzy algorithms, and heuristic programs is well worth noticing. An outgrowth of this observation is the newly emerging *natural language programming*, which is primarily concerned with the organization of information in store for question answering purposes.

5 SIMULATION LANGUAGES

Whenever a digital computer is being considered as a prospective tool for simulation, it is also necessary to give careful thought to the proper selection of software. Of particular importance in this selection is the language to be used in communicating with the machine. This section delves on describing some of the more widely used *simulation languages* and discusses ways of deciding the best language in which to program for computer implementation of a model.

During the early days of digital computer development, all programs had to be written in what is called a machine language. In a *machine language*, all instructions and data are represented by strings of zeros and ones as ultimately this is the only way one can communicate with a computer from the external world. Writing programs in machine language is tedious and error prone. For instance, a segment of a program coded in machine language (of a particular machine) would look like this:

```
010111101000110101101000
010111110000110101101001
000100011000111100001101
010010000000000111010000
000100011000111011000000
000111010000111010101101
000101100000111010101110
000101001000001001010011
```

To avoid the tedium of machine language, such codes are usually written in *assembly language*. The program written in assembly language is then translated by the computer into machine language by using another program called the *assembler*. The assembly language allows the programmer to write his code in a *mnemonic* or *symbolic* language rather than in strings of zeros and ones. A *mnemonic* is a means to aid memorization. For example, the string 101100 may be the code for a subtract operation. In an assembly language one writes SUB instead of 101100 and each time the string of letters SUB appears in a code the assembler replaces that by 101100. For instance, an assembly language version of the above cited machine language code is shown below:

Operation code	Address
LDB	06550
LDA	06551
BRM	07415
NOP	00720
BRM	07600
SKA	07255
SUB	07256
BRR	01123

It is also possible to use a mnemonic symbol for the address portion of an instruction. In such a case one says that *symbolic addressing* is used. In any event, the important thing to remember is that an assembly language has a one-to-one correspondence with machine language.

In order to translate an assembly language into machine language, the assembler must replace each mnemonic instruction with its equivalent binary code. If symbolic addressing is used, the assembler should also replace the symbolic address by its numerical equivalent. Replacement of instruction mnemonics by binary code can be done by referring to a list of all valid mnemonics and their binary codes. To do the same for the symbolic address is a little difficult because it is the programmer, not the machine designer, who usually chooses the symbolic addresses. To cope with this situation, an assembler scans the source code at least twice. During the first scan or *pass* a *symbol table* is constructed in the store giving equivalent address for each symbol. On the second pass a substitution is made to translate the code into binary language.

The assembly languages relieved a significant amount of drudgery from the task of programming a problem. However, they still retain a one-to-one correspondence with the machine language. Because of this, programming in assembly language still requires meticulous care and attention. Program modification and debugging still remain as tedious tasks. Furthermore, programs coded in the assembly language of one computer are practically useless for use in conjunction with another type of computer. In addition to these constraints, a program coded in an assembly language bears little or no resemblance to the mathematical operations it performs. Stated differently, assembly language is machine oriented, not user oriented. To combat these and other problems, the so-called *procedure oriented languages* were developed. FORTRAN, ALGOL and PL/1 are typical examples of procedure oriented languages. These languages are somewhat machine independent and therefore, once a program is written, it becomes theoretically possible to use it in conjunction with a variety of computers. Furthermore, instructions written in a language such as FORTRAN or PL/1 have a strong resemblance to mathematical equations, and as a result computer programming no longer remains alien to the mathematically oriented scientific community.

A program coded in a procedure-oriented language can be translated into an assembly language of any computer by means of another program called a *compiler*. The compiler takes the user-written program, called the *source program*, as an input and produces an assembly language code as an output, which in turn constitutes an input to the assembler whose output is a machine language code.

Developments such as the one just described always involve compromises. For instance, an assembly language code generated by a compiler is, in general, less efficient than a code written by an expert in assembly language. However, this kind of argument lost its appeal as computers became faster. Widespread use of computers for simulation further emphasized the point that there are significant advantages, from the user viewpoint in relegating more and more responsibilities to the computer at the expense of a few microseconds of additional computer time.

Early experience with simulation, however, revealed that many of the programs written—even in widely differing application areas—had functionally similar processes. Some of the more commonly occurring functions are:

(1) Record data for outputs.
(2) Arrange outputs in specified formats.
(3) Invert matrices.
(4) Perform integration.
(5) Calculate trigonometric functions.
(6) Generate pseudorandom numbers.
(7) Simulate a queue.

All of the above mentioned and several other standard operations are relatively simple in concept. Nevertheless, the details of programming and debugging are rather complicated and time consuming even with the help of procedure-oriented languages. It thus became apparent that some form of generalized approach was necessary to make simulation a practical technique. Such generalization was meant to serve the following vital functions:

(a) Reduction of the programming task. This need is not essentially different from that for other programming languages.
(b) Allowing flexibility for change. A major advantage of simulation is that it is much easier to change a representation of a system than to change the system itself. Adequate generalization of the technique is essential to assure that this advantage can be realized.
(c) Good error diagnostics.
(d) Applicability to a wide range of problems.
(e) Facilitation of model formulation.

These criteria require that a language be *problem oriented.* That is, the command and data designators in the language should contain key words and phrases that are indigenous to the standard vocabulary of the prospective user. This "language" may well be English for a business executive, a set of block diagrams for an analog computer user, or a language with such key words as ADD, INVERT, SOLVE to a user primarily interested in matrix operations.

Thus, a problem oriented or *simulation language* may be viewed as a program that allows flexibility in input operations. For instance, if a number of programs are available to solve polynomial equations under a variety of conditions, they could all be collected under one package so that a user merely feeds his input data (usually the coefficients of the polynomial) observing some syntax rules and requests a solution. When this happens, a user really does not have to possess any knowledge of programming or numerical analysis. Thus, in a way, a simulation language brings the computer to a "layman." Because of this, obviously, the scope of any given simulation language is limited to a restricted class of problems. Some of the more widely used simulation languages are listed below:

(1) To duplicate the behavior of an analog computer on a digital computer by preserving the block diagram approach to problem solving: PACTOLUS, 1130 CSMP and MIDAS.
(2) To simulate continuous systems characterized by ordinary differential equations: MIMIC, 360 CSMP, CSSL, and DYNAMO.
(3) For simulating systems characterized by partial differential equations and integral equations (still in the developmental stage).
(4) To simulate discrete or event oriented systems: GPSS, SIMSCRIPT, and SIMULA.
(5) To simulate network type of structures: GASP, GERT.
(6) For electrical circuit analysis and design: ECAP and NASAP

(7) For civil engineering applications: LOGO and STRESS.
(8) For information retrieval: DATAPLUS, and DIALOG.
(9) For simulating the behavior of one computer on another computer.
(10) For writing compilers: METAS and FSL.
(11) For machine tool control: APT.
(12) For formula manipulation: FORMAC
(13) For string manipulation: SNOBOL
(14) For list processing: IPL-V and LISP.

Most of the above languages can generally be categorized as either continuous or discrete depending upon their applicability to the characteristics of the phenomenon to be represented. *Continuous simulation languages* are appropriate when the subject is considered to consist of a continuous flow of information or material counted in the aggregate rather than by individual items. *Discrete simulation languages*, on the other hand, are used when individual items are of interest. For instance, if one is interested in simulating the growth of world population, a continuous simulation language is an appropriate choice because the number of people can be regarded as a continuous variable in time. If one is interested in modeling the growth of customers waiting in line at a supermarket checkout counter, a discrete simulation is more appropriate.

The basis for a continuous simulation language is a "continuous model," which is generally represented mathematically by a differential equation. Systems of this type were first modeled on analog computers and many are still solved using analog and hybrid computers. However, because of the limitations an analog computer places in problem size, the desire for more accuracy, and in many cases the fact that only a digital computer was available, some of the analog users have turned to digital systems. Thus, whenever a digital computer is used to simulate a continuous system, the model is necessarily approximated using discrete approximations. Thus, whenever digital computers are used, even a "continuous" simulation language is implemented in a "discrete" fashion. This subtle, nevertheless important, point should be kept in mind all the time.

EXERCISES

1. In Fig. 3.6 explain the reason for
(a) The test for $n = 3$ at the beginning.
(b) Branching on $i^2 > n$ rather than $i^2 \geq n$.

2. Write a simple algorithm to determine the greatest common divisor of two positive integers. What is the greatest common divisor of 2166 and 6099?

3. Given a set of n numbers, say $a(1), a(2), \ldots, a(n)$, how many comparisons are needed to sort them in a nondecreasing order using the following procedure: Scan the numbers from $a(1)$ to $a(n)$ for the smallest, interchange that with $a(2)$, and repeat the procedure with the remaining numbers.

4. How many comparisons are needed to sort n numbers using bubble sort? Can you suggest a way to improve this procedure, especially when the a's are partially sorted to begin with.

5. Suppose two lists $a(1), a(2), \ldots, a(n)$ and $b(1), b(2), \ldots, b(n)$ are available. Both are in a nondecreasing order to start with. It is required to produce a new list $c(1), c(2), \ldots, c(2n)$ by merging the two old lists. The new list is also required to be in a nondecreasing order. How many comparisons are required for this *merge–sort* procedure?

6. A method of sorting which sometimes results in increased speed uses an alternating direction scan. On the first pass, the largest number is bubbled down to the last position n. On the second pass, the smallest of the first $(n-1)$ numbers is bubbled up to the first position and on the third pass the largest of the numbers in positions 2 to $n-1$ is moved to position $n-1$, and so on. Write a flow chart and program to implement this method.

7. Find the real root of the equation

$$xe^x - 2 = 0$$

8. If one wishes to apply the Newton–Raphson method to find the cube root of a number a, that is, to find the positive solution of $x^3 = a$, then show that the relations corresponding to Eqs. (8) and (9) of Section 4 are

$$x_2 = (2x_1/3) + (a/3x_1{}^2), \qquad x_3 = (2x_2/3) + (a/3x_3{}^2)$$

9. One is interested in finding a root of $f(x) = 0$. Derive the recursive relation

$$x_n = x_{n-1} - f(x_n)/f'(x_n)$$

10. Using the Newton–Raphson method find to within four significant figures the root of

$$\sin x - (x + 1)/(x - 1) = 0$$

SUGGESTIONS FOR FURTHER READING

1. Further discussion on modeling and the role of optimization in modeling can be found in
   R. L. Ackoff, *Scientific Method: Optimizing Applied Research Decisions.* Wiley, New York, 1962.
2. For a comprehensive treatment of analog simulation as it is applied to a wide variety of engineering problems, see
   W. J. Karplus, *Analog Simulation: Solution of Field Problems.* McGraw-Hill, New York, 1958.

3. Other useful references to simulation are

P. S. Greenlaw, L. W. Herron, and R. W. Rawdon, *Business Simulation in Industrial and University Education*, Prentice-Hall, Englewood Cliffs, New Jersey, 1962.

H. Guetzkow *et al.*, *Simulation in Industrial Relations.* Prentice-Hall, Englewood Cliffs, New Jersey, 1963.

4. For further elaboration on Karl Popper's definition on page 86 consult

K. Popper, *The Logic of Scientific Discovery*, 3rd ed., Chapter 10, Hutchinson, London, 1968.

5. For a comprehensive treatment of the subject of algorithms, refer to

D. E. Knuth, *The Art of Computer Programming*, Vol. I: *Fundamental Algorithms.* Addison Wesley, Reading, Massachusetts, 1975.

6. More information on fuzzy sets, fuzzy algorithms and their application to the modeling of complex systems can be found in a host of papers published by Prof. L. A. Zadeh. For a sample, see

L. A. Zadeh, Toward a theory of fuzzy systems, *in Aspects of network and system theory* (R. E. Kalman and N. Declaris, eds.). Holt, New York, 1971.

R. E. Bellman and L. A. Zadeh, Decision making in a fuzzy environment, *Management Sci.* **17B**, 141–163 (1970).

7. For a mathematical introduction to the theory of fuzzy sets, see

A. Kaufmann, *Introduction to the Theory of Fuzzy Subsets*, Vol. I. Academic Press, New York, 1975.

8. It is practically impossible to list references to all the simulation languages currently available. An adequate description of the more important languages can be found in

T. H. Naylor *et al.*, *Computer Simulation Techniques.* Wiley, New York, 1968.

9. For a theoretical approach to modeling and simulation at an advanced mathematical level, consult

B. P. Zeigler, *Theory of Modeling and Simulation.* Wiley, New York, 1976.

4

Primitive Models

INTRODUCTION

In the preceding three chapters, an attempt was made to capture the salient features of the modeling methodology. No mathematics was used; only ideas were presented. In the following six chapters, an attempt will be made to illustrate the role of mathematical techniques in the modeling process. Before embarking on this mission, it is useful to pause for a moment and see how one goes through this transition, that is, from descriptive to normative models. This chapter is intended to fill this need.

One of the elegant methods of developing a model is to follow the Leibnizian approach, that is, to invoke the well established laws of nature and develop mathematical relations from them. This idea is briefly demonstrated in Section 1. Alternatively, one can follow the Lockean approach. When a well-tested law is not readily apparent, this is a very useful and powerful approach to follow. Use of this approach is described at some length in Section 2. The purpose of Section 3 is to demonstrate the interdependence of model building and data gathering phases in the study of complex phenomena. The example used in this section is chosen to demonstrate a synthesis of several ideas. First, the time element is introduced. Second, the methodology of using intuition, heuristic arguments, graphical procedures, and systematic computational algorithms is briefly demonstrated. Finally, the example illustrates how the need arises to refine and modify a model after it is initially determined. Section 4 of this chapter presents problems of yet another type. The two examples discussed in this section are chosen to illustrate methods of formalizing the ever-exciting problems, such as puzzles, on the one hand, and the equally challenging problems of building models of social systems on the other hand.

Thus, a primary purpose of this chapter is to provide motivation for the use of formal methods in model building.

1 ESTABLISHING RELATIONS USING PHYSICAL LAWS

In Chapter 3, a reference was made to the role of physical laws in the model building process. Using physical laws to derive mathematical relations is straightforward if a considerable amount of knowledge about the nature of the underlying physical process is available. With this knowledge, it is generally possible to identify the set of independent variables from the set of dependent variables. At times, it is even possible to write down equations relating these sets of variables. These equations may take the shape of an algebraic equation, such as $y = \alpha_0 + \alpha_1 x + \alpha_2 x^2 + \cdots + \alpha_n x^n$, or a differential equation, such as $dy/dx = y$, or some other complicated equation. This is precisely the procedure we normally use to describe simple phenomena such as the flow of electrical current in a circuit, the deflection of a beam under load, or the cooling of a hot body. This approach fails, however, if we attempt to develop simple equations to relate, say, rainfall to runoff. That is, the method suggested in this section is useful primarily to characterize component parts of a system; it has limited ability in characterizing truly complex systems. Therefore, the procedure of using physical laws is demonstrated with a relatively simple example from electrical circuit theory.

In electric circuit theory the basic laws derived from the principles of conservation and continuity are Kirchhoff's laws. Parenthetically it should be remarked that these laws can also be derived directly from field theory and Maxwell's equations. *Kirchhoff's voltage law* states that *the algebraic sum of the voltage differences across all the elements and sources comprising a closed path (called a loop) is equal to zero.* Kirchhoff's current law states that the

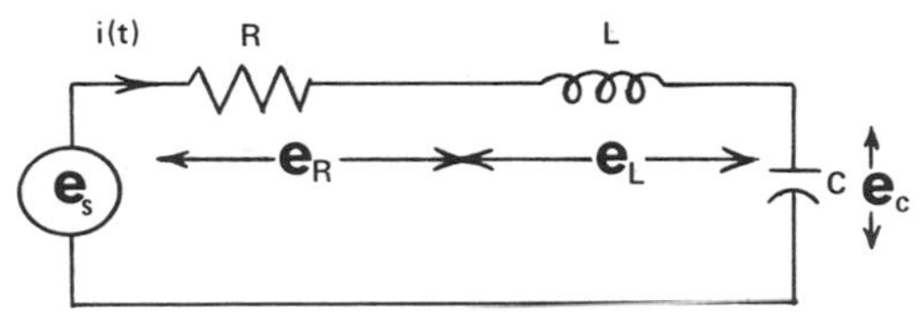

Fig. 4.1 A simple electrical circuit.

sum of the currents leaving a junction (or *node*) is equal to the sum of the currents entering the junction, or in other words: *the algebraic sum of the currents leaving a node is equal to zero.*

Example 1. Conside a simple electrical system such as the one shown in Fig. 4.1. It is desired to find the current i flowing through the circuit when an ac voltage e_s is applied. According to Kirchhoff's voltage law, the algebraic sum of all the voltage differences across all the elements is zero. If $e_R(t)$, $e_L(t)$, and $e_C(t)$ represent the voltage drops respectively across R, L, and C, and e_S represents the voltage of the source at time t then one can write, using Kirchhoff's laws,

$$-e_R(t) - e_L(t) - e_C(t) + e_S(t) = 0 \tag{1}$$

Now the values of $e_R(t)$, $e_L(t)$, and $e_C(t)$ can be obtained by invoking some elementary laws of physics as follows:

$$e_R(t) = Ri, \qquad e_L(t) = L\frac{di}{dt}, \qquad e_C(t) = \frac{1}{C}\int i\,dt \tag{2}$$

Substituting Eq. (2) in Eq. (1), and rearranging,

$$Ri + L\frac{di}{dt} + \frac{1}{C}\int i\,dt = e_S$$

This integrodifferential equation can now be regarded as a mathematical model to the system shown in Fig. 4.1. ■

2 ESTABLISHING RELATIONS VIA CURVE FITTING

While modeling many complex systems it is often difficult to identify a well-defined physical law from which mathematical relations can be developed. In some cases it may not be either prudent or practical to look for a fundamental law. Suppose that one is interested in establishing a relation between crop yield and a particular irrigation technique used. Here the variables influencing the crop yield are too many, and too little is known about how they interact. In such a case one may have to depend on statistical methods. For simplicity, consider a simple situation in which one wishes to establish a relation between two variables, say, x and y. Suppose there is reason to believe that the relation between x and y is a linear one such as $y = \alpha_0 + \alpha_1 x$. How much information about x and y is required to do this?

An answer to this question can be obtained from various points of view. For instance, as the preceding equation contains two unknown parameters, namely, a and b, two data points (x_1, y_1) and (x_2, y_2) are required to establish two equations from which α_0 and α_1 can be recovered. Viewed differently, the preceding equation describes a straight line in the xy plane, and at least two points are required to define a straight line. This argument can be generalized. To establish a relation such as $y = \alpha_0 + \alpha_1 x + \alpha_2 x^2$ it is necessary to have at least three data points and to establish an nth order polynomial-type relation, such as, $y = \alpha_0 + \alpha_1 x + \alpha_2 x^2 + \cdots + \alpha_n x^n$, at least $(n + 1)$ data points are required.

If a polynomial-type relation appears to exist between x and y, the next step would be to determine the degree n of the polynomial to be used. In some very special cases, this can be done by using the following rule. *Rule*: *If the values of x are in arithmetic progression (equidistant) and if the nth differences of the y's are constant, then the last term of the required polynomial is x^n.*

Here the nth difference of y is defined as

$$\Delta_n y = y_{n+1} - y_n; \qquad n = 1, 2, \ldots \tag{1}$$

Application of the above rule is illustrated by means of the following example.

Example 2. The temperature y is recorded at equal time intervals as shown below. One wants to fit a polynomial to this data. What would be an appropriate choice of the degree of the polynomial?

t:	0	0.1	0.2	0.3	0.4	0.5	0.6	0.7	0.8	0.9	1.0
y:	0	0.212	0.463	0.772	1.153	1.625	2.207	2.917	3.776	4.798	6.001

To answer the above question, a difference table (see Table 4.1) is constructed as shown below by using Eq. (1). For instance, the first few entries in column 3 are obtained by using

$$(\Delta_1 y)_1 = y_2 - y_1 = 0.212 - 0.0 = 0.212$$
$$(\Delta_1 y)_2 = y_3 - y_2 = 0.463 - 0.212 = 0.251$$

Similarly

$$(\Delta_2 y)_1 = (\Delta_1 y)_2 - (\Delta_1 y)_1 = 0.251 - 0.212 = 0.039$$

Inspection of the table reveals that all the third differences are approximately of the same value. According to the above rule a third degree polynomial such as

$$y = \alpha_0 + \alpha_1 x + \alpha_2 x^2 + \alpha_3 x^3$$

TABLE 4.1

A Difference Table for the Data Given in Example 2

t	y	$\Delta_1 y$	$\Delta_2 y$	$\Delta_3 y$
0	0			
		0.212		
0.1	0.212		0.039	
		0.251		0.019
0.2	0.463		0.058	
		0.309		0.014
0.3	0.772		0.072	
		0.381		0.019
0.4	1.153		0.091	
		0.472		0.019
0.5	1.625		0.110	
		0.582		0.018
0.6	2.207		0.128	
		0.710		0.021
0.7	2.917		0.149	
		0.859		0.014
0.8	3.776		0.163	
		1.022		0.018
0.9	4.798		0.181	
		1.203		
1.0	6.001			

would be a suitable choice. This difference table method tells nothing about how one determines the α's. It should be remembered that the above rule applies only when the t's are measured at *equal intervals*. Equally important is the fact that it rarely pays to take more than three or four terms in a polynomial formula on account of the labor involved in solving the equations to extract the coefficients. ■

Simple Linear Regression

Consider a problem in which one wishes to establish a relation between the height x and the weight y for all 18-year-old students. From the preceding discussion it is clear that such a relation can easily be established by measuring the height x and weight y of any two 18-year-old students. However, would such a relation be realistic? As the relation between height and weight is influenced by such factors as sex, nutrition, environment, genetic traits, and so forth, some variability is inevitable. If the relation is to be of any use, it is necessary that it hold good, for any individual from the *population* of

18-year-old students, in a statistical sense. Therefore, it is necessary to measure the heights and weights of a representative *sample* of N students, where N is generally much larger than 2. Using this data one can make a statement such as "weight has a general tendency to increase with increasing height," which is true but it only conveys qualitative information. To be more precise, it is necessary to rearrange or display the data in a format that is more suitable for an intelligent interpretation. One easy way to accomplish this task is to plot the data as a *scatter diagram* as shown in Fig. 4.2. In this figure, each point represents the ordered pair (height, weight) for one individual. Inspection of the figure reveals a possible type of relation between height and weight. Merely looking at the points on a scatter diagram is not a precise way of defining relations. A more useful method is to express the relation as a mathematical equation. This could be a linear or nonlinear equation. For simplicity let us assume that we are interested in establishing a linear relation such as $y = \alpha + \beta x$, where α and β are unknown parameters characterizing the entire 18-year-old population. However, as x and y for only a sample of size N is observed, it is possible only to make an estimate of the population *regression constant* α and the population *regression coefficient* β. In order to make this estimate one starts with a model such as

$$\hat{y} = a + bx \tag{2}$$

where a is an *estimate* of the parameter α, b is an estimate of the parameter β, and $\hat{y}$ (read this as y hat) is the estimated value of y for any given value of x. In Eq. (2), which represents a straight line, b is the slope of the line and a the magnitude of the intercept on the y axis. By changing the value of a one can get a family of straight lines parallel to each other. Similarly, changing the values of b, while keeping a at a fixed value, yields another family of straight

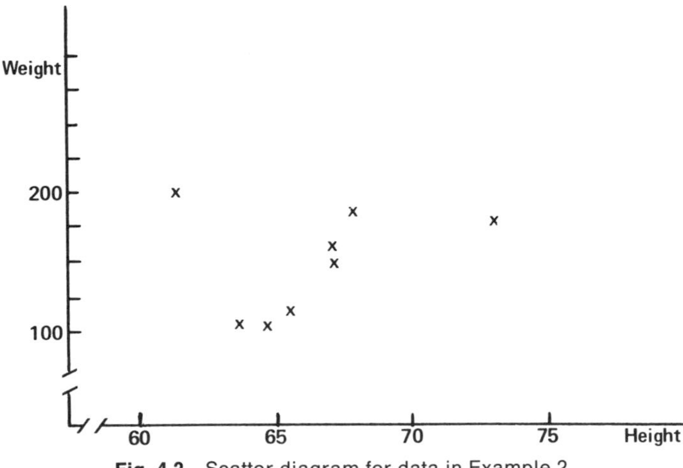

Fig. 4.2 Scatter diagram for data in Example 2.

lines all having the same intercept on the y axis. Thus, by choosing different values of a and b, it is possible to draw several straight lines through a scatter diagram. The technique of selecting one straight line, out of the multitude, that "best" describes the relation revealed by a scatter diagram is called fitting a straight line to a set of observed data.

A useful first step in developing objective methods is to define the meaning of terms like "good fit." Toward this end, it is important to note that the sample tested came from a very large population. Each person in this sample of size N has an associated pair of numbers (x_i, y_i), where x_i is the height and y_i is the weight of the ith person. The task is to build a mathematical model that, given the height of any other person from the population, would predict the associated weight. It is intuitively obvious that this prediction cannot be made exactly for any particular individual chosen from the population. However, one can make a guess by choosing a number $\hat{y}$ for the weight such that the average error committed after making a large number of such guesses would be a minimum. For certain computational and theoretical reasons a definition for this average error r is chosen to be

$$r = (1/N)(e_1^2 + e_2^2 + \cdots + e_N^2) = (1/N) \sum_{j=1}^{N} (e_j)^2 \tag{3}$$

where

$$\begin{aligned} e_j &= \left\{ \begin{matrix} \text{weight of } j\text{th individual as predicted by model} - \text{weight of} \\ \text{the same individual obtained from actual measurement} \end{matrix} \right\} \\ &= (\hat{y}_j - y_j) = (a + bx_j - y_j) \end{aligned} \tag{4}$$

Substituting Eq. (4) into Eq. (3),

$$r = (1/N) \sum_{j=1}^{N} (a + bx_j - y_j)^2 \tag{5}$$

In this model building exercise, the objective is to choose the parameters a and b such that the residual error r is minimized. It is well known (see any elementary calculus book) that the necessary conditions for r to be a minimum are

$$\frac{\partial r}{\partial a} = \frac{\partial r}{\partial b} = 0 \tag{6}$$

These conditions when applied to Eq. (5) yield

$$(1/N) \sum_{j=1}^{N} (y_j - a - bx_j) = 0, \qquad (1/N) \sum_{j=1}^{N} (y_j - a - bx_j)x_j = 0 \tag{7}$$

Solution of these two *normal equations* yields the values of the parameters a and b. It should be remembered that the values of a and b obtained by solving Eq. (7) are only *estimates* of the true values α and β which characterize the population. Before actually solving Eq. (7), the following simplifying notation is introduced: a bar over a variable is used to denote the averaging operation such as

$$\overline{y} = (1/N) \sum_{j=1}^{N} y_j \qquad \text{and} \qquad \overline{x} = (1/N) \sum_{j=1}^{N} x_j \tag{8}$$

With this notation, solution of Eq. (7) can be written as

$$b = \frac{\overline{xy} - \overline{x}\,\overline{y}}{\overline{x^2} - (\overline{x})^2} \qquad \text{and} \qquad a = \overline{y} - b\overline{x} \tag{9}$$

Thus, an estimator of weight is

$$\hat{y}_j = a + bx_j = \overline{y} + (x_j - \overline{x}) \frac{\overline{xy} - \overline{x}\,\overline{y}}{\overline{x^2} - (\overline{x})^2} \tag{10}$$

That is, the weight of an individual of a given height x_j can be estimated using Eq. (10). For computational convenience, Eq. (10) can be rewritten as

$$\begin{aligned} b &= N \sum_{j=1}^{N} x_j y_j - \left(\sum_{j=1}^{N} x_j\right)\left(\sum_{j=1}^{N} y_j\right) \bigg/ \left(N \sum_{j=1}^{N} x_j^2\right) - \left(\sum_{j=1}^{N} x_j\right)^2 \\ a &= (1/N)\left(\sum_{j=1}^{N} y_j - b \sum_{j=1}^{N} x_j\right) \end{aligned} \tag{11}$$

Example 3. A doctor wishes to determine a simple relationship between the height and weight of 18-year-old men. For this purpose several 18-year-old men were invited and the height and weight of each individual is experimentally determined as recorded below in the second and third columns.

TABLE 4.2

A Useful Computational Arrangement for Regression Calculations

No.	x	y	x^2	y^2	xy
1	66	130	4,356	16,900	8,580
2	73	180	5,329	32,400	13,140
3	67	155	4,489	24,025	10,385
4	65	110	4,225	12,100	7,150
5	64	110	4,096	12,100	7,040
6	67	161	4,489	24,921	10,787
7	68	180	4,624	32,400	12,240
8	62	200	3,844	40,000	12,400
$N = 8$	$\sum x_i = 532$	$\sum y_i = 1{,}226$	$\sum x_i^2 = 35{,}452$	$\sum y_i^2 = 195{,}845$	$\sum x_i y_i = 81{,}722$
	$\overline{x} = 66.5$	$\overline{y} = 153.25$	$\overline{x^2} = 4431.5$	$\overline{y^2} = 24480.6$	$\overline{xy} = 10215.25$

The remaining columns show the necessary computations. (See Table 4.2.)

Using the values from the last row of the above table, the values of a and b are $a = 130$ and $b = 0.65$. Therefore, the model is

$$\hat{y} = 130 + (0.65)x \qquad \blacksquare$$

The Correlation Coefficient

Another quantitative measure of the quality of the estimator can be obtained by calculating the average error committed by the estimator. This is obtained by substituting Eq. (9) into Eq. (5):

$$r = (1/N)\sum_{j=1}^{N}(a + bx_j - y_j)^2 = \overline{y^2} - a\overline{y} - b\overline{xy} \tag{12}$$

$$= [y^2 - (\overline{y})^2](1 - \rho^2)$$

where

$$\rho^2 = \frac{(\overline{xy} - \overline{x}\,\overline{y})^2}{[\overline{y^2} - (\overline{y})^2][\overline{x^2} - (\overline{x})^2]} \tag{13}$$

$$= \frac{[N\sum_j x_j y_j - (\sum_j x_j)(\sum_j y_j)]^2}{[N\sum_j x_j^2 - (\sum_j x_j)^2]}[N\sum_j y_j^2 - (\sum_j y_j)^2] \tag{14}$$

Note that the notation $\sum_j$ is an abbreviation for $\sum_{j=1}^{N}$ if there is no ambiguity about N. The term ρ^2 is called the *coefficient of determination* and ρ is called the *coefficient of correlation.* If $\rho = \pm 1$, the residual error r vanishes and a linear algebraic relation is said to exist between x and y. If $\rho \equiv 0$, then by virtue of Eqs. (13) and (11), we have $\hat{y}_j = \overline{y}$, which implies that observations on x do not aid in predicting y.

One must be extremely cautious in interpreting the meaning of correlation and what it says about a relation between two variables. A little reflection indicates that the above type of regression is indeed an attempt to mathematically relate two variables when indeed there is no physical or logical basis to believe that such a relation exists. In other words, no one can prevent one from conducting a correlation analysis between the temperature in New Delhi and political activity in Washington if one so wishes. That is, any two seemingly unrelated variables can be assigned to the x and y axes of a graph to get a line or curve to pass through the points of the scatter diagram in the least-square sense. To illustrate the point consider the values of x and y:

x:	0	5	19	28	0	39	41	42	46	52	60	61	64	66	84	97
y:	35	57	41	42	42	31	49	37	51	45	55	58	34	36	50	35

A scatter diagram for this data is shown in Fig. 4.3 (see the points denoted by crosses). No evidence of any type of relation is discernible from the scatter diagram. However, suppose one has reason to believe that y ought to be a function of x. Maybe x is related to y through a third variable, say, z, via

$$y = z(1/x) \qquad \text{or} \qquad z = xy$$

Now z is plotted against x, and suddenly an apparent linear relationship appears (see the points denoted by circles in Fig. 4.3). After all, x and y are related! This is all pure nonsense because the table of values for x and y are obtained from a table of random numbers. Precisely for reasons like this, one must be careful in detecting *artificial* or *spurious correlations*.

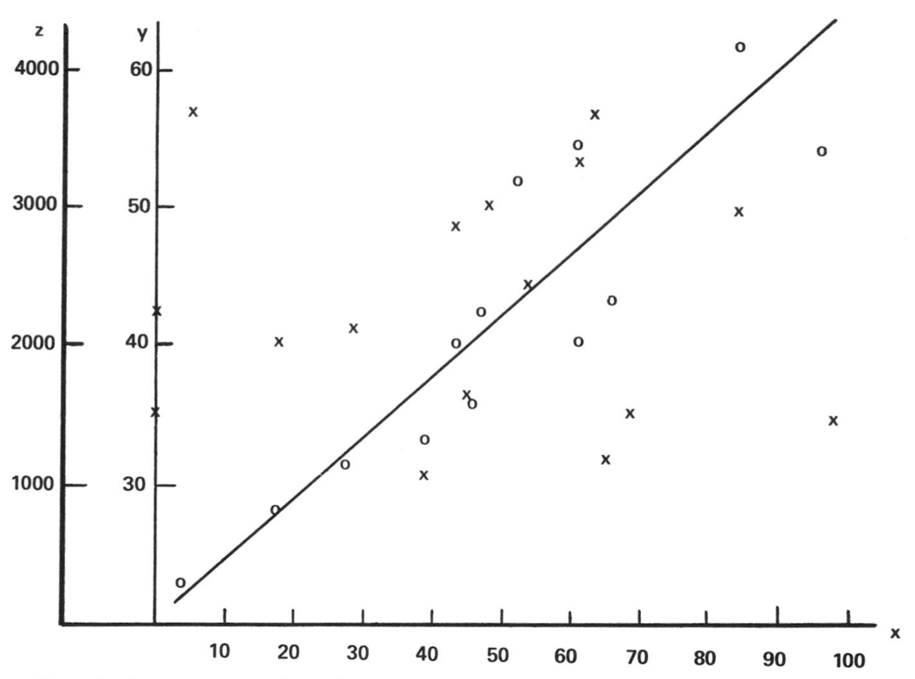

Fig. 4.3 Demonstration of spurious correlation. The crosses are y's. The circles are z's.

Homoscedasticity

A basic assumption of the least squares regression technique is that the deviation of the error terms is constant; it is independent of the magnitude of the variables. This is the *homoscedastic* assumption. When the assumption is violated, that is, when the data are *heteroscedastic*, measures of the accuracy of the parameters will be misleading. Heteroscedasticity can be detected easily by plotting the residual against the experimentally observed data, as

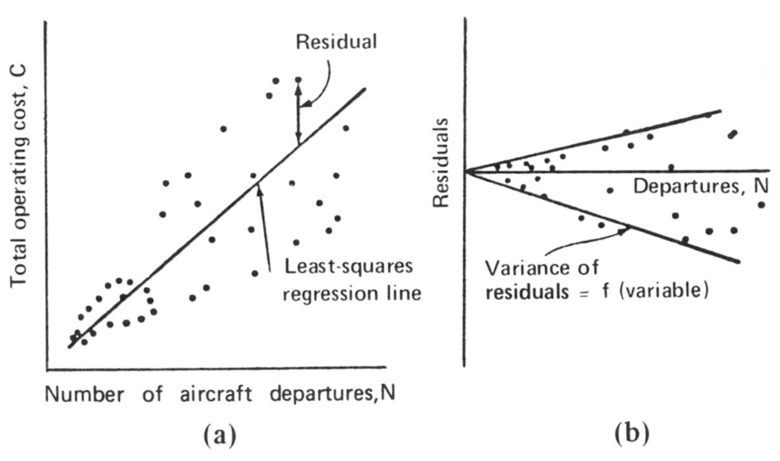

Fig. 4.4 Example of heteroscedastic data. (Reproduced with permission from R. deNeufville and J. H. Stafford, *System Analysis for Engineers and Managers*, copyright © 1971, McGraw-Hill, New York.)

shown in Fig. 4.4. The cure for heteroscedasticity is to transform the data so that the variance of the residuals will become more nearly constant. If the size of the residual increases directly with the absolute magnitude of the variable, heteroscedasticity can be cured quickly by performing regression on the logarithm of the appropriate variable.

Multiple Linear Regression

One method of improving the quality of a model is to include as many factors as possible in the set of independent variables. If the weight y depends upon the height x_1, sex x_2, nutrition $x_3, \ldots,$ then one can try a model such as

$$y = a_0 + a_1x_1 + a_2x_2 + \cdots + a_Mx_M = a_0 + \sum_{i=1}^{M} a_ix_i \tag{15}$$

where $x_1, x_2, \ldots, x_M$ are factors influencing the weight. To establish a relation such as Eq. (15), first, it is necessary to have data on y as well as on $x_1, x_2, \ldots, x_M$. Assume, as before, that a sample of size N has been chosen and the values of all variables involved are recorded. (Note the difference between M and N.) Proceeding as before, one can define r, equate its partial derivatives with respect to $a_0, a_1, \ldots, a_M$ to zero, and arrive at a set of M normal equations:

$$\begin{aligned}
a_0N + a_1\sum x_{1i} + a_2\sum x_{2i} + \cdots + a_M\sum x_{Mi} &= \sum y_i \\
a_0\sum x_{1i} + a_1\sum x_{1i}^2 + a_2\sum x_{1i}x_{2i} + \cdots + a_M\sum x_{1i}x_{Mi} &= \sum x_{1i}y_i \\
&\vdots \\
a_0\sum x_{Mi} + a_2\sum x_{Mi}x_{1i} + a_2\sum x_{Mi}x_{2i} + \cdots + a_M\sum x_{Mi}^2 &= \sum x_{Mi}y_i
\end{aligned} \tag{16}$$

where all the summations are from $i = 1$ to N. From the first equation of this set

$$a_0 = \left(\sum_{i=1}^{N} y_i - \sum_{j=1}^{M} a_j \sum_{i=1}^{N} x_{ij} \right) \Big/ N \tag{17}$$

Substituting for a_0 in the remaining equations of the set, and making use of the notation

$$\begin{aligned} X_1X_2 &= \sum_{i=1}^{N} x_{1i}x_{2i} - \left(\sum_{i=1}^{N} x_{1i} \sum_{i=1}^{N} x_{2i} \right) \Big/ N \\ X_1Y &= \sum_{i=1}^{N} x_{1i}y_i - \left(\sum_{i=1}^{N} x_{1i} \sum_{i=1}^{N} y_i \right) \Big/ N \end{aligned} \tag{18}$$

and so on, the normal equations appear as

$$\begin{aligned} a_1X_1{}^2 + a_2X_1X_2 + \cdots + a_MX_1X_M &= X_1Y \\ a_1X_1X_2 + a_2X_2{}^2 + \cdots + a_MX_2X_M &= X_2Y \\ &\vdots \\ a_1X_MX_1 + a_2X_MX_2 + \cdots + a_MX_M{}^2 &= X_MY \end{aligned} \tag{19}$$

In matrix notation, the system of equations to be solved for $a_i, i = 1, 2, \ldots, M$ can be written as

$$\begin{pmatrix} g_{11} & g_{12} & \cdots & g_{1M} \\ g_{21} & g_{22} & \cdots & g_{2M} \\ & & & \vdots \\ g_{M1} & g_{M2} & \cdots & g_{MM} \end{pmatrix} \begin{pmatrix} a_1 \\ a_2 \\ \vdots \\ a_M \end{pmatrix} = \begin{pmatrix} c_1 \\ c_2 \\ \vdots \\ c_M \end{pmatrix} \tag{20}$$

or more compactly as

$$\mathbf{Ga} = \mathbf{c}$$

where

$$\begin{aligned} g_{ll} &= \sum_{i=1}^{N} x_{li}{}^2 - \left(\sum_{i=1}^{N} x_{li} \right)^2 \Big/ N \qquad \text{for} \quad l = 1, 2, \ldots, M \\ g_{kl} = g_{lk} &= \sum_{i=1}^{N} x_{ki}x_{li} - \left(\sum_{i=1}^{N} x_{ki} \sum_{i=1}^{N} x_{li} \right) \Big/ N \qquad \text{for} \quad l \neq k \\ c_l &= \sum_{i=1}^{N} x_{li}y_i - \left(\sum_{i=1}^{N} x_{li} \sum y_i \right) \Big/ N \qquad \text{for} \quad l = 1, 2, \ldots, M \end{aligned} \tag{21}$$

Solution of Eq. (20) can be written compactly as

$$\mathbf{a} = \mathbf{G}^{-1}\mathbf{c} \tag{22}$$

where $\mathbf{G}^{-1}$ is the inverse of the matrix $\mathbf{G}$.

Multicollinearity

From the above discussion it is evident that, computationally, solving the regression equation is tantamount to inverting a matrix. Even though this appears to be a straightforward task at first sight, deeper analysis shows that obtaining $\mathbf{G}^{-1}$ is beset with difficulties. The reason for these difficulties can be explained from many viewpoints. In practice the independent variables x_i are related to each other in some fashion. For instance, suppose that one is attempting to find a relation between precipitation (y) and temperature (x_1) and humidity (x_2). However, humidity and temperature are related to each other. This situation, called *multicollinearity*, makes it very difficult to estimate the parameters. From an information-content point of view, the set of equations (19) is not generally independent or the matrix **G** in Eq. (20) is generally *ill conditioned.*

This difficulty suggests that there may be something basically unrealistic in the use of multiple linear regression to describe highly complex nonlinear systems. What is really needed is a method of describing the effect of a natural group of factors that expresses how they *actually interact* rather than being forced by an artifical mathematical relation with a linear additive structure. Physical or biological processes exhibiting saturation phenomena are abundant. Linear regression models can never truly duplicate the saturation effects, asymptotic behaviors, and so forth. The differential equation approach is a powerful tool to handle these types of problems.

Finding the Best Type of Formula

There is nothing in the method of least squares that restricts the use of a straight line

$$\hat{y} = a + bx$$

as a model to an observed phenomenon. The preceding analysis can be easily extended to a variety of simple nonlinear models. Typical and elementary nonlinear models are:

the power relation	$y = bx^{\alpha}$, $y = ab^{x}$
the exponential relation	$y = ae^{\beta x}$
the double exponential relation	$y = ae^{\alpha x} + be^{\beta x}$
the inverse type relation	$y = a + b/x$; $y = 1/(a + bx)$
the polynomial relation	$y = a + bx + cx^2 + dx^3$
the logistic relation	$y = k/(1 + be^{-\alpha x})$

If the scatter diagram indicates a fairly smooth curve without sharp turns or bends, it is likely that a nonlinear relation exists. As an aid in determining which of the nonlinear relations to use in any given problem, the following suggestions are offered:

(1) If the observed data give a straight line graph when plotted on a *logarithmic paper*, try formulas of the type

$$y = bx^{\alpha}$$

(2) If the data give a straight line when plotted on a *semilogarithmic paper*, the proper choice would be

$$y = ae^{\beta x} \qquad \text{or} \qquad y = ab^{x}$$

(3) If the points $(1/x, y)$ or $(x, 1/y)$ lie on a straight line when plotted on an ordinary rectangular coordinate paper, the proper formula is

$$y = a + b/x$$

in the first case and

$$y = 1/(a + bx)$$

in the second case. If none of the above give satisfactory results, a polynomial model

$$y = a_0 + a_1 x + a_2 x^2 + \cdots + a_n x^n$$

can be used to fit any data by taking a sufficient number of terms in the series.

(4) Many growth processes such as those occurring in biological systems are known to be governed by the logistic law of growth. Regression methods, such as those described so far, have been found to be ill suited to this kind of nonlinear problem exhibiting saturation effects. This problem can be overcome by using the *differential equation* for logistic growth. This topic will be discussed briefly in Section 3 and at length in Chapter 10.

3 MORE COMPLEX PARAMETER ESTIMATION PROBLEMS

In many scientific investigations, the use of mathematics is dictated not so much by a desire to solve equations as by the desire to test hypotheses. An important problem in this context is that of deriving or defining an appropriate model from observed data. This, in its various aspects, is the inverse problem. The nature and various facets of an inverse problem were already discussed in Chapter 2. The conceptual difficulty in solving an inverse problem stems from several angles. In the design of an experiment the choice of data to be collected is determined in part by prior knowledge about the system and this prior knowledge is usually obtained from some type of a model, even a descriptive model. Thus, data and theory are closely intertwined.

The question of uniqueness is very important in the context of solving an inverse problem. Because data gathering itself is an error prone operation, there is really no way of knowing how good a model is because the full set of possible alternatives is seldom available for comparison. Therefore, a useful criterion for selecting a good model is physical plausibility. That is, a good model should (a) fit the data accurately, (b) be theoretically consistent with realities, (c) have parameters that have physical meaning and that can be measured independently of each other, and finally (d) prove useful in making predictions.

A Trial and Error Procedure

Consider the problem of population growth. Suppose that there exists a means of counting the number of cells in a test tube at various times. For concreteness, let the population be N_0 at time $t = t_0$. The measured values of population at times $t_1, t_2, t_3, \ldots$ are plotted as shown in Fig. 4.5. Now using these observed data it is proposed to test the hypothesis that the rate of growth is directly proportional to the population size.

What is meant by "test of a hypothesis"? If the population N indeed grows at a rate proportional to population size, then N must satisfy the equation

$$\frac{N(t + \Delta t) - N(t)}{\Delta t} = kN(t); \qquad N(t = 0) = N_0 \tag{1}$$

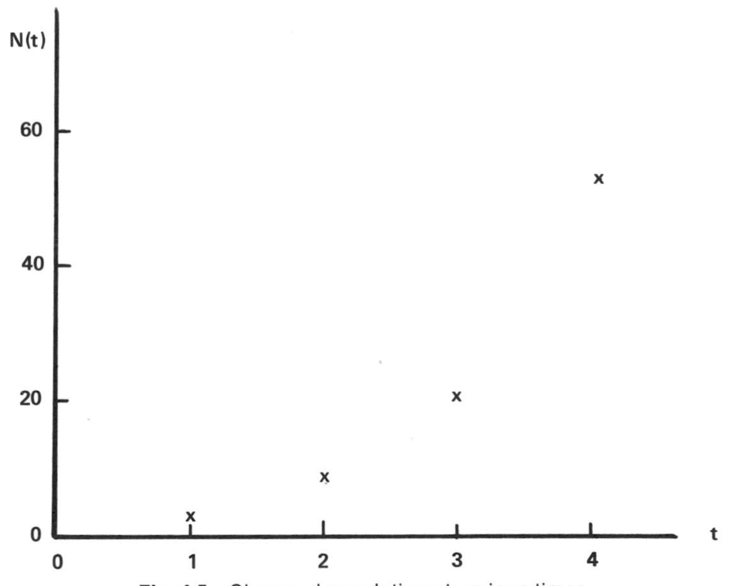

Fig. 4.5 Observed population at various times.

As the variables N and t of Eq. (1) are observed and plotted (Fig. 4.5), the only unknown is the constant k. Our goal is to determine that value of k which when substituted in Eq. (1) yields a solution curve that passes through as many observed points as possible.

In the present case, determination of k is quite simple. One can proceed to solve Eq. (1), by first writing it as

$$N(t + \Delta t) = (1 + k(\Delta t))N(t)$$

Therefore,

$$N(\Delta t) = (1 + k(\Delta t))N_0$$
$$N(2\,\Delta t) = (1 + k(\Delta t))N(\Delta t) = (1 + k(\Delta t))^2 N_0$$

and, using induction,

$$N(n\,\Delta t) = (1 + k(\Delta t))^n N_0 \qquad \text{for} \quad n = 1, 2, \ldots \tag{2}$$

Since N_0 is known, one can calculate $N(n\,\Delta t)$, $n = 1, 2, \ldots$, etc., for any chosen value of k. Thus, one way of estimating k is to solve Eq. (2) for various values of k, and compare the results with the data given in Fig. 4.5. The value of k that best fits the data is the required answer.

A Graphical Procedure

Alternatively, one can quickly recognize that Eq. (1) as $\Delta t \to 0$ is the same as the differential equation

$$\frac{dN}{dt} = kN; \qquad N(0) = N_0 \tag{3}$$

whose solution can be written as

$$N(t) = N_0 e^{kt} \tag{4}$$

or

$$\ln N(t) = \ln N_0 + kt \tag{5}$$

Hence, ln $N(t)$ plotted against t should be a straight line with a slope k, as shown in Fig. 4.6. Thus, k is obtained by simply measuring the slope of the line in Fig. 4.6. Actual plotting of this graph can be considerably simplified by using a specially marked graph paper called *semilog paper*. In practice, as a result of either experimental error or deviations from ideal behavior, all the points may not fall on a straight line. In such cases, an approximate value of k can be obtained by using one of several approximating procedures such as the least-squares method.

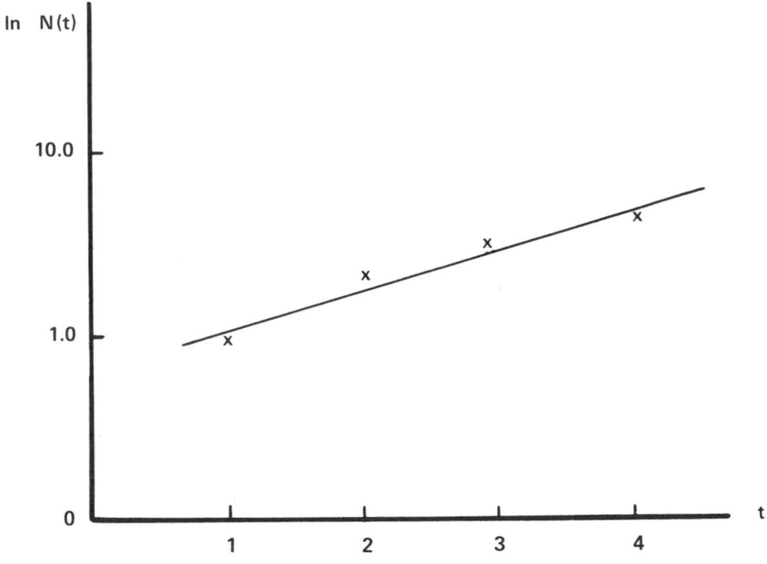

Fig. 4.6 A semilog plot of population growth.

Estimation of the unknown parameter k was found to be a simple matter in this case primarily because the observed growth process, more or less, followed an exponential law. If the observations were continued over prolonged periods of time, competition for food increased and it transpired that under such circumstances the exponential growth curve tends to level off. It would be useful if the graphical method developed above could be adapted to this situation. Consider once again the problem of population growth. Let us also assume that the size of a cell colony, measured at hourly intervals, yields a set of six points of the type shown in Fig. 4.7. The task here is to determine the equation whose solution curve fits the six observed points. From the experience gained so far, it is tempting to try $N = e^{bt}$ as a possibility. Indeed, this hunch is pretty good, and $N = e^{bt}$ gives a reasonable fit over the first four hours of growth (see the dotted line in Fig. 4.7). Afterward the growth levels off, and $N = e^{bt}$ fails to fit. This in turn means that $N = e^{bt}$ is only a good approximation for the initial stages of growth, and $(dN/dt) = bN$ is an acceptable model for small values of t.

For large values of t, the leveling off process suggests that the population is not growing as fast as expected. That is, some nonlinearity is putting a lid on the growth process. Probably members of the colony are perishing as N increases. This is possible because overcrowding increases the death rate. Therefore, one can possibly try

$$k = b - mN$$

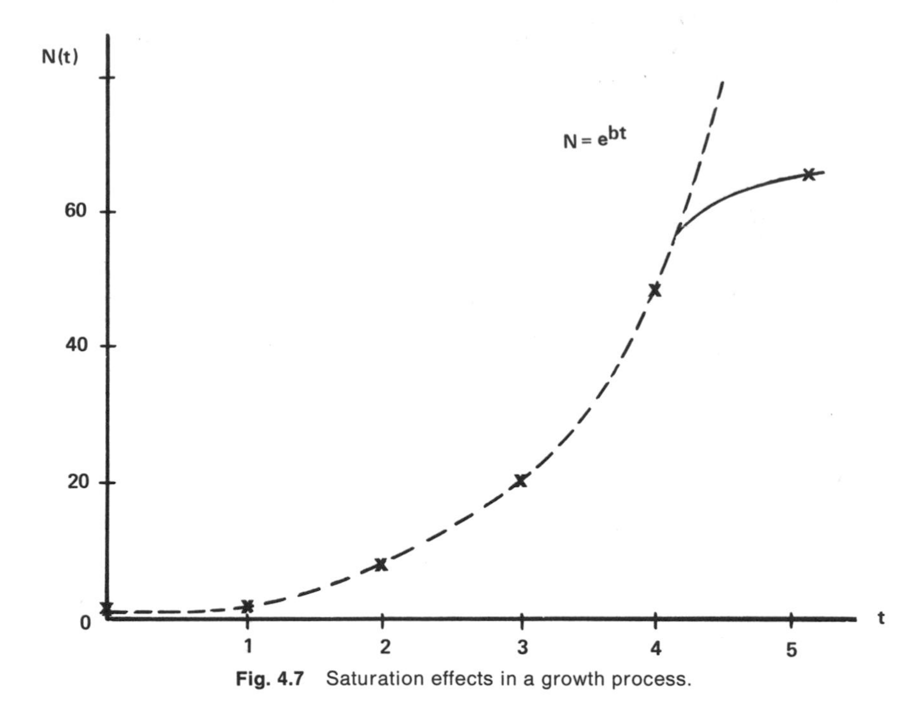

Fig. 4.7 Saturation effects in a growth process.

suggesting that the mortality is increasing with N. Now Eq. (3) can be modified as

$$\frac{dN}{dt} = bN - mN^2; \qquad N(0) = N_0 \tag{6}$$

Perhaps solution of Eq. (6) might yield a better fit.

The model described by Eq. (6) is more complicated than the one in Eq. (3) due to several reasons. First, Eq. (6) is now nonlinear and therefore more difficult to solve. Second, Eq. (6) contains two parameters, b, the birth rate and m, the mortality rate, whereas Eq. (3) contains only k, the "fertility" rate. Thus, if Eq. (6) describes the observed data, it would be far superior to Eq. (3).

A convenient way of roughly estimating b and m without having to solve the differential equation is as follows. Let $N(t)$ denote the solution of Eq. (6) at any time t and let N_k be the experimentally observed size of the population at the kth hour. As the first four experimentally observed values fit closely with $N = e^{bt}$, an estimate of b can be obtained, as before, by plotting the first four points on a semilog graph paper. To obtain an estimate of m, it is necessary to keep the colony under observation until a steady population has been reached. Steady population level implies that the rate of growth of population is zero. That is, $(dN/dt) = 0$, which in turn means

$$bN - mN^2 = 0 \tag{7}$$

where the N used in Eq. (7) is the steady state population. To emphasize this point, Eq. (7) is rewritten as

$$bN_s - mN_s^2 = 0 \tag{8}$$

where N_s is the steady state population. As long as this steady state population is not zero, one can write

$$m = b/N_s \tag{9}$$

Thus, estimates of b and m are obtained without too much difficulty.

Application of this simple technique often encounters difficulties in practice. For example, it may be quite difficult to measure the size of the cell colony when it is very small. Second, the differential equation that assumes N as a continuous variable may not be a good model when N is very very small. Third, it could be very time consuming to wait for the population to reach a steady state.

Parameter Estimation by a Search Technique

The heuristic technique of estimating the parameter values, presented earlier, has its own merits and should always be used whenever feasible. For equations that are a little more complex a good deal of intuition, experience, and insight into the problem is necessary for any meaningful estimation of parameters. Furthermore, if the unknown parameters are more than a few, as is the case with large systems, heuristic techniques may not work all the time. What is needed is a general and systematic procedure that yields fast results. One such method is to conduct an exhaustive experiment by solving Eq. (6) again and again with various sets of b, m, and N_0, if N_0 is also unknown, until the solution with the required characteristic is found. This may be an awkward approach if hand calculations are contemplated, but it is a logical alternative if high-speed computers are used.

For simplicity of argument let us assume that both b and N_0 in Eq. (6) are known to start with or have been determined according to some method or other. Now it is required to estimate the value of the only other unknown left, namely m. The solution of Eq. (6) therefore depends on the value of m and time t. This can be denoted by writing the solution as $N(m, t)$—thus explicitly showing the dependence of N on both m and t. Now, a simple-minded procedure is to pick a value for m arbitrarily, say m_1, substitute it in Eq. (6), solve it and compare this solution $N(m_1, t)$ with the observed data N_k, $k = 1, 2, \ldots, K$. Then, using the knowledge gained so far, the value of

m_1 can be altered to m_2 and to m_3 and so on until a *satisfactory* solution to Eq. (6) is obtained. How to measure the degree of satisfaction attained is once again a choice left to the systems analyst. One way, which is already familiar to us, is to attempt a least-squares fit between the solution $N(m, t)$ and the observed data N_k, $k = 0, 1, 2, \ldots, K$. For instance, an error E can be defined as

$$E = \sum_{k_1=1}^{K} (N(m, t_k) - N_k)^2 \tag{10}$$

As m assumes values $m_1, m_2, \ldots, m_M$, Eq. (6) is solved repeatedly, and the corresponding value of E is computed. The value of m that yields the smallest value of E thus gives an estimate of the unknown.

Curve-Peeling Techniques

The graphical procedure described earlier is characterized by the fact that there is only one exponential term in the solution of the underlying differential equation. In many problems, the solution is characterized by the presence of several exponential terms. For concreteness, consider a system which, under observation, produced the data, shown in Table 4.3, as it evolved in time. From a firsthand knowledge about the system, there is reason to believe that the above data probably represents a model that can be represented by a mathematical structure such as

$$x(t) = A_1 \exp(-\lambda_1 t) + A_2 \exp(-\lambda_2 t) + A_3 \exp(-\lambda_3 t) \tag{11}$$

where $\lambda_1 > \lambda_2 > \lambda_3 \geq 0$. It is required to find $A_1, A_2, A_3, \lambda_1, \lambda_2$, and λ_3.

The first step of the peeling procedure is to rearrange Eq. (11) as

$$\begin{aligned} x(t) = A_3 \exp(-\lambda_3 t)[1 &+ (A_1/A_3) \exp[-(\lambda_1 - \lambda_3)t] \\ &+ (A_2/A_3) \exp[-(\lambda_2 - \lambda_3)t]] \end{aligned} \tag{12}$$

TABLE 4.3

Data for the Peeling Procedure

t	$x(t)$	$y(t)$	$z(t)$	t	$x(t)$	$y(t)$	$z(t)$
0	100.00	70.00	50.00	7	8.00	0.61	—
1	47.85	23.28	11.15	8	6.42	0.37	—
2	29.96	9.84	2.49	9	5.18	0.22	—
3	21.49	5.01	0.55	10	4.19	0.13	—
4	16.31	2.83	0.12	11	3.40	0.08	—
5	12.70	1.67	0.03	12	2.77	0.05	—
6	10.04	1.00	0.006	13	2.25	0.03	—
				14	1.84	0.02	—
				15	1.50	0.01	—

If $|\lambda_1|$ and $|\lambda_2|$ are sufficiently larger than $|\lambda_3|$, then for sufficiently large t, the second and third terms in the parenthesis on the right side of Eq. (12) are small enough when compared to unity to be negligible. Therefore, for large t, Eq. (12) can be approximated by

$$x(t) \simeq A_3 \exp(-\lambda_3 t) \tag{13}$$

or

$$\ln x(t) = \ln A_3 - \lambda_3 t \tag{14}$$

Thus, a plot of $\ln x(t)$ versus t, for large t, should give a straight line with intercept $\ln A_3$ on the ordinate and a slope $-\lambda_3$. Therefore, the data given above is plotted on a semilog paper (curve (a) in Fig. 4.8) and the value of $\ln A_3$ and $-\lambda_3$ are obtained from the tangent drawn to the curve at large values of t. From the figure $A_3 = 30$ and $\lambda_3 = 0 \cdot 2$.

Having determined A_3 and λ_3, the second step of the peeling procedure is to subtract $A_3 \exp(-\lambda_3 t) = 30 \exp(-0.5t)$ from the data in Table 4.3 to

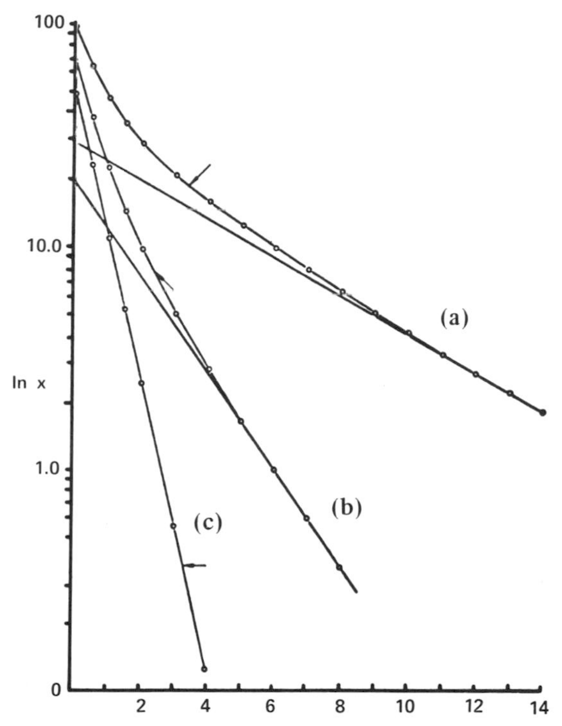

Fig. 4.8 Graphical peeling method applied to function (a) $x(t) = 50 \exp(-1.5t) + 20 \exp(-0.5t) + 30 \exp(-0.2t)$. (b) $y(t) = 50 \exp(-1.5t) + 20 \exp(-0.5t)$; (c) $z(t) = 50 \exp(-1.5t)$.

get $y(t)$ as shown in the third column of Table 4.3. Now

$$y(t) = A_1 \exp(-\lambda_1 t) + A_2 \exp(-\lambda_2 t)$$
$$= A_2 \exp(-\lambda_2 t)[1 + (A_1/A_2) \exp[-(\lambda_1 - \lambda_2)t]] \tag{15}$$

Again, if $|\lambda_1|$ is sufficiently larger than $|\lambda_2|$, the above procedure can be repeated and $y(t)$ can be approximated by

$$y(t) \simeq A_2 \exp(-\lambda_2 t) \qquad \text{or} \qquad \ln y(t) = \ln A_2 - \lambda_2 t \tag{16}$$

The values of A_2 and λ_2 are obtained as before by measuring the intercept and the slope of the tangent of a plot of ln $y(t)$ versus t (curve b in Fig. 4.8). From the figure, $A_2 = 20$ and $\lambda_2 = 0.5$.

By repeating the procedure once again the values of A_1 and λ_1 were found to be respectively 50 and 1.5. Thus,

$$x(t) = 50 \exp(-1.5t) + 20 \exp(-0.5t) + 30 \exp(-0.2t) \tag{17}$$

4 ELEMENTARY STATE TRANSITION MODELS

Consider the problem of finding the shortest route between home and office of an individual in a city. In this problem, the collection {streets, cars, drivers, etc.} may be regarded as the elements of a system and the location of the individual's car at any time may be thought of as the state of the system. The task of finding the shortest route may be stated as the problem of finding the least time in which the system can be transferred from an initial state (home) to a final state (office). Indeed, the sequence of state transitions is a continuum. However, in this problem as it is considered, it is sufficient to distinguish a finite number of states: S_1 may refer to the car at street corner 1, S_2 at corner 2, and S_N at corner N. From this viewpoint, a graph of the road map may itself be regarded as a *state transition graph* or *state diagram*. Indeed, state diagrams can be profitably used to solve a variety of other problems.

Example 4. Consider the problem of three cannibals and three edible missionaries seeking to cross a river in a boat that can hold at most two persons and can be navigated by any combination of two individuals. Under no circumstances can the missionaries be outnumbered by the cannibals anywhere.

This problem has been solved time and again in various guises. One can give a simple verbal description as a solution to the problem:

(1) Send two cannibals across the river.
(2) Return one cannibal.
(3) Send two cannibals across the river.

(4) Return one cannibal.
(5) Send two missionaries across the river.
⋮

This problem can also be viewed as a discrete system changing its states. Let the ordered pair (m, c) denote the number of missionaries and cannibals on the first bank. The state of the system can be completely described by this ordered pair. Then, the above transitions can be written abstractly as

(1) $(3, 3) \to (3, 1)$
(2) $(3, 1) \to (3, 2)$
(3) $(3, 2) \to (3, 0)$
(4) $(3, 0) \to (3, 1)$
(5) $(3, 1) \to (1, 1)$
⋮

These states and transitions can also be represented pictorially as a state diagram. To build this diagram, the following convention is adopted. Every possible pair (m, c) is denoted by a point in the mc plane. As both m and c can take any of the four values 0, 1, 2, and 3, there are 16 possible points (or states) in the figure (see Fig. 4.9a). However, all these states are not admissible. As the cannibals can never outnumber the missionaries anywhere, the points for which $c > m > 0$ represent inadmissible states. Note that the states for which $c > m = 0$ are admissible as there are no missionaries to be eaten anyway. To emphasize this point Fig. 4.9a is redrawn as in

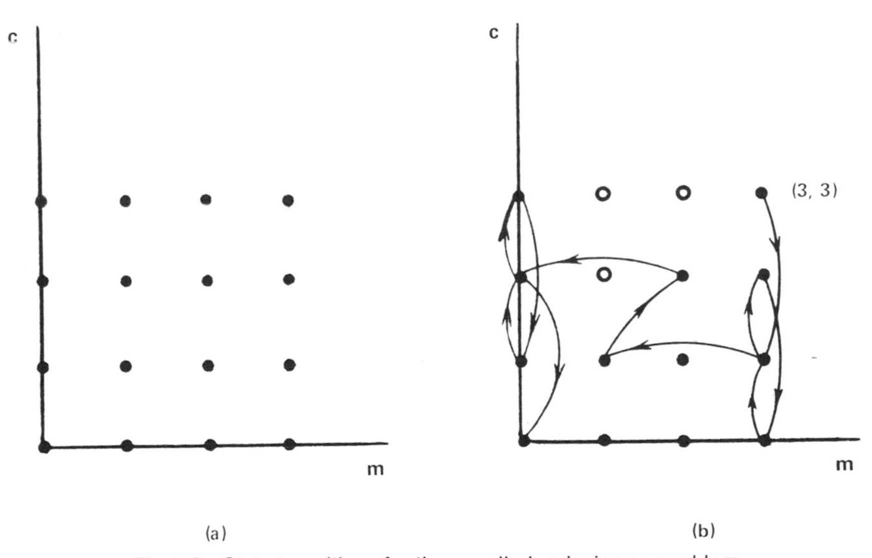

Fig. 4.9 State transitions for the cannibal–missionary problem.

Fig. 4.9b, showing the inadmissible states as little circles. With this notation, every passage of the boat represents a transition from one admissible point to another one in the diagram.

Note also that a passage of the boat from the first bank to the second bank corresponds either to a downward motion or a right to left motion in the state diagram. Similarly, a passage from the second bank back to the first bank corresponds to an upward motion or a left–right motion. Thus, starting from the given initial state, namely (3, 3), it is conceivable to reach the desired state, namely (0, 0), by a sequence of transitions by observing the simple rule that travel along the arcs leading down or to the right must alternate with travel leading to the left or upward. There is one more item to be considered. As the boat cannot hold more than two individuals, all admissible transitions are not reachable from any given state. For instance, one cannot go from state (3, 3) to (0, 3) in one step as it requires the boat to carry three. Stated differently, denote the set of all admissible states which are on the right-most column in Fig. 4.9b by R, on the left-most column by L and along the diagonal by D. Since the boat holds only two, it is impossible to go directly from a type R state to type L state and vice versa. That is, type L states are not reachable from a type R state. Any solution to the puzzle must therefore include a type D state as an intermediate step.

Observing these rules, a solution to the puzzle is shown in Fig. 4.9b via a sequence of arcs. By inspection, it is possible to tell whether there are any other solutions to the problem. ■

Example 5. Consider the task of building a model to the life cycle of a nuclear American family. The term "nuclear family" means a collection consisting of a man, a woman, and their children, excluding all other relatives and in-laws. The components of the system are obviously men and women of marriageable age and unmarried children. (At the time of this writing, the traditional monogamous union of a couple in wedlock is *not* regarded as the only way of defining marriage, at least by some.) In view of this, the states of a life cycle model are defined as follows:

Let M denote a married couple, $\bar{M}$ denote an unmarried couple, Φ denote no children, C denote children living at home, and $\bar{C}$ denote children living away from home.

State S_1: $(\bar{M}, \Phi)$, stands for an unmarried couple living together and the couple does not have any children.

State S_2: $(\bar{M}, C)$, stands for an unmarried couple and their children living at home.

State S_3: $(\bar{M}, \bar{C})$, stands for an unmarried couple and their children living away from home.

States S_4 (M, Φ); S_5 (M, C); and S_6 $(M, \bar{C})$ are similarly defined.

It is not difficult to see the kind of transitions that can take place between these states. Unlike the cannibal–missionary problem, it is now necessary to assign some probabilities to these transitions to lend reality to the dynamics of the social process. Let

$P(M)$ = probability of a couple getting married.
$P(C)$ = probability of having a child (not necessarily their own genetic offspring).
$P(L)$ = probability of a child leaving home (not necessarily to college).
$P(S)$ = probability of a spouse leaving home

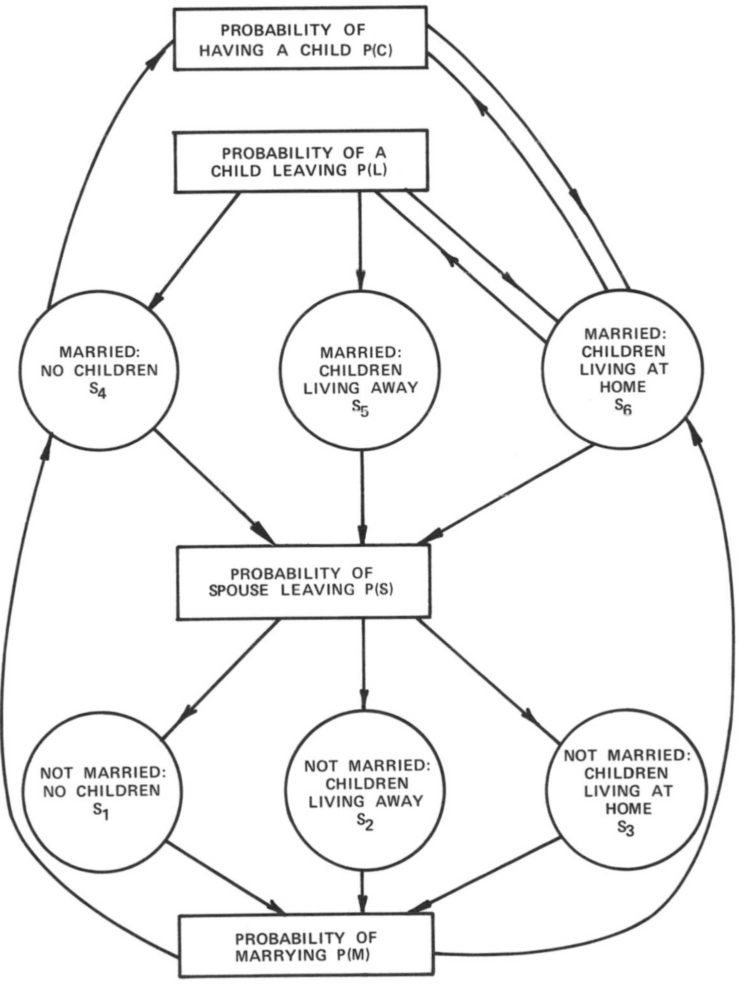

Fig. 4.10 A block diagram of a family life cycle.

A state graph of this problem can be constructed by treating each vertex as a state and each weighted directed edge as a transition from one state to another, the weights being the transition probabilities defined above. A graph describing this situation is drawn in two different ways in Figs. 4.10 and 4.11. Notice that although Fig. 4.10 is highly descriptive, it is not a state graph.

Using the state transition graph as a starting point, one can either simulate the dynamics of a family using a computer or answer many interesting questions. The intent in presenting this example is merely to show how one can adapt graph theoretic techniques to attack problems arising in a purely social context.

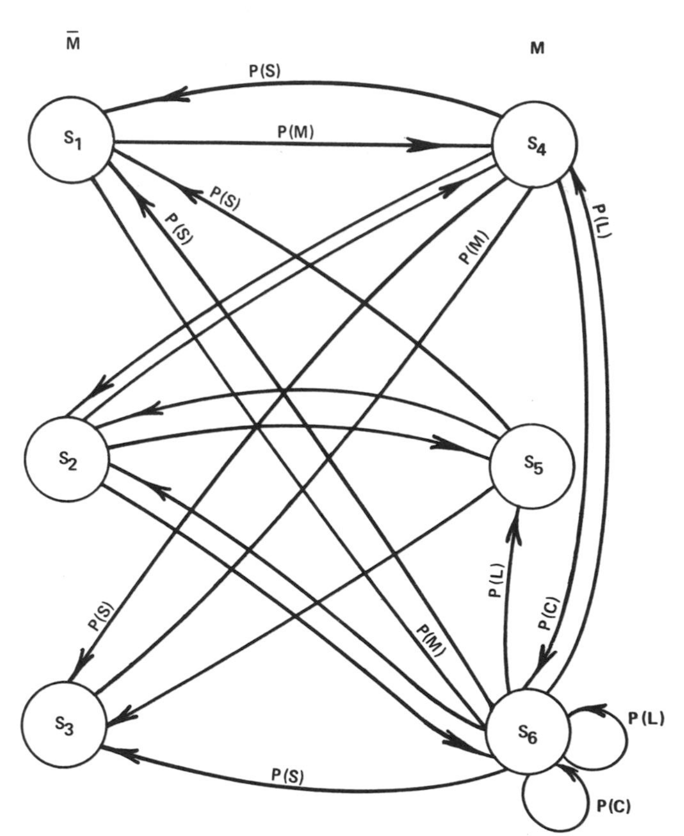

Fig. 4.11 A state transition model for the family life cycle. (Adapted from R. Werner and J. J. Verner, *A Pragmatic Approach to Social Systems Modeling.* Social Systems Simulation Group, San Diego, California.)

EXERCISES

1. Show that the equation of the straight line that fits the following data is $y = 4.75 + 0.395x$.

x:	2.0	4.0	6.0	8.0	10.0
y:	5.5	6.3	7.2	8.0	8.6

2. Repeat the problem in Example 3 by plotting the weight on the x axis and height on the y axis. Are x and y related by the same equation as before? Explain your answer.

3. Generalize the results of linear regression discussed in the text to the problem of establishing a relationship between a variable y and a set of variables x_1, x_2, x_3, x_n by using an equation of the form

$$y = b_0 + b_1 x_1 + b_2 x_2 + \cdots + b_n x_n$$

This procedure is called multiple linear regression.

4. (a) If the points on a scatter diagram fail to present a linear trend, a better fit to the data may be obtained by using the parabola, $y = b_0 + b_1 x + b_2 x^2$ and the method of least squares. Show that the parameters, a, b, and c can be obtained by solving

$$b_2\left(\sum_{i=1}^{N} x_i^2\right) + b_1\left(\sum_{i=1}^{N} x_i\right) + b_0 N = \sum_{i=1}^{N} y_i$$

$$b_2\left(\sum_{i=1}^{N} x_i^3\right) + b_1\left(\sum_{i=1}^{N} x_i^2\right) + b_0\left(\sum_{i=1}^{N} x_i\right) = \sum_{i=1}^{N} x_i y_i$$

$$b_2\left(\sum_{i=1}^{N} x_i^4\right) + b_1\left(\sum_{i=1}^{N} x_i^3\right) + b_0\left(\sum_{i=1}^{N} x_i^2\right) = \sum_{i=1}^{N} x_i^2 y_i$$

(b) Show that the following data fit an equation of the form $y = 3 + 0.4x - 0.01x^2$:

x:	2.00	4.00	6.00	8.00	10.00	12.00	14.00
y:	3.76	4.44	5.04	5.56	6.00	6.36	6.64

5. Generalize the above procedure when one wishes to fit an mth degree polynomial

$$y = b_0 + b_1 x + b_2 x^2 + \cdots + b_m x^m$$

to the data points.

6. (a) If the data points on a scatter diagram do not appear to have a linear trend, a power curve such as $y = b_0 x^m$ may be tried. Show that b_0 and m can be estimated by the formulas

$$\log b_0 = \frac{1}{N}\left(\sum_{i=1}^{N} \log y - m \sum_{i=1}^{N} \log x\right)$$

$$m = \frac{\sum \log x \sum \log y - N\sum \log x \log y}{(\sum \log x)^2 - N\sum(\log x)^2}$$

(b) For the following data, show that

$$b_0 = 3 \text{ and } m = 2$$

x:	1	2	3	4	5	6
y:	3	12	27	48	75	108

7. The number of individuals N from a given population whose income *exceeds* x is given by the law

$$N = A/x^b$$

This is called *Pareto's law* of income distributions. Find the parameters of Pareto's law from the following data. Also estimate the number of millionaires in the population.

Income class ($)	Number of people in that class
1000–5000	32,000,000
5000–25,000	1,100,000
25,000–250,000	110,000
Over 250,000	4,000

8. (a) Find expressions for b_0 and m in order to establish a relation such as $y = b_0 e^{mx}$ between variables x and y.

(b) Find an equation of the above form from the following data:

x:	2.0	3.0	4.0	5.0	6.0	7.0	8.0	9.0
y:	3.5	5.0	6.2	9.0	13.0	16.0	23.0	30.0

9. Construct a difference table for the following data:

x_1	x_2	x_3	x_4	x_5	x_6	x_7	x_8	x_9
y_1	y_2	y_3	y_4	$y_5 + \varepsilon$	y_6	y_7	y_8	y_9

where ε is an error committed in observing y_5. Construct the table up to the fifth difference and discuss the nature of propagation of ε in the table. If the location of where the error was committed is not known in advance, suggest a procedure for locating and correcting that error.

10. One value of y in the following data was recorded erroneously. Locate and correct that error

x:	1	2	3	4	5	6	7	8
y:	5	8	17	38	77	140	233	362

11. Consider a dynamical system represented by the system of differential equations, written in vector–matrix form,

$$\dot{\mathbf{x}} = \mathbf{A}\mathbf{x} + \mathbf{b}u$$

Determine the parameter matrices of the system. That is, determine the entries of the entities **A** and **b** using the following measurements and the regression technique. The relevant vectors and matrices are defined as follows:

$$\mathbf{x} = \begin{bmatrix} x_1 \\ x_2 \end{bmatrix}, \qquad \mathbf{A} = \begin{bmatrix} a_{11} & a_{12} \\ a_{21} & a_{22} \end{bmatrix}, \qquad \mathbf{b} = \begin{bmatrix} b_1 \\ b_2 \end{bmatrix}$$

The observed measurements are as follows:

t	0	1	2	3	4	5	6	7	8	9
x_1	1.0	0.99	0.97	0.96	0.95	0.945	0.94	0.935	0.93	0.925
x_2	0	−0.1	−0.19	−0.23	−0.28	−0.25	−0.22	−0.18	−0.14	−0.07
u	1.0	1.25	1.5	1.75	2.0	2.25	2.5	2.75	3.0	3.25

t	10	11	12	13	14	15	16	17	18
x_1	0.92	0.925	0.93	0.94	0.95	0.97	0.99	1.02	1.05
x_2	0	0.09	0.17	0.26	0.36	0.45	0.55	0.65	0.76
u	3.5	3.75	4.0	4.25	4.5	4.75	5.0	5.25	5.5

t	19	20	21	22	23	24	25	26	27
x_1	1.1	1.15	1.22	1.28	1.32	1.36	1.44	1.52	1.58
x_2	0.86	0.96	1.05	1.13	1.21	1.3	1.38	1.45	1.51
u	5.75	6.0	6.25	6.5	6.75	7.0	7.25	7.5	7.75

t	28	29	30	31	32	33	34	35	36
x_1	1.66	1.74	1.83	1.92	2.02	2.1	2.18	2.27	2.37
x_2	1.57	1.63	1.69	1.74	1.78	1.82	1.86	1.89	1.91
u	8.0	8.25	8.5	8.75	9.0	9.25	9.5	9.75	10.0

t	37	38	39	40	41	42
x_1	2.47	2.57	2.66	2.76	2.86	2.97
x_2	1.93	1.95	1.97	1.98	2.02	2.05
u	10.25	10.5	10.75	11.0	11.25	11.5

Hint: Approximate the derivative by a difference approximation such as

$$\dot{x} = [x(k+1) - x(k)]/\Delta t$$

where $x(k) = x(t_k)$. Now convert the equation into the discrete form

$$x(k+1) = Fx(k) + gu(k)$$

Now define $\mathbf{w}(k) = (x_1(k), x_2(k), u(k))^T = (w_1(k), w_2(k), w_3(k))^T$ and

$$\mathbf{\Phi} = \begin{bmatrix} a_{11} & a_{12} & b_1 \\ a_{21} & a_{22} & b_2 \end{bmatrix}$$

Now the discrete form of the given equation can be rewritten as

$$\mathbf{x}(k+1) = \mathbf{\Phi}\mathbf{w}(k)$$

Determine the parameters using multiple linear regression. (Answer: $a_{11} = 0$, $a_{21} = 1$, $a_{21} = -4$, $a_{22} = -2.6$; $b_1 = 0$, $b_2 = 1.5$.)

SUGGESTIONS FOR FURTHER READING

1. For a rigorous discussion of regression analysis, see

   C. Daniel and F. S. Wood, *Fitting Equations to Data*. Wiley, New York, 1971.

   N. R. Draper and H. Smith, *Applied Regression Analysis*. Wiley, New York, 1967.

2. For further discussion of difference tables, see any elementary book on numerical analysis, for example,

   C. F. Gerald, *Applied Numerical Analysis*. Addison-Wesley, Reading, Massachusetts, 1970.

3. More problems, such as the cannibal–missionary problem, can be found in

   R. E. Bellman, K. L. Cooke, and J. A. Lockett, *Algorithms, Graphs, and Computers*. Academic Press, New York, 1970.

   T. L. Saaty and R. G. Busacker, *Finite Graphs and Networks*. McGraw-Hill, New York, 1965.

5

Forecasting

INTRODUCTION

Forecasting, or prediction of the probable future course of events, is the most pervasive and important element in efficient planning. The ultimate effect of a decision generally depends on the outcome of factors that cannot be foreseen at the time a decision is made. The role of forecasting cuts across many fields of endeavor—management, operations research, economics, and systems analysis, to mention a few. In the great majority of day to day decisions, forecasting is done using a heuristic and intuitive approach based on subjective evaluations. While dealing with complex systems involving large investments of money and manpower, mathematical and statistical methods would be more useful in imparting quantitative attributes to the quality of forecasts made. As no forecast can be made in a complete vacuum, an essential prerequisite for a good forecast is a good set of data. Precisely because of this, no discussion on modeling and simulation is complete without an adequate discussion of the procedures for measurement and subsequent handling of data.

1 THE NATURE OF DATA

In modeling circles there is a notion that can be roughly stated as follows: No meaningful data gathering effort can be launched without a specific reference to a model and no model can be meaningfully constructed without some preliminary data. This notion is more or less a truism and adequate attention should therefore be paid in coordinating data collection and

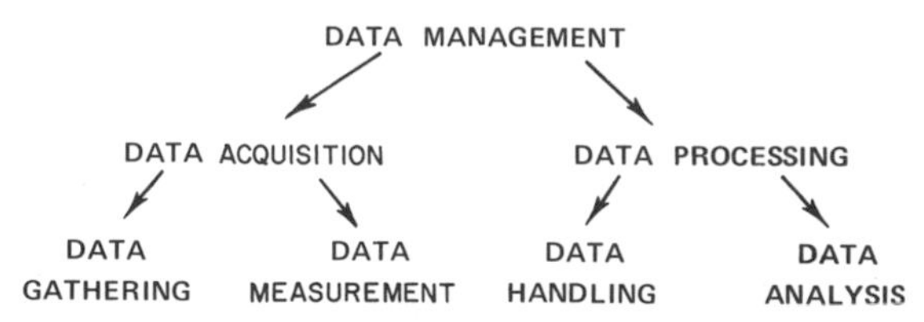

Fig. 5.1 Aspects of data management.

modeling efforts. Management of data, therefore, involves several operations. Some of the more important aspects of data management are shown in Fig. 5.1.

Data Measurement

Of all the activities shown in the lower echelons of Fig. 5.1, the problem of data measurement is perhaps the most difficult and least understood. The question of measurability and subsequent numerical representation of any particular attribute of an object is not an easy one to resolve. How can one measure attributes such as atmospheric pollution, social welfare or attitude of a decision maker? The questions that can be raised and the problems to be resolved in this context were already discussed at some length in Chapter 2.

Data Gathering

Data gathering methods as well as the means and devices of data gathering are strongly influenced by the nature of data required by a particular problem. Perhaps the easiest way of gathering data is to look for sources of data that are already collected.

Data Sources

Typical sources for data that have already been collected are local, state, and federal government agencies. In most countries the equivalents of the U.S. Geological Survey, Weather Bureau, and the Bureau of the Census do exist, and they serve as excellent sources of data. Some data that may not fall within the jurisdiction of a country can be obtained from the appropriate United Nations agency such as the World Health Organization or the World Meteorological Organization. Foundations, institutions, and other nonprofit organizations such as the National Science Foundation in the United States or its equivalent in other countries also serve as potential sources of data.

Professional associations such as the Institute of Electrical and Electronic Engineers (IEEE), the American Medical Association (AMA), insurance companies, legal associations, libraries, and newspapers are also potential sources of data.

A disadvantage of this approach is that the data supplied by these agencies may not be in the format required by the person seeking the data. In this event, one may have to embark on a well-planned data-gathering project, even though this is an expensive operation. While planning experiments to gather data, it is useful to understand the types of data that one can collect.

Kinematic data deal with the motion of bodies in space and time. Thus, kinematic data are invariably collected with reference to a coordinate system and a time reference. Data about satellite orbits and movement of vehicles such as ships and aircraft are examples of this kind. Kinematic data are usually gathered on line.

Dynamic data provide information about temporal variation of an attribute. Variations in daily temperatures at a recording station, changes in land values, and fluctuations in stock market over a period of time are typical examples of dynamic data.

Static data provide information about attributes that do not change. The height of a building, length of a vehicle, and width of a traffic lane are examples of static data.

Statistical data provide information about the distribution or arrangement of measurements. Data such as heights of children in a school are typical of this kind.

Behavioral data refer to information about the nature of behavior of individuals or groups of individuals. Attitudes of individuals to political and public issues are typical examples of behavioral data. These types of data are usually collected by surveys, questionnaires, or other types of public opinion polls.

Statistical sampling is an essential part of any data collection activity. Some of the more important methods of sampling are: straightforward sampling, importance sampling, random sampling, stratified random sampling, two-stage sampling, randomized block, ratio estimate, systematic sampling, quota sampling, Russian roulette, split plot, fractional replication, nested hypercubes, compacted hypercubes, orthogonal designs, sequential bifurcation, Latin squares, and Youden squares.

Once a sampling strategy is adopted, the next step is to decide on the techniques and tools for obtaining samples. Samples can be obtained from field crews who regularly visit field stations, or they can be monitored from remote control devices, or samples can also be obtained from large scale remote sensing experiments.

Data Handling

The data handling activity consists of two aspects: data reduction and data storage. The data reduction activity essentially consists of converting ungrouped, unorganized data into regrouped, reorganized data. The goal of data reduction is to create order out of disorder. One should remember that what is organized data for one particular application may turn out to be unorganized data for another application. From this viewpoint, data reduction is a sorting problem. Data reduction problems can also be visualized as consisting of operations such as formating, editing, analog-to-digital conversion, or any other like operation involving data manipulation.

Data Storage

Data storage is an important operation and should be done with a view to retrieving segments of the data efficiently. Central to the data storage task is organization. Although some of the most potentially valuable information exists in the minds of individuals as a by-product of their experience, nevertheless, in terms of quantity, there is an abundance of recorded knowledge. Its bulk and heterogeneity is what makes organization imperative. Organization implies classifying, indexing, cataloguing, or otherwise identifying information content within a file, a library, or any other collection. The data storage activity should aim at practical methods of making recorded information available to a wide class of users.

Data Analysis

Data analysis is the process of extracting meaningful results from a mass of data. Typically, data analysis implies construction and use of algorithms for such activities as regression analysis, curve fitting, goodness of fit, hypothesis testing, and similar statistical techniques.

In practice, since routine operations done with data do not fall into any one particular category as described above, a combination of them are required. Operations such as pattern recognition for the identification of geometric and structural features from remote sensing measurements and synthetic generation of data belong to this category.

2 STATISTICAL ATTRIBUTES OF DATA

Statistical techniques play a fundamental role in several aspects of data management and model building. A thorough study of techniques for the treatment of data would take us too far afield. Only a brief discussion of the more important aspects of the problem will be presented in the sequel.

The terms *population* and *sample* were defined in the context of regression analysis. Two more basic concepts would be useful in modeling complex systems: random variable and random process. Random variables are discussed here. Random processes will be treated in Section 4.

A random variable, which is generally denoted by the upper case Latin character such as X, is a real number associated with some observable phenomenon. Yet, if the phenomenon were to be observed repeatedly under essentially uniform and identical conditions, the same real number would not necessarily result. As a simple example of a *random variable*, consider an experiment consisting of several trials of tossing a coin. Suppose that we say that variable X assumes the value $+1$ whenever heads appear and -1 if tails appear. Then the value that X is going to assume depends upon the outcome of a particular trial. If the outcome is "heads," the value of X is $+1$, if the outcome is "tails," the value of X is -1. Thus, X defines a mapping from the set of all possible outcomes of the experiment to the set $\{+1, -1\}$. This function X is called a random variable. (See also Appendix 2.)

What is the probability that "heads," for example, will appear on a trial? Intuitively we know that this probability is one-half, if the coin is a fair one. This does not mean that heads and tails alternate in a sequence of trials. All it says is that after a large number of trials, about one-half of the outcomes will be heads and one-half will be tails. Thus, probability is a quantitative measure of the chance or likelihood of an event occurring. Probability is expressed by a quantity P, given by

$$P = m/n$$

where m is the number of favorable outcomes and n is the total number of equally likely events.

Associated with every random variable, one can define a *probability distribution.* In simple terms, a probability distribution is a systematic arrangement of numerical data. Consider an experiment of measuring daily rainfall data in a given month. Assume that 0.1 inch of rainfall was recorded on one day, 0.2 inch on two days, 0.3 inch on three days, 0.4 inch on three days, etc. This information can be plotted on a graph as shown in Fig. 5.2. A smooth curve of the plot can be thought of as a *probability density function* $f(x)$ of the random variable X.

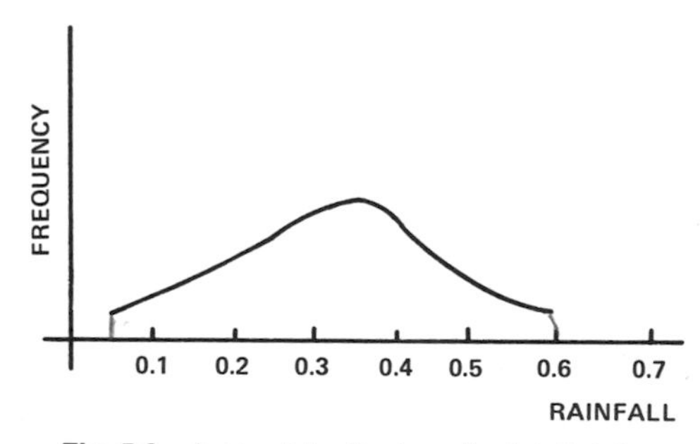

Fig. 5.2 A possible display of rainfall data.

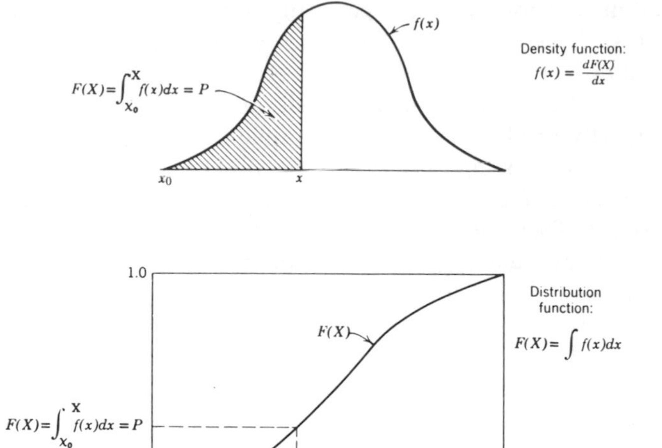

Fig. 5.3 Density and distribution functions. (Reproduced with permission from F. F. Martin, *Computer Modeling and Simulation*, Copyright © 1968, Wiley, New York.)

The integral of the density function is the *probability distribution* function $F_X(x)$:

$$F_X(x) = \int_{\infty}^{x} f_X(x)\, dx \tag{1}$$

The subscript merely denotes that the random variable under consideration is X. The total area below the density function is always represented as unity. The area under a portion of the probability density curve enclosed by two ordinates x and $x + dx$ is the probability that a random measurement x_i lies between x and $x + dx$. The relation between $f_X(x)$ and $F_X(x)$ is illustrated in Fig. 5.3. From this figure and Eq. (1) one can deduce that $F(x_1)$ is nothing but the probability that the random variable X assumes a value less than or equal to any specific real number x_1. This statement can be written

$$F(x_1) = P[X \leq x_1] \tag{2}$$

Attributes of Density Functions

A random variable is completely characterized by a specification of its density function which can be *continuous* or *discrete* and can take different shapes under different circumstances. It is convenient to concisely characterize a random variable X by one, two or more parameters rather than specifying the entire density function by displaying the curve.

The *expected value* of the random variable X denoted by $E(x)$ or μ, is defined as

$$\mu = E(x) = \int_{-\infty}^{\infty} x f_X(x)\, dx, \qquad \text{continuous case} \tag{3a}$$

$$= \sum x_i f_X(x_i), \qquad \text{discrete case} \tag{3b}$$

This is nothing but the average value or mean value of a random variable. This is also called the *first moment*.

The *second moment* is a measure of the dispersion or spread of the density curve about the mean. The second moment, also known as *variance*, for continuous and discrete cases, is given by

$$\sigma^2 = \int_{-\infty}^{\infty} (x - \mu)^2 f_X(x)\, dx, \qquad \text{continuous case} \tag{4a}$$

$$= \sum (x_i - \mu)^2 f_X(x_i), \qquad \text{discrete case} \tag{4b}$$

In either case, it can be proved that

$$\sigma^2 = E(x^2) - [E(x)]^2 = E(x^2) - \mu^2 \tag{5}$$

The square root of the variance σ is called the *standard deviation*. The third moment is a measure of symmetry of the density curve. The third moment, also known as *skewness*, is given by

$$g = \int_{-\infty}^{\infty} (x - \mu)^3 f_X(x)\, dx, \qquad \text{continuous case} \tag{6a}$$

$$= \sum (x_i - \mu)^3 f_x(x_i), \qquad \text{discrete case} \tag{6b}$$

Typical shapes of a density function for three different values of g are shown in Fig. 5.4.

The fourth moment, called *kurtosis*, is a measure of the peakedness of a curve and is given by

$$k = \int_{-\infty}^{\infty} (x - \mu)^4 f_X(x)\, dx \tag{7}$$

An analogous formula holds good for discrete case. If $\sigma = 1$, and $k > 3$,

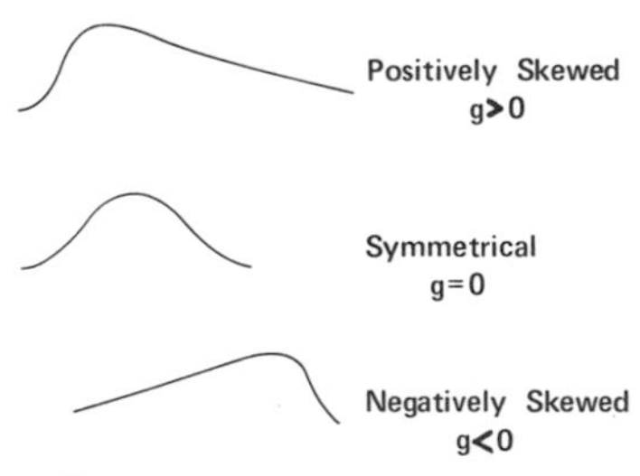

Fig. 5.4 Skewness of probability density function.

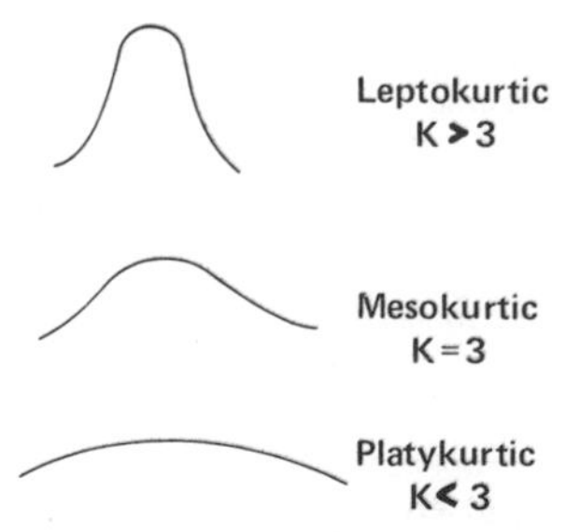

Fig. 5.5 Kurtosis of probability density function ($\sigma = 1$).

then the density curve is leptokurtic (tall and thin). For $k = 3$, it is mesokurtic (medium) and for $k < 3$, it is platykurtic (flat) (see Fig. 5.5).

It is important to remember that statistical attributes such as mean, variance, skewness, kurtosis, etc. can be defined with respect to a particular sample under consideration or with respect to the entire population. The utility of statistical methods stems from the fact that they enable one to draw certain inferences about the properties of the entire population by studying some of the properties of a relatively small sample. For example, the *sample mean* or sample average, defined by

$$\bar{x} = (1/N) \sum_{i=1}^{N} x_i \tag{8}$$

can be used as an estimate of the *population mean* μ defined in Eq. (3). Similarly, the *sample variance*, defined by

$$S^2 = [1/(N-1)] \sum_{i=1}^{N} x_i^2 - [N/(N-1)](\bar{x})^2 \tag{9}$$

can be used as an estimate of the *population* variance σ^2, defined in Eq. (4).

In an analogous fashion, a computationally useful formula for *sample skewness* G can be written as follows:

$$G = \frac{(1/N) \sum_{i=1}^{N} x_i^3 - (3/N^2)\left(\sum_{i=1}^{N} x_i^2\right)\left(\sum_{i=1}^{N} x_i\right) + (2/N^3)\left(\sum_{i=1}^{N} x_i\right)^3}{\left[(1/N) \sum_{i=1}^{N} x_i^2 - \bar{x}^2\right]^{1.5}} \tag{10}$$

3 PROBABILITY DISTRIBUTIONS AND THEIR UNDERLYING MECHANISMS

Many types of probability distributions are of interest in modeling and simulation. It is useful to understand, at least in a qualitative way, the circumstances under which a particular distribution arises in practice.

Some Important Distributions

Some of the more important distributions and their characteristics are briefly described here.

Normal Distribution

Most important of the distributions is the normal (or Gaussian) distribution whose density function

$$f(x) = [(2\pi)^{-1/2}/\sigma] \exp[-(x-\mu)^2/2\sigma^2] \tag{1}$$

has the familiar symmetrical bell shape. This distribution is characterized by two parameters: the mean μ and the standard deviation σ. The importance of normal distribution stems from the fact that it arises in a natural way in many problems of practical interest. Furthermore, a good part of mathematical and statistical analysis becomes relatively simple, if the underlying distributions are normal. Normal distributions appear in problems where a particular effect y is the result of a number of independent additive causes x_i, $i = 1, 2, \ldots, n$, regardless of the nature in which the individual x_i are distributed. For example, the quantity of water flowing in the lower reaches of a river tend to be normally distributed because this quantity is the sum of flows from a large number of tributaries.

Exponential Distribution

Consider a sequence of similar events starting at times $t_1, t_2, \ldots$. For example, at time t_1, one receives a telephone call. A second telephone call arrives at time t_2, and so on. It is relatively easy to show that these times are exponentially distributed with a density function

$$f(t) = \lambda e^{-\lambda t} \tag{2}$$

where λ, the parameter of the distribution, represents the average number of calls arriving per unit time interval. Indeed, exponential distribution can be found in any situation in which events occur in a random manner, independent of each other, and the chance of that event occurring in any interval is proportional to the duration of that interval. Births, deaths, accidents, and world conflicts are examples of this category.

Poisson Distribution

Consider a long time interval T and divide it into n small intervals and let $\Delta = T/n$. What is the probability that 0, 1, 2, 3, 4, ... events occur in the time interval Δ? Poisson distribution occurs as an answer to this question.

For example, the number of telephone calls coming during a 24-hr period can be very large. Yet, the probability of a call coming during a particular second is quite small. In a situation like this, we might expect the probability of getting 0, 1, 2, 3, 4, . . . telephone calls to follow the Poisson distribution, whose density function is defined by

$$f(x) = e^{-\lambda}(\lambda^x/x!); \qquad x = 0, 1, 2, \ldots \tag{3}$$

where $\lambda > 0$. Other examples following Poisson distribution are: the number of cars passing a preassigned point on a single lane road during a fixed time interval, the number of aircraft arriving on a runway in a fixed time interval, and the number of wars occurring in a specified time period. One should remember, however, that the events should occur independently. In heavy traffic, for example, cars tend to travel in platoons and aircraft get stacked before landing. Then, cumulative effects come into play and the distribution tends to become normal.

There is an interesting and well-known relation between exponential and Poisson distributions. It can be shown that if (1) the total number of events occuring during any given time interval is independent of the number of events that have already occurred and (2) the probability of an event occurring in the interval t to $t + \Delta t$ is approximately $\lambda(\Delta t)$ for all values of t, then (a) the density function of the interval t between the occurrence of consecutive events is exponential, and (b) the probability of x events occurring during time t is Poisson distributed.

Pareto Distribution

Attributes such as weights, heights, and IQs have the property that they tend to cluster around an average value with a relatively narrow spread. That is, it is relatively difficult to find people weighing as much as 300 kg or measuring 5 m tall or having an IQ of 300 because these values are too far removed from the average. However, it is not unusual to find people with extreme degrees of wealth or poverty. Attributes such as wealth are characterized by the Pareto distribution, whose density function is

$$f(x) = 1 + ax^{-v} \tag{4}$$

Pareto distribution can be quickly detected by plotting the data on a log–log paper because

$$\log(1 - f(x)) = \log a - v \log x$$

is a straight line with a negative slope. Pareto distribution was also found to be valid for the scientific productivity of people measured in terms of papers published.

Log–Normal Distribution

The log–normal distribution is defined by

$$f(x) = x^{-1}(2\pi\sigma^2)^{-1/2} \exp\{-(\log x - \mu)^2/2\sigma^2\} \tag{5}$$

where μ is the average value of log x. That is, if $y = \log x$ is normally distributed, then x is said to be log-normally distributed, Note that

$$\log f(x) = -\tfrac{1}{2}\log(2\pi\sigma^2) - (\log x - \mu)^2/2\sigma^2 - \log x \tag{6}$$

Therefore, the data plotted on a log–log paper is a parabola.

Both Pareto and log–normal distributions are characterized by long tails. William Shockley attempted to provide a phenomenological explanation for these long tails. He claims that a distribution with long tails appears if the successful completion of a number of tasks is a prerequisite for the occurrence of an event. Consider the event "the publication of a technical paper." The process of publishing a paper can be visualized as consisting of various tasks: (1) ability to think of a good problem, (2) ability to work on it, (3) ability to recognize an unusual and worthwhile result, (4) ability to recognize when to stop experimenting and start writing up, (5) ability to write precisely and to the point, and (6) ability to use the reviewer's criticism constructively and profit from it. If the probabilities of successfully performing each of the above tasks are $p_1, p_2, \ldots, p_6$, then the probability of successfully publishing a paper is the product

$$p = p_1 p_2 \cdots p_6 \tag{7}$$

The log–normal aspect of this becomes apparent by taking the logarithm of Eq. (7):

$$\log p = \log p_1 + \log p_2 + \cdots + \log p_6 \tag{8}$$

Even if the individual log p_i's are not normally distributed, the sum tends to be normally distributed. This observation, made once before, is based on a very fundamental theorem in statistics called the "central limit theorem."

Testing Goodness of Fit

In practice, an analyst is often faced with the problem of determining whether a set of data may be looked upon as values of a random variable having a given distribution. To illustrate this procedure, we consider the following problem. Observations were made on a single lane of a fairly busy highway, counting the frequency with which 0, 1, 2, 3, ... cars passed a marker during fixed time intervals of 5-min duration. This information is shown in the first two columns of Table 5.1. It is desired to test whether the

TABLE 5.1

Testing for Poisson Distribution

$N_5 = x_i$	Measured frequencies f_i	Poisson probabilities with $\lambda = 3$	Expected frequencies e_i
0	18	0.0498	21.9
1	53	0.1494	65.7
2	103	0.2240	98.6
3	107	0.2240	98.6
4	82	0.1680	73.9
5	46	0.1008	44.0
6	18	0.0504	22.2
7	10 }	0.0216	9.5 }
8	2 } 13	0.0081	3.6 } 14.3
9	1 }	0.0027	1.2 }

number of cars passing the marker in the given time period is a random variable having a Poisson distribution.

In Table 5.1, N_5 is the number of cars passing the marker in the 5-min interval. The measured frequency f_i tells the number of times the event described in column 1 occurred. For instance, on 53 different occasions only one car passed the marker during a 5-min interval. On 103 different occasions, two cars passed the marker, and so forth. Therefore, the entire test period took a total of 440 test intervals or $440 \times 5 = 2200$ min. During the entire test period, a total of $(0 \times 18) + (1 \times 53) + (\cdots) + (9 \times 1) =$ 1341 cars passed the marker. Therefore, on the average $(1341)/440 = 3.05 \approx$ 3.0 cars passed the marker in a unit (i.e., 5-min) interval. Using a value of $\lambda = 3.0$, Eq. (3) is evaluated for values of $x = 0, 1, 2, \ldots, 9$, and these results are shown in column 3 of the table. Multiplying these probabilities by the total frequency, 440, we get the theoretically expected frequency e_i, and this is shown in column 4 of the table. Comparing the measured frequency f_i with the expected frequency f_i, the agreement appears to be reasonably good.

A more formal method of testing the agreement between e_i and f_i is the chi-square test for the goodness of fit. In this test the quantity χ^2, defined by

$$\chi^2 = \sum_{i=1}^{m} \frac{(f_i - e_i)^2}{e_i} \tag{9}$$

is compared with $\chi^2_{\alpha,m-2}$ which is generally available in standard statistical tables. If $\chi^2 < \chi^2_{\alpha,s-t-1}$, then the hypothesis that the observed data comes from a Poisson distribution is accepted. The subscript α is a small number, typically chosen to be in the interval (0.005, 0.1), and reflects the degree of accuracy desired. The subscript $s - t - 1$ represents the *degrees of freedom*,

TABLE 5.2 Table of Values of χ^2, for the Chi-Square Test[a]

ν	$\alpha = .995$	$\alpha = .99$	$\alpha = .975$	$\alpha = .95$	$\alpha = .05$	$\alpha = .025$	$\alpha = .01$	$\alpha = .005$	ν
1	.0000393	.000157	.000982	.00393	3.841	5.024	6.635	7.879	1
2	.0100	.0201	.0506	.103	5.991	7.378	9.210	10.597	2
3	.0717	.115	.216	.352	7.815	9.348	11.345	12.838	3
4	.207	.297	.484	.711	9.488	11.143	13.277	14.860	4
5	.412	.554	.831	1.145	11.070	12.832	15.086	16.750	5
6	.676	.872	1.237	1.635	12.592	14.449	16.812	18.548	6
7	.989	1.239	1.690	2.167	14.067	16.013	18.475	20.278	7
8	1.344	1.646	2.180	2.733	15.507	17.535	20.090	21.955	8
9	1.735	2.088	2.700	3.325	16.919	19.023	21.666	23.589	9
10	2.156	2.558	3.247	3.940	18.307	20.483	23.209	25.188	10
11	2.603	3.053	3.816	4.575	19.675	21.920	24.725	26.757	11
12	3.074	3.571	4.404	5.226	21.026	23.337	26.217	28.300	12
13	3.565	4.107	5.009	5.892	22.362	24.736	27.688	29.819	13
14	4.075	4.660	5.629	6.571	23.685	26.119	29.141	31.319	14
15	4.601	5.229	6.262	7.261	24.996	27.488	30.578	32.801	15
16	5.142	5.812	6.908	7.962	26.296	28.845	32.000	34.267	16
17	5.697	6.408	7.564	8.672	27.587	30.191	33.409	35.718	17
18	6.265	7.015	8.231	9.390	28.869	31.526	34.805	37.156	18
19	6.844	7.633	8.907	10.117	30.144	32.852	36.191	38.582	19
20	7.434	8.260	9.591	10.851	31.410	34.170	37.566	39.997	20
21	8.034	8.897	10.283	11.591	32.671	35.479	38.932	41.401	21
22	8.643	9.542	10.982	12.338	33.924	36.781	40.289	42.796	22
23	9.260	10.196	11.689	13.091	35.172	38.076	41.638	44.181	23
24	9.886	10.856	12.401	13.848	36.415	39.364	42.980	45.558	24
25	10.520	11.524	13.120	14.611	37.652	40.646	44.314	46.928	25
26	11.160	12.198	13.844	15.379	38.885	41.923	45.642	48.290	26
27	11.808	12.879	14.573	16.151	40.113	43.194	46.963	49.645	27
28	12.461	13.565	15.308	16.928	41.337	44.461	48.278	50.993	28
29	13.121	14.256	16.047	17.708	42.557	45.722	49.588	52.336	29
30	13.787	14.953	16.791	18.493	43.773	46.979	50.892	53.672	30

[a] This table is based on Table 8 of *Biometrika Tables for Statisticians*, Vol. I, by permission of the *Biometrika* trustees.

in which s is the number of terms added (here, $s = m$) and t is the number of parameters in the distribution under consideration (here, $t = 1$, because Poisson distribution has only one parameter). Therefore, in the present problem, $s - t - 1 = m - 1 - 1 = m - 2$.

In order to apply the chi-square test to the given data, first we have to "pool" (or combine) the last three entries of the table to make sure that each expected frequency is at least 5. Thus, we get

$$\chi^2 = \frac{(18 - 21.9)^2}{21.9} + \frac{(53 - 65.7)^2}{65.7} + \frac{(103 - 98.6)^2}{98.6} + \frac{(107 - 98.6)^2}{98.6}$$
$$+ \frac{(82 - 73.9)^2}{73.9} + \frac{(46 - 44.0)^2}{44.0} + \frac{(18 - 22.2)^2}{22.2} + \frac{(13 - 14.3)^2}{14.3}$$
$$= 5.95$$

Let us choose $\alpha = 0.05$. The number of terms is $v = m - 2 = 6$ because $m = 8$ after "pooling" the last three entries in the table. From Table 5.2, the value of $\chi^2_{0.05,6} = 12.592$. Because $\chi^2 < \chi^2_{0.05,6}$, the hypothesis that the given data comes from a population with a Poisson distribution is accepted.

How to Guess the Distribution

In many applications one is confronted with experimental data, such as the one given in Table 5.1, but has no idea about the distribution of the population from which it has been sampled. An intuitive approach would be a trial and error procedure in which one hypothesizes a distribution, applies the chi-square test, and checks whether the hypothesis is correct. More formal methods are available to handle this problem. A complete discussion of this topic is analytically too complicated to be considered here. It suffices to say that considerable insight can be obtained by inspecting the various moments of the data. A heuristic approach as to how one can use some of the sample moments to determine the population distribution is presented in Section 6.

Probability Graph Paper

On several earlier occasions, it was seen that the use of a special graph paper, such as a semilog or log–log paper, often simplifies analysis. Several other types of graph paper can be advantageously used in statistical studies. For example, one can profitably use the so called *arithmetic probability graph paper*, or simply the *normal probability paper*, to check whether a sample of data can be considered to be drawn from a normal population. This paper, which is commercially available, has a special vertical scale called the *probability scale* and the horizontal scale is the ordinary scale. Before using the probability paper, certain preliminary processing of the data is usually advantageous. In this context, one can usually work either with ungrouped data or grouped data. To illustrate these procedures a set of 100 data points is considered:

0.78	0.38	0.72	0.65	0.72	0.92	0.78	0.65	0.92	0.78
1.36	1.43	0.65	0.48	0.83	0.48	0.72	0.48	0.65	0.78
0.65	1.00	0.78	0.78	1.03	1.26	0.48	0.48	1.06	0.96
0.65	0.92	0.72	0.78	0.78	0.48	0.28	0.36	0.83	0.48
0.78	0.49	0.36	0.78	0.78	0.83	0.88	0.96	1.03	1.21
0.88	0.57	0.72	1.03	0.92	0.96	0.78	1.09	0.92	1.12
0.65	0.65	0.83	0.72	0.72	0.78	0.72	1.09	0.83	0.83
0.83	1.06	0.57	0.78	1.23	1.09	1.03	0.18	0.65	1.34
0.96	0.65	0.48	1.18	1.12	0.18	0.48	0.72	0.57	0.55
0.96	0.65	0.96	0.51	0.65	1.21	1.48	0.96	0.96	1.40

Ungrouped Data

In this procedure, the above numbers are first arranged in an increasing (or decreasing) order and labeled as $z_1, z_2, \ldots, z_n$. Now the z_i are plotted on the x axis and $[(i - \frac{1}{2}) \times 100]/n$ % on the y axis of the probability graph paper. If n is large, it is not necessary to plot every point; every fifth or tenth point will generally suffice. Such a graph with $z_1, z_{11}, \ldots, z_{91}$ and $z_{\max}$ plotted on the x axis and $(100i - 50)/n$ on the y axis is shown in Fig. 5.6.

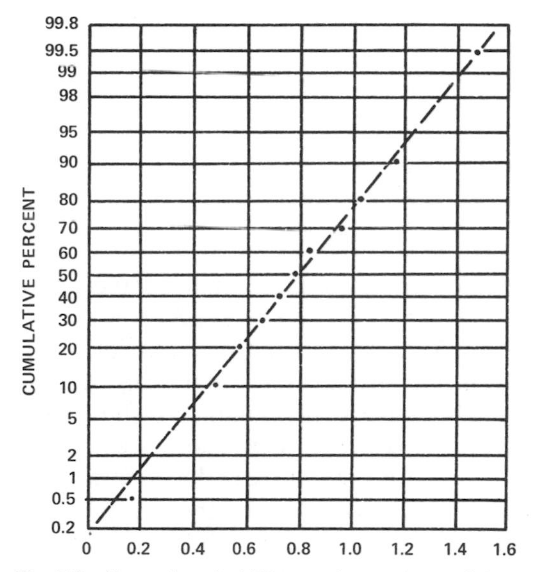

Fig. 5.6 Normal probability graph—ungrouped data.

Grouped Data

Alternatively, one can group the data into classes before plotting. Seven classes are arbitrarily chosen, and the 100 data points are sorted into one of the classes as shown below.

Class limits	Frequency
0.10–0.29	3
0.30–0.49	13
0.50–0.69	17
0.70–0.89	32
0.90–1.09	23
1.10–1.29	7
1.30–1.49	5

Note that all the classes are of the same size and the classes do not overlap. The above type of disjoint class limits are used in preference to overlapping limits such as 0.10–0.30, 0.30–0.50, . . . , because we would have difficulty in deciding where to put 0.30, 0.50, . . . , if these values had occurred in the data. Now a cumulative frequency distribution can be obtained by rearranging the above information as

Data points	No. of observations
Less than 0.095	0
Less than 0.295	3
Less than 0.495	16
Less than 0.695	33
Less than 0.895	65
Less than 1.095	88
Less than 1.295	95
Less than 1.495	100

The above information can now be plotted on the probability paper as shown in Fig. 5.7.

From Figs. 5.6 and 5.7 it is clear that the data fall roughly on a straight line, which in turn means that the given data can be considered as being drawn from a normally distributed population. It should be understood that a probability graph paper is a crude device for checking whether a distribution follows the pattern of a normal curve and therefore it should be used with caution. Other probability papers besides the normal paper are also available. For example, the log–normal probability paper, which is also commercially available, can be used to detect a log–normal distribution.

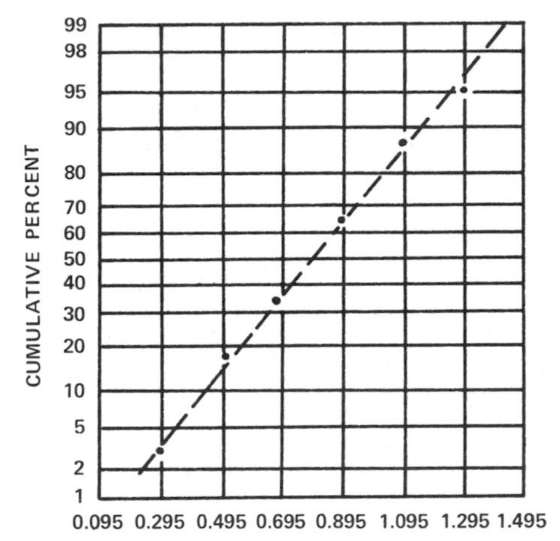

Fig. 5.7 Normal probability graph—grouped data.

4 GENERATION OF RANDOM NUMBERS

Before the computer era, techniques for generating random numbers were based on outputs of physical devices; for example, a balanced roulette wheel, a die, a coin, a radioactive device, noise from a vacuum tube, and many others yield outputs that pass tests for randomness. In the context of modeling and simulation, random numbers, generated via the use of physical devices, suffer from two fundamental drawbacks. First, it is difficult to connect a physical device to a computer. Second, numbers generated with the help of physical devices are not reproducible and are therefore, unsuitable for conducting simulation experiments under controlled conditions.

Random numbers generated with the help of mathematical relations are not truly random, since the sequence of numbers is completely deterministic. However, these sequences, called *pseudorandom numbers*, pass most statistical tests for randomness. Nowadays several computer programs for generating uniformly distributed pseudorandom numbers are available in many computer libraries. Common methods used in generating uniformly distributed pseudorandom numbers are the mid-square, mid-product, and the congruential techniques. Of these the congruential method is more popular and will be described.

The *multiplicative congruential method*, or *power residue* method, is based on the recursive formula

$$U_k = aU_{k-1} (\text{mod } m); \qquad k = 1, 2, \ldots \tag{1}$$

where the relation "$= x(\text{mod } m)$" denotes the selection of the remainder from the aforementioned division of x by m. A fundamental requirement for the use of this method is an initial, user-supplied, number U_0, called the *seed* from which $U_1, U_2, \ldots$ can be generated. This seed should be selected as randomly as possible from among an admissible set of seeds. Once the seed is selected, the sequence defined by $U_1 = aU_0 (\text{mod } m)$, $U_2 = aU_1 (\text{mod } m) = a^2 U_0 (\text{mod } m)$ or, in general

$$U_k = a^k U_0 (\text{mod } m) \tag{2}$$

can be easily generated on a digital computer. If the modulus m is selected in accordance with the computer's arithmetic unit, then divisions can be performed by mere shifting operations. Specifically, one selects $m = p^e$, where p is the number of numerals in the number system of the computer and e is the number of such digits in the standard word size of the computer. For a binary computer, $m = 2^b$, where b is the number of bits per word. While using a binary computer, the seed U_0 is chosen to be any odd integer and a is chosen to be of the form

$$a = 8t \pm 3; \qquad t > 0, \quad \text{an integer} \tag{3}$$

The pseudorandom numbers thus generated lie in the interval $[0, 2^b]$. By placing a radix point before each number in the above sequence, one gets pseudorandom numbers in the interval $[0, 1]$.

Random Numbers with Nonuniform Distributions

A nonuniform random number is a number that has been randomly selected from a collection of random numbers with any specified probability distribution. There are three commonly used methods of generating these: the *inverse transform method*, the *rejection method*, and the *method of rectangular approximation*. In all cases, one usually starts by generating uniformly distributed random numbers.

Inverse Transformation Method

In this method, the problem is to solve $U = F(x)$ for x where U is uniformly distributed between 0 and 1 and where F is the required distribution function. For instance, if one wishes to generate exponentially distributed random numbers with an expected value μ and variance σ, then one starts with a definition of the density function

$$f(x) = \alpha e^{-\alpha x}; \qquad x \geq 0, \quad \alpha > 0 \tag{4}$$

The distribution function $F(x)$ can now be written, using its definition as

$$F(x) = \int_{-\infty}^{x} f(t)\, dt = \int_{-\infty}^{x} \alpha e^{-\alpha t}\, dt = 1 - e^{-\alpha x} \tag{5}$$

The expected value and variance of X are given by

$$\mu = E\{X\} = \int_0^\infty x\alpha e^{-\alpha x}\, dx = 1/\alpha \tag{6}$$

$$\sigma^2 = \text{var}\{X\} = \int_0^\infty (x - 1/\alpha)^2 \alpha e^{-\alpha x}\, dx = 1/\alpha^2 = \mu^2 \tag{7}$$

Notice that the exponential distribution is characterized only by one parameter, namely α, because $\mu = \sigma$ as evidenced by the above two equations.

Now, to generate exponential random variates, one sets $U = F(x)$ and solves for x. Because the uniform distribution U is symmetrical, $F(x)$ and $1 - F(x)$ are interchangeable. Therefore, equating $1 - F(x)$ to Eq. (5), one gets

$$U = F(x) = e^{-\alpha x} \tag{8}$$

Consequently,

$$x = -(1/\alpha) \ln U = -\mu \ln U \tag{9}$$

There are many distributions for which the inverse transformation method cannot be used because it is impossible to analytically invert the distribution function. The normal distribution is such a case. Whenever inversion is possible, the inverse transformation method is by far the most efficient.

Rejection Method

This method can be applied to generate any distribution as long as the range of the random variable is finite. In theory, this restriction eliminates infinite-tail distributions such as normal and exponential from consideration. However, finite limits can be set beyond which an observation rarely occurs.

To see how the method works, suppose that the desired density function is $f(x)$. It is easy to scale x and f such that $0 \le x$ and $f \le 1$. Now, two uniformly distributed random numbers U_1 and U_2 are generated. One chooses $x = U_1$ if $U_2 \le f(U_1)$. Otherwise, one rejects these values U_1 and U_2 and generates a new pair.

Other Methods

Several other specialized techniques are available. As an example of one of these methods, consider the problem of generating random variates that exhibit gamma distribution with an expected value μ and variance σ^2. Once again, one starts with a definition of the density function for a gamma distribution:

$$f(x) = \alpha^k x^{(k-1)} e^{-\alpha x}/(k-1)! \tag{10}$$

where $\alpha > 0$, $k > 0$, and x is nonnegative. This problem differs from the exponential distribution described earlier because an explicit form for the distribution function $F(x)$ does not exist in this case. However,

$$\mu = E\{X\} = k/\alpha \tag{11}$$

$$\sigma^2 = \text{var}\{X\} = k/\alpha^2 \tag{12}$$

Notice that the gamma distribution becomes identical to the exponential distribution for $k = 1$ and to the Erlang distribution if k is a postive integer. As k increases, the gamma distribution tends to the normal distribution in an asymptotical manner. In a majority of practical situations, the need for generating gamma variates for nonintegral values of k does not arise. Therefore, the discussion in this section is limited to generating the so-called Erlang gamma variates.

Generating of Erlang gamma variates (i.e., for k = integer) can be accomplished by taking the sum of k exponential variates $x_1, x_2, \ldots, x_k$ with identical expected values $1/\alpha$. Therefore, the formula for generating Erlang gamma variates x is

$$x = \sum_{i=1}^{k} x_i = -\frac{1}{\alpha} \sum_{i=1}^{k} \ln U_i \tag{13}$$

An equivalent, but computationally faster, formula is

$$x = -1/\alpha \left(\ln \prod_{i=1}^{k} U_i \right) \tag{14}$$

As another example of a special method, consider the problem of generating normally distributed random variates. A very popular method (based on the central limit theorem) of generating normally distributed variates with mean μ_x and variance ${\sigma_x}^2$ is to use

$$x = \sigma_x (12/K)^{1/2} \left(\sum_{i=1}^{K} U_i - K/2 \right) + \mu_x \tag{15}$$

where U_i are uniformly distributed variates in the interval (0, 1). The value of K that should be used is usually determined by balancing computational efficiency against accuracy. Generally one chooses $K = 12$ to take advantage of the obvious computational simplicity.

5 TIME SERIES

In Section 2 the concept of a random variable was introduced. In general each random variable arises as an outcome, or a measured response of a conceptual experiment. Each trial of this conceptual experiment when completed yields, in a sense, a random variable. In several contexts, the conceptual experiment will provide observations serially. In such situations, it is useful to associate a *function*, rather than a number, with each outcome of a trial. As several trials constitute an experiment, a family of functions would result at the conclusion of the experiment. This family is called a *stochastic* or *random process*.

A crude method of generating a random process is the following. Let a variable y have a nominal value of, say, 100. We wish to change this value by, say, ± 1, and plot the history of the values of y as a graph. In order to determine how y should change, a person is asked to toss a coin several times. As per the rules of the experiment, the current value of y is changed by $+1$ whenever "heads" occur and by -1 whenever "tails" occur. During the course of several tosses, which in this case constitutes a trial, the path the variable y takes is plotted as shown in Fig. 5.8 (the dotted line). This "curve," thus generated, is one realization of a random process. If the coin is tossed once again, say, by a different person, the resulting curve in all likelihood will be different (the dashed line of Fig. 5.8). Similarly each trial results in a different curve. What we have in Fig. 5.8 is a set of realizations out of infinitely many possible realizations of a random process.

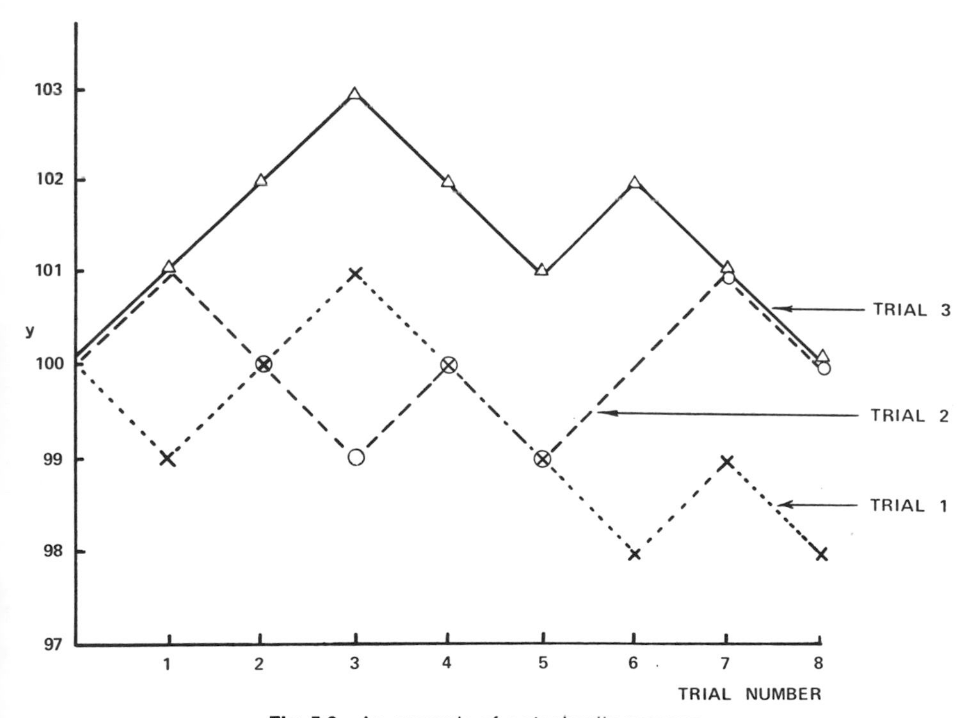

Fig. 5.8 An example of a stochastic process.

More concretely, any one curve of Fig. 5.8 can be thought of as the fluctuation of the closing price of a particular stock as the days progress. In this particular example, the abscissa represents time, and for a given stock for a given period, we get only one curve—all other realizations are only possibilities. Thus, a time series is a realization of a particular random process under study.

Of the various types, data that has the structure of a time series plays an especially important role in modeling and forecasting. A *time series* is nothing but a sequence of values a variable assumes as time progresses. The daily stock market prices of a commodity is an example of a time series. Monthly or yearly stream flow data is another example. Indeed, any sequence of observations made on a variable, in a dynamic system, can be regarded as a time series. Central to the definition of time series is the assumption that the sequence of recorded observations on a given variable is viewed as one particular realization of a random process.

Relevant Attributes of Time Series

Consider a plotting of the quarterly sales of an established line of machine tools (Fig. 5.9). This is a time series x_i. Comparison of Fig. 5.9 with Fig. 5.8 reveals their basic similarity; they both represent particular realizations of some underlying random processes. Indeed, the daily stock prices of a vigorously growing company, shown in Fig. 5.10, is a member of another random

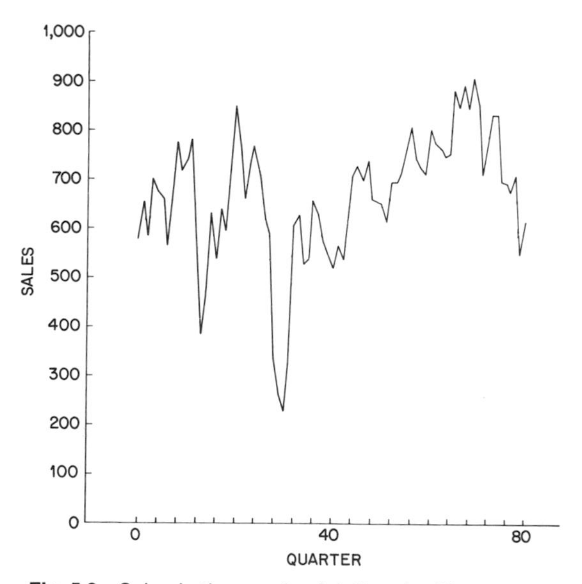

Fig. 5.9 Sales in thousands of dollars for 80 quarters.

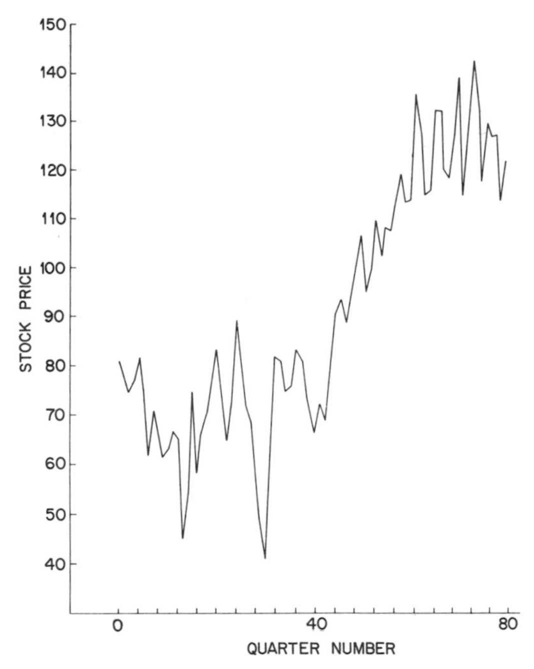

Fig. 5.10 Stock prices for over 80 quarters.

process. To distinguish one from another, it is useful to characterize each process by means of certain parameters.

Inspection of Fig. 5.9 reveals that the sale of machine tools, even though fluctuating in an unpredictable manner, exhibits no tendency to wander too far from its historical average or mean of about $666,000. A time series possessing this property is said to be *stationary*. That is, a stationary time series exhibits no *trend*. On the other hand, the time series in Fig. 5.10 exhibits what might be called a distinct change in level over the period of observation. The behavior of this time series is *nonstationary*. Certain time series exhibit no trend but possess a *periodicity* or seasonal dependence. Average monthly temperatures of a town plotted over a period of several years is likely to exhibit a seasonal periodicity. Similarly, civilian employment in the United States exhibits, fairly consistently, a low point at mid-winter and a high point in mid-summer.

The presence or absence of stationarity in a time series can be determined either by visual inspection of data points on a graph or by a simple regression test. This test is performed by fitting a straight line $x = a + bt$ to the set of points $\{(t_i, x(t_i))\}$. If $b \approx 0$ in this test, then the regression line is approximately parallel to the t axis, indicating the absence of a trend. The periodic components of a time series, if any, can be detected either by visual inspection of a graph or by means of a harmonic (or Fourier) analysis. In this test, one attempts to describe the function $x(t)$ as a weighted sum of a series of sines and cosines. Detailed description of these procedures is beyond the scope of this work. However, there are standard program packages, such as the BMD programs of the University of California, that are capable of performing a number of statistical tests.

The trend and periodic components in a time series are often referred to as the deterministic portion of a time series. Removal of the deterministic part leaves a residual series that is random in nature. This random portion can be characterized by such statistical attributes as probability density function, *autocorrelation* or *autocovariance function* and the *power spectral density function.*

For a given population, the auto, or serial, or lag-k correlation coefficient is defined as

$$\rho_k = \frac{E\{[x(t) - \mu][x(t+k) - \mu]\}}{E\{[x(t) - \mu]^2\}E\{[x(t+k) - \mu]^2\}^{0.5}} \tag{1}$$

If the time series is stationary, then ${\sigma_x}^2$ is the same at time $t + k$ as well as at time t. For this case, ρ_k simplifies to

$$\rho_k = \frac{E\{[x(t) - \mu][x(t+k) - \mu]\}}{{\sigma_x}^2} \tag{2}$$

where ${\sigma_x}^2$ is the variance of the random variable x(t). Also, the covariance function is defined as

$$\operatorname{cov}\{x(t), x(t+\mu)\} = E\{[x(t) - \mu][x(t-k) - \mu]\} \tag{3}$$

Hence, the lag-k serial correlation coefficient for a given population is

$$\rho_k = \frac{\operatorname{cov}\{x(t), x(t+k)\}}{\operatorname{var}\{x(t)\}} \tag{4}$$

As before, an estimate of r_k of ρ_k can be obtained from a sample via the formula

$$r_k = \frac{\sum_{i=1}^{N-K} x_i x_{i+k} - [1/(N-K)][\sum_{i=1}^{N-K} x_i][\sum_{i=k+1}^{N} x_i]}{([\sum_{i=1}^{N-K} {x_i}^2 - [1/(N-K)](\sum_{i=1}^{N-K} x_i)^2]^{0.5} \times [\sum_{i=k+1}^{N} {x_i}^2 - [1/(N-K)](\sum_{i=k+1}^{N} x_i)^2]^{0.5})}$$

where $x_i = x(t_i)$ and N is the total number of points i in the series.

The autocorrelation function gives a measure of persistence in a time series. Consider a time series showing the volume of water flowing in a river. It has been observed, from experience, that low flow is more likely to be followed by a low flow than by a high flow. Similarly, a high flow is more likely to be followed by a high flow. That is, river flows have a tendency to persist. In statistical parlance, one describes such a phenomenon by saying that successive flows are *positively correlated.*

Thus, if one's goal is to analyze a time series, some of the relevant attributes are: average value, trend, periodicity, and autocorrelation.

6 A MODEL FOR GENERATING RIVER FLOW DATA

In many simulation studies, the need for time series data stretching over long intervals of time is very real. For instance, it is generally difficult or impossible to get data on rainfall, weather conditions, flow in rivers, etc. for long stretches of time. When the required amount of data are not available from any of the traditional data sources, then one has to conduct new experiments—either under laboratory conditions or under field conditions. Biological experiments on animals are problems of this type. Sometimes, it may not be possible to conduct experiments for the sake of getting data. For example, it is practically impossible to conduct field studies and experiments in order to get data on environmental pollution levels during historical times; the pollution problem, as we perceive it now, was recognized only in the early 1960s. Under circumstances such as these, a widely popular approach is to generate data on a computer. This process of generating data is variously referred to as synthetic generation of data, synthesis of time series or simulation of a stochastic process. A first step in this effort would be to obtain a characterization of the data to be generated. If a sample of actual data is available as a record of finite length, then one can attempt to simulate this record on the computer such that the simulated data resembles the available data record.

At the beginning of Chapter 1, the idea of using simulation as a planning and decision making tool in the management of water resource systems was introduced. Some of the macromodels one would use in such a problem were displayed in Fig. 1.1. The present discussion on synthetic generation of time series adds a new dimension to the planning process by increasing the scope and sensitivity of the evaluations. For convenience in discussing specifics, the forthcoming discussion will be focused on the computer generation of data pertaining to the flow of water in a stream. With suitable modifications the same ideas can be applied to other time series as well.

Early civil engineers realized that the flow patterns of different streams vary considerably and that the history of flows in a particular stream provides a very valuable clue to the future behavior of a stream. Once the future behavior can be predicted, one can plan water use wisely despite inflow fluctuations. A simple minded forecasting tool in this context is to inspect the historical record of flows and ascertain features such as the frequency of floods and the duration of droughts. However, use of historic records alone involves several drawbacks. Historical records are apt to be quite short. The exact pattern of flows during the historical period is very unlikely to recur during the period prescribed by the planning horizon. Historical records alone are not sufficient to predict risks such as the likelihood of a reservoir running dry. A remedy to this problem is to generate data in a synthetic manner. The goal of such an activity is not always the pursuit of a phenomenological explanation of flows; the real flows are most likely generated in nature by a complex process that depends on a host of hydrological, meteorological, and topographical characteristics. The justification is primarily operational and pragmatic.

A Crude Queuing Model

As an aid to visualize the context in which synthetic stream flows would be useful, assume that a reservoir is being tested for its performance. The size of the reservoir is assumed to be 10 units. A stream, whose mean annual flow is 6 units, feeds the reservoir. It is further assumed that 6 units of water are required to be released annually from the reservoir for various uses. Assume that the rules by which water is released are set as follows:

(1) If fewer than 6 units are available in the reservoir, release all the water.

(2) If the total available water (i.e., current storage plus current inflow) is more than 16 units, then keep the reservoir full (at 10 units) and release all additional water.

(3) For all other intermediate cases, 6 units are released and the reservoir is left at some intermediate level.

A problem such as this can be conveniently studied using simulation. The relevant attributes of this simulation are the outflow and current available water. The release rules are well specified and can be used as an operating rule in determining the outflow or output. The current available water depends on two variables: current storage and current inflow. Current storage in turn depends upon current inflow and outflow. The inflow, however, is a random quantity. Therefore, if the pattern of inflows can be predicted, in a statistical sense, the entire reservoir operation can be simulated. An extremely crude simulation is to prepare an accounting of water as shown in Table 5.3. In this simulation, the reservoir is assumed to have 9 units of water at the beginning of the first period. This is shown in row 1, column 2. The entries in column 3 are quantities selected at random with the stipulation that their average value is 6. The entries in column 4 are deter-

TABLE 5.3

An Accounting of Flows in and out of a Reservoir[a]

Period	Initial Storage	Inflow	Release	Final Storage	Deficit	Surplus Release
...	...	...	...	...	...	...
...	...	...	...	...	...	...
...	...	...	...	9	...	...
1	9	12	11	10	0	5
2	10	11	11	10	0	5
3	10	12	12	10	0	6
4	10	5	6	9	0	0
5	9	4	6	7	0	0
6	7	2	6	3	0	0
7	3	1	4	0	2	0
8	0	3	3	0	3	0
9	0	5	5	0	1	0
10	0	8	6	2	0	0
11	2	5	6	1	0	0
12	1	4	5	0	1	0
13	0	5	5	0	1	0
14	0	7	6	1	0	0
15	1	8	6	3	0	0
16	3	2	5	0	1	0
17	0	2	2	0	4	0
18	0	1	1	0	5	0
19	0	7	6	1	0	0
20	1	11	6	6	0	0
...					18	

[a] Adapted with permission from Fiering and Jackson, *Synthetic Streamflows*, Water Resources Monogr. I, p. 8, American Geophysical Union, Washington, D.C., 1971.

mined by the release rules. At the end of each time period there is either a surplus or deficit of water. This knowledge can be profitably used to ascertain the quality of design of a dam. Thus, if we can simulate the inflows, we have a useful model to predict the fluctuations of water in a reservoir. Parenthetically, it is instructive to observe that Table 5.3 is essentially a simulation of a queue.

Simulation of Inflows

Simulation of inflows is tantamount to generating a sequence of numbers corresponding to the flow in a stream. This sequence should exhibit the same statistical properties as the data obtained from a stream. Therefore, a first step in a simulation of this kind is to characterize the actual data obtained from the stream, which is itself a time series. In general, the ith member of a time series, denoted by x_i, can be visualized as consisting of two parts and therefore can be written as

$$x_i = d_i + e_i; \qquad i = 1, 2, \ldots \tag{1}$$

where d_i is the deterministic part and e_i is the random part. The deterministic part, in turn, can be visualized as consisting of a trend T_i and one or more oscillatory effects P_i about the trend. That is, Eq. (1) can be rewritten as

$$x_i = T_i + P_i + e_i; \qquad i = 1, 2, \ldots \tag{2}$$

The trend has been defined loosely as a slow, long-term movement in a time series. The trend can be linear, quadratic, or some such function of time. It is generally preferable to work with time series that exhibit no trend. That is, it is preferable to remove any trend and render the time series stationary. In terms of intuitive ideas, removal of the trend is equivalent to making the time series have a constant mean and allowing it to fluctuate about the mean.

Removal of Linear Trend

As stated earlier, one way of detecting a trend is to conduct a simple regression test. Assuming the presence of a linear trend, one can represent the time series as

$$x_i = a + bt_i + P_i + e_i; \qquad i = 1, 2, \ldots \tag{3a}$$

or equivalently as

$$x(t) = a + bt + P(t) + e(t) \quad \text{at} \quad t = t_i; \qquad i = 1, 2, \ldots \tag{3b}$$

where $a + bt$ represents the linear trend, $P(t)$ the periodic part, and $e(t)$ the random component. For a fixed duration of time (Δt), the incremental time series assumes the form

$$x^*(t) = x(t + \Delta t) - x(t) = b(\Delta t) + P(t + \Delta t) - P(t) + e(t + \Delta t) - e(t) \tag{4}$$

Thus, the time series defined by $x^*(t)$ has no trend. Therefore, a linear trend can be removed by defining a new time series $x^*(t)$. In practice, the incremental series can be obtained easily by constructing a difference table.

Removal of Quadratic Trend

If a quadratic trend exists in a time series, then it can be written as

$$x(t) = a + bt + ct^2 + P(t) + e(t) \tag{5}$$

where $a + bt + ct^2$ represents the quadratic trend. This can be removed by defining a new series defined by

$$x^{**}(t) = x(t + 2(\Delta t)) - 2x(t + \Delta t) + x(t) \tag{6a}$$

$$= k + P(t + 2(\Delta t)) - 2P(t + \Delta t) + P(t)$$
$$+ e(t + 2(\Delta t)) - 2e(t + \Delta t) + e(t) \tag{6b}$$

where k is a constant. Thus, it can be seen that a relatively simple linear combination of terms in the time series produced a new time series that is free of both linear and quadratic trends. A special case of $x^{**}(t)$ arises if (Δt) is chosen to be unity. Then Eq. (6a) becomes

$$x^{**}(t) = x(t + 2) - 2x(t + 1) + x(t) \tag{7}$$

This quantity is called the *moving average.*

There are several pros and cons to the process of rendering a time series stationary by removing the trend. Trend removal, though straightforward, has to be done with caution. Unless a record of sufficient length is available for examination, certain low-frequency oscillatory components may appear, like trend, and the methods described above may remove such periodic components. Second, an attempt at removing the trend from a random series of uncorrelated observations may introduce spurious oscillations in the residual series. On the credit side, there are some advantages in working with a trendless, that is, stationary, time series. For instance, the autocorrelation function of a stationary time series depends only on the time lag τ between, not the absolute location of, the times for which the autocorrelation function is evaluated. This feature has a practical significance because the autocorrelation function of a stationary time series can be used, sometimes, to determine the underlying mechanism that may be generating the time series. Furthermore, a stationary time series has a *spectral distribution function,* which is nothing but the Fourier transform of the autocorrelation function. Therefore, a spectral analysis (i.e., decomposition of the time series into its frequency components) can be easily conducted to reveal any oscillatory effects present in the time series. Several computational simplifications arise if a stationary time series is also normally distributed. Thus, there is considerable merit in isolating and removing the trend.

Once the trend is removed, the resulting time series contains the periodic component $P(t)$, if any, and the random component $e(t)$. For the purpose of simplifying the present discussion, the periodic component is assumed to be nonexistent, thus leaving us only with the random component, which is now stationary. This stationary series $\{e_i\}$ can be either correlated (i.e., consequent flows are interdependent) or uncorrelated. Simulation of an uncorrelated series is the easiest because it involves a generation of a sequence of random numbers $\{e_i\}$ with the prescribed distribution. It is generally difficult to generate $\{e_i\}$ with an arbitrary autocorrelation function. However, several heuristic procedures are available to generate time series with any autocorrelation function exhibiting any specified distribution. A strong theoretical base for these methods is nonexistent and any attempt at establishing a reasonable theoretical base is possible only if the random component of the time series is normally distributed with a zero mean.

Heuristic Generation of Correlated Time Series

If the stationary series $\{e_i\}$ exhibits a correlation, then one has to use a more complicated scheme in characterizing it. Correlation effects can be included by writing a general expression for $x(t)$ as

$$x(t) = b_0 + b_1 x(t-1) + \cdots + b_m x(t-m) + e(t) \tag{8}$$

That is, $x(t)$ is a linear combination of the m previous flows. Schemes such as the one above are called *autoregressive*. The simplest of the autoregressive schemes is the Markov scheme defined by

$$x(t) = b_0 + b_1 x(t-1) + e(t) \tag{9a}$$

or, in discrete notation,

$$x_i = b_0 + b_1 x_{i-1} + e_i \tag{9b}$$

A phenomenological explanation to this equation is that the present flow x_i depends upon the previous flow x_{i-1} only. Due to this fact, Eq. (1) is often called the *lag-one Markov model*. The task of building a lag-one Markov model, therefore, boils down to a specification of b_0, b_1 and the nature of the distribution of e_i. A heuristic method of assigning values to b_0 and b_1 is to write

$$x_i = \mu + \rho_1(x_{i-1} - \mu) + e_i \tag{10}$$

where μ is the mean and ρ_1 is the lag-one serial correlation coefficient. That is, we assume the flow x_i to be the sum of the mean flow, a proportion of the departure of the previous flow from the mean and a random component. Notice that the mean and variance of x_i are influenced by the mean and variance of e_i through Eq. (10). Therefore, it is necessary to understand the nature of the distribution of e_i. Let us assume that the e_i are normally distributed with zero mean and a constant variance σ_e. Now the mean and variance of x_i can be calculated as follows:

$$E\{x_i\} = \mu + \rho_1 E\{x_{i-1}\} - \rho\mu + 0$$

The last term is zero because $\{e_i\}$ has zero mean. Writing $E\{x_i\} = E\{x_{i-1}\}$, the above equation yields $E\{x_i\} = \mu$. Thus, the sequence generated by Eq. (10) has the same mean value as the population mean. The variance of x_i can also be calculated in a similar manner:

$$\begin{aligned} \operatorname{var}\{x_i\} &= \operatorname{var}\{\mu + \rho_1(x_{i-1} - \mu) + e_i\} \\ &= E[\mu + \rho_1(x_{i-1} - \mu) + e_i]^2 - \mu^2 \end{aligned}$$

Denoting the variance of x_i by σ^2 and the variance of e_i by $\sigma_e{}^2$, the above equation becomes

$$\sigma^2 = \rho^2\sigma^2 + \sigma_e{}^2, \qquad \text{or} \qquad \sigma_e{}^2 = (1 - \rho^2)\sigma^2 \tag{11}$$

That is, the variance of $\{x_i\}$ is influenced by the variance of $\{e_i\}$. If $\{e_i\}$ has a variance of $\sigma_e{}^2$, then $\{x_i\}$ has a variance of $(1/1 - \rho^2)\sigma_e{}^2$. Therefore, if $\{x_i\}$ has to exhibit a variance of σ^2 (viz., the population variance), then $\{e_i\}$ has to exhibit a variance of $(1 - \rho^2)\sigma^2$. How shall we make $\{e_i\}$ exhibit a variance of $(1 - \rho^2)\sigma^2$? It is known that if a random variable t_i is normally distributed and serially independent with zero mean and unit standard deviation, then $t_i\sigma\sqrt{1 - \rho_1{}^2}$ is once again a normally distributed, serially independent, random variable with zero mean as before, but now with a variance $= \sigma^2(1 - \rho_1{}^2)$. Therefore, if the desire is to preserve not only the mean and lag-one correlation coefficient, but also the variance, then Eq. (10) should be modified to read

$$x_i = \mu + \rho_1(x_{i-1} - \mu) + t_i\sigma\sqrt{1 - \rho_1{}^2} \tag{12}$$

The ideas presented so far will now be illustrated by means of a specific example. Consider the data shown in the first two columns of Table 5.4. They show the annual flows for the Selway River in Idaho in hundreds of acre-feet (haf). The task now is to generate synthetic time series data $\{q_i\}$ that exhibit the same statistical properties as the given data $\{x_i\}$.

The solution process can be visualized as consisting of two stages. The first stage is the analysis stage. At this stage the statistical properties of the given data are calculated. During the second stage, an attempt is made to synthesize $\{q_i\}$ using various assumptions. A step by step procedure of this process is given below.

Step 1 A regression test on the given data shows that there is no significant trend, so there is no need to worry about trend removal. Inspection of the data suggests the absence of a periodic component. Had the given data been monthly flows, rather than annual flows, we would have observed some periodicity reflecting the seasonal variation of flows, and so no serious attempt was made to detect periodicities.

TABLE 5.4

Annual Flows, Selway River, Idaho in Hundreds of Acre Feet[a]

i	x_i	x_i^2	x_i^3	$x_i x_{i+1}$	$x_i x_{i+2}$	$x_i - \bar{x}$
1	429	184041	78953589	161304	220935	-159.8
2	376	141376	53157376	193640	135360	-212.8
3	515	265225	136590875	185400	300760	-73.8
4	360	129600	46656000	210240	162360	-228.8
5	584	341056	199176704	263384	254040	-4.8
6	451	203401	91733851	196185	217833	-137.8
7	435	189225	82312875	210105	321030	-153.8
8	483	233289	112678587	356454	393645	-105.8
9	738	544644	401947272	601470	329886	149.2
10	815	664225	541343375	364305	463735	226.2
11	447	199809	89314623	254343	245850	-141.8
12	569	323761	184220009	312950	400576	-19.8
13	550	302500	166375000	387200	585750	-38.8
14	704	495616	348913664	749760	481536	115.2
15	1065	1134225	1207949625	728460	814725	476.2
16	684	467856	320013504	523260	416556	95.2
17	765	585225	447697125	465885	505665	176.2
18	609	370881	225866529	402549	326424	20.2
19	661	436921	288804781	354296		72.2
20	536	287296	153990656			-52.8
Σ	11776	7500172	5177696020	6921190	6576666	0.0

[a] Reproduced with permission from M. B. Fiering and B. B. Jackson, *Synthetic Streamflows*. Water Resources Monogr. I, p. 44, American Geophysical Union, Washington, D.C., 1971.

Step 2 Using the formulas given in earlier sections, the first three moments are calculated:

$$\text{mean} = \bar{x} = 1/20 \sum_{i=1}^{20} x_i = 11{,}726/20 = 588.8 \quad \text{haf}$$

$$\text{variance} = S^2 = 1/19(7{,}500{,}172 - 20(588.8)^2) = 29{,}813.85 \quad (\text{haf})^2$$

$$\text{standard deviation} = S = (29{,}813.85)^{1/2} = 172.667 \quad \text{haf}$$

$$\text{skewness} = G = 0.9916$$

Similarly, the first two correlation coefficients are calculated.

$$\text{lag-one correlation coefficient} = r_1 = 0.3782$$

$$\text{lag-two correlation coefficient} = r_2 = 0.2467$$

Calculation of the above statistical parameters is facilitated by performing some preliminary calculations as shown in Table 5.4.

Step 3 Now the analyst should decide on the degree of complexity required in the model. Inspection of the statistical parameters suggest the following. The positive value of the skewness suggests (see Fig. 5.4) that x_i is probably not normally distributed because a normal distribution has zero skewness. Conceivably, x_i has a log–normal distribution; that is, $\log(x_i)$ is probably normally distributed because a log–normal distribution exhibits a positive skewness. Second, the lag-two serial correlation coefficient is not negligible. This means that x_i probably depends not only on x_{i-1} but also on x_{i-2}.

Step 4 To simplify the present discussion, it is useful to make some assumptions. For instance, it is assumed that the current interest is really to preserve only the mean, variance and the lag-one serial correlation coefficient in the synthetic data. Therefore, the following generating scheme, suggested by Eq. (12), is used:

$$q_i = \bar{x} + r_1(q_{i-1} - \bar{x}) + t_i S\sqrt{1 - r^2} \tag{13}$$

where $\bar{x}$, r_1, and S are obtained from Step 2. The values of t_i are obtained from a random number generator. For the purpose of this example, some of the t_i values are shown in Table 5.5. Now the results obtained from Eq. (13) are summarized in Table 5.6. Entries in the last column of this table are the synthetic flows. In simulations of this kind it is customary to ignore the first several (say, 50) values of q_{i+1} to minimize any possible errors in the start-up procedure. Once the sequence $\{q_i\}$ of the required length is available, it is advisable to verify whether q_i indeed resembles the original sequence $\{x_i\}$. This can be readily done by calculating all the relevant statistical parameters of $\{q_i\}$ and comparing them with those obtained for $\{x_i\}$. This task is left as an exercise to the student.

It is important to note that a normal distribution has been chosen in spite of the skewness present in the data. If there is a strong desire to preserve the skewness also in $\{q_i\}$, then one should consider the possibility of generating a log–normal distribution. The procedure for this is very similar to the one described above, except for the fact that now one works with the logarithm of x_i. The importance of normal and log–normal distributions stems from the fact that many streams exhibit these distributions. The basic reason for the frequent appearance of normally distributed flows is the central limit theorem. The gist of this theorem says that a variable which is the sum of several identically distributed random variables derived from *any* distribution with a finite mean and variance is distributed approximately normally. Thus, if the flow in a stream in a given time period is the result of several additive factors, then the flow can be considered as the sum of several

TABLE 5.5

Standard Normal Random Sampling Deviates

	0	1	2	3	4	5	6	7	8	9
0	0.132	0.736	0.795	0.799	0.165	0.640	0.314	0.420	0.915	0.867
1	0.104	0.319	0.930	0.100	0.731	0.393	0.855	0.720	0.581	0.399
2	0.491	0.434	0.390	0.649	0.638	0.089	0.978	0.228	0.075	0.810
3	0.582	0.834	0.074	0.581	0.137	0.385	0.807	0.400	0.901	0.497
4	0.024	0.125	0.788	0.982	0.386	0.145	0.788	0.472	0.817	0.992
5	0.847	0.302	0.254	0.024	0.583	0.776	0.175	0.307	0.212	0.942
6	0.268	0.814	0.065	0.996	0.633	0.069	0.058	0.451	0.873	0.362
7	0.506	0.181	0.464	0.662	0.638	0.475	0.772	0.346	0.213	0.230
8	0.096	0.875	0.577	0.393	0.102	0.542	0.577	0.788	0.039	0.414
9	0.668	0.459	0.100	0.591	0.619	0.631	0.303	0.918	0.171	0.444
10	0.279	0.261	0.038	0.559	0.322	0.479	0.611	0.776	0.099	0.219
11	0.757	0.787	0.375	0.652	0.070	0.180	0.139	0.288	0.968	0.835
12	0.507	0.973	0.990	0.819	0.768	0.220	0.685	0.920	0.954	0.975
13	0.463	0.935	0.317	0.958	0.782	0.843	0.063	0.703	0.037	0.988
14	0.752	0.405	0.548	0.385	0.710	0.730	0.002	0.299	0.844	0.682
15	0.729	0.192	0.230	0.673	0.196	0.189	0.845	0.905	0.077	0.169
16	0.832	0.207	0.297	0.620	0.654	0.438	0.397	0.371	0.524	0.583
17	0.622	0.337	0.491	0.671	0.090	0.392	0.535	0.641	0.421	0.506
18	0.073	0.275	0.206	0.256	0.234	0.327	0.680	0.856	0.281	0.975
19	0.988	0.808	0.556	0.711	0.409	0.321	0.332	0.811	0.916	0.604

Reproduced with permission from M. B. Fiering and B. B. Jackson, *Synthetic Streamflows.* Water Resources Monograph. I, p. 20, American Geophysical Union, Washington, D.C., 1971.

TABLE 5.6

Synthetically Generated Streamflows

i	q_i	$q_i - \bar{x}$	$r_1(q_i - \bar{x})$	$\bar{x} + r_1(q_i - \bar{x})$	t_i	$t_i s\sqrt{1-r_1^2}$	q_{i+1}
0	588.80	0.00	0.00	588.80	-0.523	-83.60	505.20
1	505.20	-83.60	-31.62	557.18	0.611	97.66	654.85
2	654.85	66.05	24.98	613.78	-0.359	-57.38	556.40
3	556.40	-32.40	-12.26	570.54	-0.393	-62.82	513.73
4	513.73	-75.07	-28.39	560.41	0.084	13.43	573.83
5	573.83	-14.97	-5.66	583.14	-0.931	-148.81	434.33
6	434.33	-154.47	-58.42	530.38	-0.027	-4.32	526.06
7	526.06	-62.74	-23.73	565.07	0.798	127.55	692.63
8	692.63	103.83	39.27	628.07	1.672	267.26	895.32
9	895.32	306.52	115.92	704.72	-1.077	-172.15	532.57

Reproduced with permission from M. B. Fiering and B. B. Jackson, *Synthetic Streamflows*. Water Resources Monograph. I, p. 64, American Geophysical Union, Washington, D.C., 1971.

random variables and therefore will be approximately normal. There is one practical problem, however, with the use of normal distribution. Because a normal density function is symmetrical about the mean, at times a negative value for a flow will result. These values are only used to generate the subsequent flow value and discarded from further consideration.

EXERCISES

1. A random variable X is known to be uniformly distributed in $0 \leq x \leq 1$ and the density function is $f(x) = 1$. Calculate the first four moments.

2. By looking at a given graph, can you tell whether it represents a probability distribution function or probability density function? How?

3. Following is a list of variables. Which of these could be represented by discrete distributions? Which by continuous distribution functions?

Biological life span	Number of trees per acre
Mortality rate	Number of defective parts
Weights of animals	Demand at a service counter
Cloud density	Error in the speed of a car
Temperature variations	Thickness of a wire

4. The heights in inches of a certain group of children are given by: 35, 38, 43, 23, 30, 33, 41, 37, 50, 36, 39, 42, 48, 44, 59, 51, 55, 46, 45, 49. Find the first four moments of this data. Study these moments and tell whether the underlying distribution function is symmetrical or not.

5. Generate 1000 uniformly distributed random numbers using a standard library routine available at your computer center
(a) in the interval (0, 1),
(b) in the interval (0, 99)

6. Generate 100 approximately normally distributed random variables of zero mean and unit variance by first generating 1200 uniformly distributed random variables U_i and computing for each of the 100 nonoverlapping dozens of numbers, the quantity

$$r_i = \left[\sum_{i=1}^{12} U_i - 6 \right]$$

7. It is desired to test whether the number of gamma rays emitted per second by a radio active substance is a random variable having a Poisson distribution with $\lambda = 2.4$. Test this hypothesis using the chi-square test with $\alpha = 0.05$ using the following results obtained in 300 1-sec intervals.

Number of gamma rays	Frequency	Number of gamma rays	Frequency
0	19	4	44
1	48	5	35
2	66	6	10
3	74	7	4

8. The following is the distribution of the number of days which an electrical light bulb functioned while in continuous use.

Number of days	Frequency	Number of days	Frequency
0–5	37	15–20	13
5–10	20	20–25	8
10–15	17	25–30	5

Using $\alpha = 0.05$, test the hypothesis that this data constitute a random sample from an exponential population.

SUGGESTIONS FOR FURTHER READING

1. For some interesting discussions on forecasting, see

O. Morgenstern, K. Knorr, and K. P. Heiss, *Long-term Projections of Power: Political, Economic and Military Forecasting*. Bellinger Pub., Cambridge, Massachusetts, 1973.

2. For a discussion of statistical foundations and simulation methodology, consult
   G. A. Mihram, *Simulations: Statistical Foundations and Methodology*. Academic Press, 1971.
3. For further discussion on simulation tools, refer to
   R. E. Shannon, *System Simulation: The Art and Science*. Prentice-Hall, Englewood Cliffs, New Jersey, 1975.
   G. Gordon, *System Simulation*. Prentice-Hall, Englewood Cliffs, New Jersey, 1969.
   G. S. Fishman, *Concepts and Methods in Discrete Event Simulation*. Wiley, New York, 1973.
4. Techniques for generating random numbers with various distributions can be found in
   T. H. Naylor, J. L. Balintfy, D. S. Burdick, and K. Chu, *Computer Simulation Techniques*. Wiley, New York, 1968.
5. Most of the material in Section 6 was drawn from an excellent and easy to read monograph by
   M. B. Fiering and B. B. Jackson, *Synthetic Streamflows*. Water Resources Monograph I, American Geophysical Union, Washington, D.C., 1971.
6. For a good discussion of the chi-square test, see
   J. E. Freund, *Mathematical Statistics*, 2nd ed., Prentice-Hall, Englewood Cliffs, New Jersey, 1971.
7. A number of statistical techniques of forecasting in the context of developing models to urban systems in general and problems such as facility location planning, and resource allocation in particular can be found in
   W. Helly, *Urban Systems Models*. Academic Press, New York, 1975.

6

Recognition of Patterns

INTRODUCTION

Recognition can be regarded as a basic attribute of living organisms. A pattern is a description of an object. Thus, recognition of patterns is a basic activity of all living organisms. Indeed, the ability to recognize patterns is a necessary part of survival. In the animal kingdom one's survival depends upon one's ability to recognize friend and foe. An infant learns to recognize its mother's face and voice at an early age. Indeed, we can recognize people from an analysis of their hand-writing, fingerprints, and voice prints.

Acts of recognition can be divided into two major types: recognition of concrete items and recognition of abstract items. Recognition of spatial and temporal patterns using one's visual and aural sensory apparatus belongs to the former type. Examples of spatial patterns are alphanumeric characters, fingerprints, and pictures. Temporal patterns include speech waveforms, electrocardiograms, time series, and target signatures. Recognition of conceptual patterns, such as the proof of a theorem, belongs to the latter type.

The subject of pattern recognition spans a number of disciplines. What is common among all these disciplines is the task of recognizing members of a given class from a conglomerate containing fuzzy elements from many classes. In diagnosing a disease, for example, a doctor observes a number of fuzzy symptoms and recognizes a particular disease.

1 FUNDAMENTAL PROBLEMS OF RECOGNITION

In order to recognize any pattern, first, one has to have data. Collecting this data is the sensing problem. If one is trying to recognize an abnormality in an electrocardiogram, then one has to have a signal $f(t)$ describing the fluctuations in the depolarizing potential of the heart muscle. One may wish to record this signal by sampling the skin potential at discrete instants of time. The resulting information is the *pattern vector* $\mathbf{x} = (x_1, \ldots, x_n)^T$ where $x_i = f(t_i)$. If the interest is to recognize the most desirable plan in a resource management problem, then the sensing problem involves a description of each plan vector (analogous to the pattern vector). Here the elements of the plan vector would be the attributes (or features) of the plan. Thus, the pattern vector contains all the measured information. When the measurements yield information in the form of real numbers, then it is often useful to think of a pattern vector as a point in an n-dimensional Euclidean space, which may be called the feature space or the attribute space.

An item closely related to the sensing problem is the question of sampling. Consider once again the problem of analyzing an electrocardiogram. The electrical skin potential can be measured at a number of points on the body. Then is it enough to study one signal? A doctor may wish to study a number of signals tapped from various parts of the body. Similarly, management looks at a number of alternative plans before choosing one. Theoretically one can collect an infinite number of cardiograms and consider an infinite number of alternative plans. However, for practical reasons, one has to limit the objects to be studied for recognition purposes. Therefore, the starting point of the recognition problem is generally a finite set A of objects. In a number of cases, the set A may represent the entire population. However, A typically represents a sample from a much larger population. If A represents a sample, then one has to apply statistical techniques to study the sampling process itself.

The second problem in recognizing patterns concerns the extraction of characteristic features or attributes from the given data. In a speech recognition problem, for example, a sample of the relevant features (or attributes) are duration of sound, ratios of energy in various frequency bands, the location of spectral peaks, and the movement of the peaks with time. In a problem involving the management of a river some of the relevant attributes are economic impact, water quality, ecological damage, and recreational advantages. To enable one to recognize one plan as being superior to the others, a capacity to discriminate one plan from another is required. As each plan is characterized by several attributes, these are the variables whose values enable us to discriminate one plan from another. If one or more of these attributes are omitted, either deliberately or inadvertently, the nature and character of a plan may completely change. When a relevant attribute is left out, there may exist no discriminatory feature and the points in the attribute space may cluster into an amorphous mass from which nothing practical can be inferred. Thus, selection of a proper set of attributes is the second important problem in recognizing patterns. This stage is generally referred to as the *feature extraction* or *preprocessing*.

The third problem in recognizing patterns involves the determination of a decision procedure. In a speech recognition problem, for instance, on what basis do we put some speech patterns in one class? In a resource management problem, on what basis do we choose a plan? For instance, one can conceivably associate an index J_i with plan P_i, and the numerical value of J_i may be used as a key in discriminating one plan from another. This stage generally involves the introduction of a set of parameters that must be estimated. This gives rise to the idea of using pattern recognition as a tool for parameter identification.

The ideas presented in the foregoing discussion are illustrated here by a simple example. Consider the problem of classifying a subset of the professional people into two classes: A_1, the class of all professional singers, and A_2, the class of all professional wrestlers. First, we need a collection of attributes to characterize singers and wrestlers. Attributes such as the color of the skin and hair are not very useful for classification as they do not possess sufficient discriminating power for the purpose of this example. Height of the individuals may or may not have significance. Attributes such as weight, size of the biceps muscle, and sweetness of voice probably will have a discriminating power. Weight can be measured on a ratio scale using, say, kilograms as a unit. Size is a fuzzy concept but it can be quantified, for example, on an ordinal scale using units such as, say, large and small. Alternatively, one can use the principle of substitution. That is, size can be measured by measuring the circumference of the muscle. Now circumference can once again be measured on the ratio scale using, say, centimeters as a unit. Sweetness is a psychophysical attribute and is generally measured subjectively on an ordinal scale. Assuming that all the relevant attributes are properly measured and quantified, the next step is to represent each person by a point in the attribute space. For convenience in plotting, only two attributes are considered at a time. For various pairs of attributes, the resulting dis-

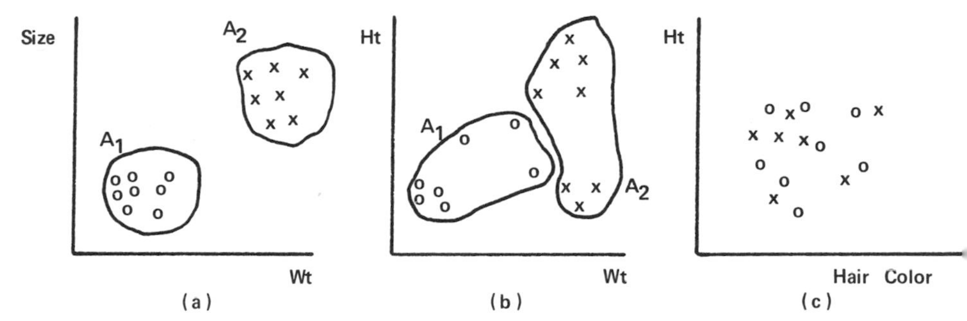

Fig. 6.1 Clustering of various attributes of wrestlers and musicians: The circles represent wrestlers and the x's represent musicians. (a) Easy to discriminate clusters; (b) difficult to discriminate clusters; (c) impossible to discriminate clusters.

tribution of points is shown in Fig. 6.1. Inspection of Fig. 6.1 reveals the role the attributes play in the discrimination of classes. In Fig. 6.1a, the discrimination boundaries are easy to see because there are two distinct clusters separated from one another. In Fig. 6.1b it is still possible to detect a discrimination boundary even though the clusters A_1 and A_2 are extremely close. In Fig. 6.1c, it is impossible to discriminate between the clusters. Thus, in this example, the concept of closeness is used intuitively as a tool for classification. Therefore, the first order of business is to study the concept of closeness a little more systematically.

2 NEIGHBORHOODS AND DISTANCES

The word "neighbor" is a common household word. Such terms as "neighbor," "neighboring block," and "the community neighborhood" convey a sense of proximity or distance. The term "distance," however, can have a broader interpretation. Usage such as "she married a distant relative," "you have come a long way," and "they are poles apart at the negotiating table" all refer to some generalized notion of distance. Thus, the term "distance" can assume different meanings in different contexts. In modeling of complex systems it is important to understand the precise meaning of the term "distance," as it occurs almost everywhere. The modeling of such diverse problems as traffic flow, pattern of telephone calls, migration and mobility of populations, marriage partner selection rules, and newspaper and magazine circulation all involve the concept of distance.

Metric Measures of Distance

In mathematical parlance there is a subtle difference between neighborhood and distance. It is useful to recognize that distance is not critical in the mathematical concept of neighborhood. For a given set of points *neighborhoods* can be defined to suit any given set of circumstances. Let R be the set of all real numbers. Any real number x is an element of this set. A neighborhood to this element or point can be defined as follows: choose an arbitrary positive number ϵ and consider as the neighboor of x, the set of all points y_i for which $|x - y_i| < \epsilon$. Thus the neighborhood of a point x is itself a set. To emphasize this, a neighborhood of a point x that satisfies $|x - y_i| < \epsilon$ is called an ϵ *neighborhood* and is denoted by $N_\epsilon(x)$.

Once neighborhoods have been defined on a set, that set is considered, *by definition*, as a *space*. That is, a space is a collection of points in which neighborhoods have been assigned to each point. Making a set into a space by assigning neighborhoods does not mean that the set ceases to be a set. It adds only a little bit of structure to an otherwise structureless set. However, the amount of structure imparted to a set by assigning neighborhoods to its points is quite minimal and is also qualitative. A quantitative structure can be imparted to a space by introducing a method of measuring the *size of neighborhoods* and *distances* between them. Then the space is called a *metric space*. Thus, the concept of *distance* is vital for the quantitative study of sets, relations in sets, and systems in general.

Distance between Points

In a one-dimensional space, distance between any two points x_1 and x_2 can be measured by $|x_1 - x_2|$, where the bars indicate absolute value. This concept can be generalized to higher dimensions. Consider two points A and B in a two-dimensional space, as shown in Fig. 6.2. From elementary geometry we know that $AB^2 = AC^2 + BC^2$. If the coordinates of A and B are respectively (x_1, y_1) and (x_2, y_2), then $AC = (x_2 - x_1)$ and $BC = (y_2 - y_1)$. Therefore, the distance between the points A and B is given by

$$d(A, B) = [(x_2 - x_1)^2 + (y_2 - y_1)^2]^{1/2} \tag{1}$$

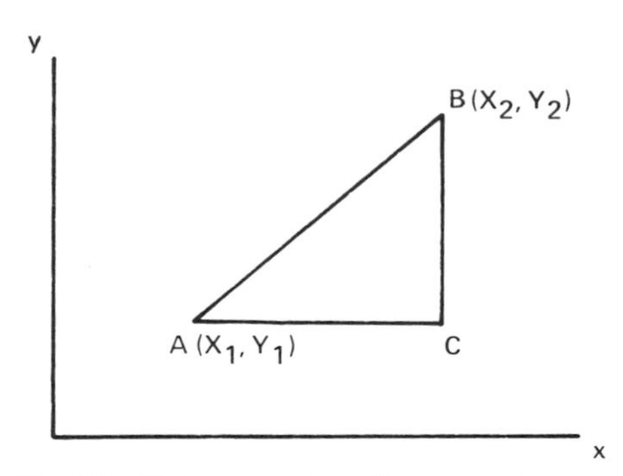

Fig. 6.2 Distance in a two-dimensional space.

This concept of distance can be formalized. Consider a set S and a pair of points (a, b) in S. Let the function $d(a, b)$ define a number denoting the distance between a and b using intuitive notions. It is easy to see that the function $d(a, b)$ satisfies the following conditions:

$$\begin{aligned} &\text{(i)} \quad d(a, b) \geq 0 \\ &\text{(ii)} \quad d(a, b) = d(b, a) \\ &\text{(iii)} \quad d(a, b) + d(b, c) \geq d(a, c) \end{aligned} \tag{2}$$

The first condition says that distances are always positive. The distance between two points is equal to zero if and only if ("iff" is an abbreviation for this) a and b represent the same point. The second condition says that any yardstick used to measure distances should be symmetrical. The third condition, called the *triangle inequality*, says that the "straight line distance" is always smaller, or at most equal, to the distance measured along any other path.

Intuitively, the distance between two points, "as a crow flies," satisfies the above definition of a distance. Such a distance is called the *Euclidean* distance. However, this intuitive notion fails in many unconventional, not necessarily complicated, cases. This failure can occur either if the geometry or topology of the field over which measurements are taken changes or if the "yardstick" to measure distances changes. An example belonging to the former category is the fact that the shortest distance from London to New York is not necessarily the path along which "a crow flies." (It is yet to be determined whether a crow ever flew this route!) Commercial planes generally use the short "polar route." As another down to earth example, the distance traveled by a person in going from his home (point A) to work (point B) is not necessarily the straight line distance between A and B. Because of one-way streets, the distance from A to B could conceivably be different than the distance from B to A. As an example belonging to the latter category, the distance between two points along a river may be defined in terms of the *time* it takes to row a boat from one point to another. Rowing downstream takes less time than rowing upstream. Thus, the "distance" from a point upstream to a point downstream is less than the "distance" between the same two points in the opposite direction.

Distance between Sets

In many applications it becomes necessary to measure distance not between two points but between two sets. Consider two sets A and B. Let n be the number of points in A and m be the number of points in B. What is the distance between A and B? As one can measure distance from any point in A to any other point in B, and taking into account the symmetry of distance measure, one can easily see nm different distances between A and B. One way to avoid this proliferation of distances is to measure the distance between the centers of distribution of A and B. With reference to a Cartesian coordinate system the *center of distribution* for any set of points can be defined as

$$\overline{x} = (1/n) \sum_{i=1}^{n} x_i, \qquad \overline{y} = (1/n) \sum_{i=1}^{n} y_i \tag{3}$$

where (x_i, y_i) are the coordinates of point i. Then distance $d(A, B)$ between the sets A and B can now be defined as

$$d(A, B) = [(\overline{x}_A - \overline{x}_B)^2 + (\overline{y}_A - \overline{y}_B)^2]^{1/2} \tag{4}$$

where $(\bar{x}_A, \bar{y}_A)$ and $(\bar{x}_B, \bar{y}_B)$ are respectively the centers of distribution of A and B.

The need to measure distances between sets is widespread. Consider a set of S variables $\{x_n, x_{n-1}, \ldots, x_2, x_1\}$. Each variable x_i takes on either the value 0 or 1. Thus, at any time the set looks like a string of zeros and ones. This string is called a *binary number*. As each x_i can stand for either of two numbers (namely, 0 or 1), the set S can represent 2^n different binary numbers. Each of these binary numbers is said to represent a *state* of the set S.

There is considerable practical interest in defining a "distance" that quantifies the separation of two different states of the set S. The number of variables in which two states differ is called the *Hamming distance*. For instance, distance between the states 011010 and 111001 is three since they differ in three variables, namely, x_1, x_2, and x_6. The concept of Hamming distance plays an important role in coding theory and analog-to-digital conversion.

A very special case of distance between two sets occurs in applied mathematical analysis. Let A be the set of all points on the straight line $y = a + bx$ and B be the set of order pairs $\{(x_1, y_1), (x_2, y_2), \ldots, (x_n, y_n)\}$. Usually, the set A corresponds to a model and the set B corresponds to observed data. One is interested to know how good the model is. Mathematically this "goodness" of the model is measured in terms of the distance between the sets A and B. There are two commonly used methods of measuring this distance. In both of these methods, one first draws lines, parallel to the y axis, from each point in B until they meet the line $y = a + bx$. The lengths of these lines are measured and labelled as $\epsilon_1, \epsilon_2, \ldots, \epsilon_n$. Then one measure of distance between A and B is

$$d(A, B) = |\epsilon_1| + |\epsilon_2| + \cdots + |\epsilon_n| = \sum_{i=1}^{n} |\epsilon_i| \tag{5}$$

Another widely used measure of distance between A and B is

$$d(A, B) = (\epsilon_1{}^2 + \epsilon_2{}^2 + \cdots + \epsilon_n{}^2)^{1/2} = \left(\sum_{i=1}^{n} \epsilon_i{}^2\right)^{1/2} \tag{6}$$

Distance between Functions

Going one step further, it is also possible to define a distance between two lines or curves. Consider two curves $f(t)$ and $g(t)$. It is desired to find the distance between $f(t)$ and $g(t)$ over an interval, say, $[0, T]$. As $f(t)$ is a set of points and $g(t)$ is another set of points, the notion of distance between sets can be generalized. An analog to Eq. (5) is

$$d(f(t), g(t)) = \int_0^T |f(t) - g(t)|\, dt \tag{7}$$

This is called the integral of the absolute value of the error (or iae) criterion for measuring distance. An analog to Eq. (6) is

$$d(f(t), g(t)) = \int_0^T (f(t) - g(t))^2 \, dt \tag{8}$$

This is called the integral of the square of the error (or ise) criterion for measuring distance. (See also page 33.)

A widely used distance measure, among statisticians, is the *Mahalnobis distance*. This distance refers to a measure of separation between groups of statistical data and is being used widely for cluster analysis and pattern recognition. Any discussion of this topic requires some background in the elementary notions of probability and statistics. A brief explanation of Mahalnobis distance can be given as follows. Consider two samples. Each sample contains n random variables. It is necessary to measure some "distance" between these samples, and it should be characterized by a single number. Let $\boldsymbol{\mu}_1$ be a vector of mathematical expectations of sample 1; $\boldsymbol{\mu}_2$ is a vector of mathematical expectation for sample 2. It is easy to see that vectors $\boldsymbol{\mu}_1 - \boldsymbol{\mu}_2$ are a generalization of the difference between averages for the one-dimensional situation. In such an instance, a covariance matrix $\mathbf{A}$ is similar to variance in the one-dimensional situation. If we want to determine a single number which is based on observation values and scaled similar to the variance, then the simplest measure is

$$D^2 = (\boldsymbol{\mu}_1 - \boldsymbol{\mu}_2)^T \mathbf{A}^{-1} (\boldsymbol{\mu}_1 - \boldsymbol{\mu}_2) \tag{9}$$

where the superscript T stands for the transpose.

Social Distances

In many applications, particularly in behavioral sciences, distances in the physical sense do not exist. Social distance is a concept long used and it refers to the degree of perceived separation between different groups. The perception of distance, in a social context, is conditioned by many socio-psychological elements and the way these elements interrelate is based to a degree on the individual's value system. Since social distance is a composite variable and is based on numerous mutually perceived characteristics, it is difficult to measure. However, sometimes it is quite possible to rank social groups in increasing order of perceived social distance such as family, friends, neighbors. This ranking is often based on the premise that individuals apparently receive the largest rewards, in terms of satisfaction, from meeting close relatives, and successively smaller rewards from meeting friends and neighbors.

Sociologists use the term "distance between occupational categories" to imply a measure of separation between two occupations. Such measures are useful when one is interested in problems such as the following:

(a) In what way does the profession of a parent group influence the profession of the offspring group? (What is the distance between parent profession and offspring profession?)

(b) What is the pattern of movement of, say, scientists with doctoral degrees from one subfield of specialty to another?

Problems of this kind invariably invoke probabilistic concepts and ideas related to what are called state transitions.

Nonmetric Measures of Distance

Measures of similarity or proximity between two sets need not be limited to distance type functions. For illustrative purposes let us consider the problem of determining the similarity between two plans X and Y. Each plan is characterized by n attributes. Let x_i be the value of the ith attribute in plan X and y_i be the corresponding value in plan Y. This information is shown below.

$$\begin{array}{lllll} \text{Plan } X: & x_1 & x_2 & \cdots & x_n \\ \text{Plan } Y: & y_1 & y_2 & \cdots & y_n \end{array}$$

Now the set $\{x_1, x_2, \ldots, x_n\}$ can be represented by the vector $\mathbf{x}$. Similarly, the vector $\mathbf{y}$ is defined. That is

$$\mathbf{x} = (x_1, x_2, \ldots, x_n)^{\mathrm{T}}, \qquad \mathbf{y} = (y_1, y_2, \ldots, x_n)^{\mathrm{T}}$$

Thus, plan X can be represented by the vector $\mathbf{x}$ and plan Y by the vector $\mathbf{y}$ in an n-dimensional space. If α is the angle between the vectors $\mathbf{x}$ and $\mathbf{y}$, then the inner product between $\mathbf{x}$ and $\mathbf{y}$ can be written, variously, as

$$(\mathbf{x}, \mathbf{y}) = \mathbf{x}^{\mathrm{T}}\mathbf{y} = \sum_{i=1}^{n} x_i y_i = \|\mathbf{x}\| \, \|\mathbf{y}\| \cos \alpha \tag{10}$$

where $\|\mathbf{x}\|$ denotes the length of the vector $\mathbf{x}$. From Fig. 6.3 it is evident that the inner product is nothing but the product of the length of the vector $\mathbf{x}$ and the length of the projection of $\mathbf{y}$ on $\mathbf{x}$. Solving for $\cos \alpha$

$$\cos \alpha = \mathbf{x}^{\mathrm{T}}\mathbf{y}/\|\mathbf{x}\| \, \|\mathbf{y}\| = \sum_{i=1}^{n} x_i y_i \Bigg/ \left[\left(\sum_{i=1}^{n} x_i^2\right)\left(\sum_{i=1}^{n} y_i^2\right)\right]^{1/2} \tag{11}$$

When both vectors are oriented in the same direction, $\cos \alpha$ is a maximum. Thus, $\cos \alpha$ can bc used as a measure of similarity between $\mathbf{x}$ and $\mathbf{y}$ and this is often denoted by the symbol $s(\mathbf{x}, \mathbf{y})$. This type of similarity function is useful only if the points representing the plans tend to cluster along some linear dimension and also when the clusters are sufficiently separated.

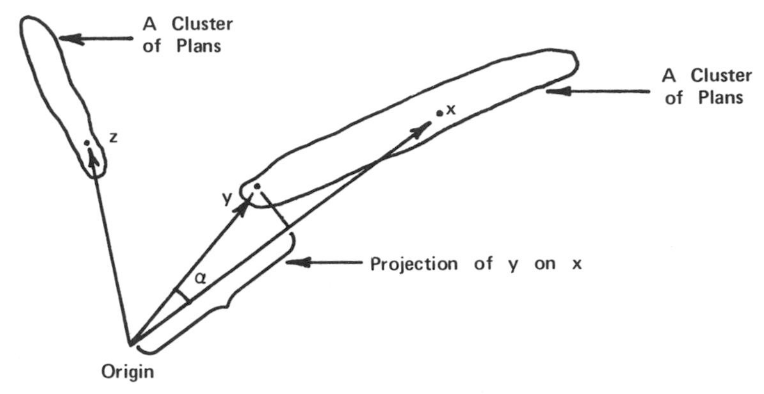

Fig. 6.3 Illustration of similarity measure.

The similarity function can be given an interesting physical interpretation if the x_i's and y_i's take on only binary values. Suppose the attribute i is capital expense, then $x_i = 1$ means that the plan X is expensive and $x_i = 0$ implies that plan X is not expensive. In other words, $x_i = 1$ if the plan X possesses the attribute i. Then $\mathbf{x}^T\mathbf{y}$ simply gives the number of attributes shared by plans X and Y. Similarly $\|\mathbf{x}\|\ \|\mathbf{y}\| = [(\mathbf{x}^T\mathbf{x})(\mathbf{y}^T\mathbf{y})]^{1/2}$ is the geometric mean of the number of attributes possessed by plan X and the number possessed by plan Y. Thus the similarity function $s(\mathbf{x}, \mathbf{y})$ is a measure of common attributes shared by both the plans. For this reason, a binary variant of Eq. (11) has received widespread attention. This is the so-called *Tanimoto measure* defined by

$$s(\mathbf{x}, \mathbf{y}) = \frac{\mathbf{x}^T\mathbf{y}}{(\mathbf{x}^T\mathbf{x} + \mathbf{y}^T\mathbf{y} - \mathbf{x}^T\mathbf{y})} \tag{12}$$

Giving an interpretation to this measure is left as an exercise to the reader. There are a number of other nonmetric measures that were found to be useful in a variety of contexts and the interested reader should consult the extensive literature available on this subject.

Notice that there is no conceptual difference between a similarity measure and a distance measure. A similarity measure gives an idea about the proximity and distance gives an idea about separation.

3 A HIERARCHICAL APPROACH TO CLUSTERING

Once the concept of proximity is available in the form of a distance (or similarity) function, one can proceed to apply this measure to study how and to what extent various objects (such as plans, fingerprints, cardiograms) are similar to each other. Consider once again the decision making problem involving m plans and n attribute, as shown in Table 6.1. Using one of the distance or similarity functions as a guide, one can now proceed to associate or cluster either the plans or the attributes in this table. That is, the clustering procedure can be applied either to the rows or to the columns. There is an important difference, however, between the problems of clustering rows and clustering columns. A column contains values all of which are in the same units. A row, on the other hand, cuts across all the attributes and the unit of measurement of each attribute is likely to be different. This heterogeneity makes it difficult to define clusters of plans within the context of a given set of attributes. The purpose of this section is to look into the problem of clustering rows; the question of clustering columns will not be discussed further.

TABLE 6.1

A Plan–Attribute Matrix

Plans \ Attributes	a_1	a_2	$\cdots$	a_j	$\cdots$	a_n
p_1	x_{11}	x_{12}	$\cdots$	x_{1j}	$\cdots$	x_{1n}
p_2	x_{21}	x_{22}	$\cdots$	x_{2j}	$\cdots$	x_{2n}
$\vdots$	$\vdots$	$\vdots$		$\vdots$		$\vdots$
p_i	x_{i1}	x_{i2}	$\cdots$	x_{ij}	$\cdots$	x_{in}
$\vdots$	$\vdots$	$\vdots$		$\vdots$		$\vdots$
p_m	x_{m1}	x_{m2}	$\cdots$	x_{mj}	$\cdots$	x_{mn}

Equalization of Attribute Scores

Let the distance between plan $\mathbf{p}_j$ and $\mathbf{p}_k$ be defined as

$$D_r(\mathbf{p}_j, \mathbf{p}_k) = \left[\sum_{i=1}^{n} |x_{ji} - x_{ki}|^r \right]^{1/r} = d_{jk} \tag{1}$$

For instance, for $r = 1, j = 1$, and $k = 2$, the above equation becomes

$$D_1(\mathbf{p}_1, \mathbf{p}_2) = |x_{11} - x_{21}| + |x_{12} - x_{22}| + \cdots + |x_{1n} - x_{2n}| = d_{12} \tag{2}$$

For $r = 2$, Eq. (1) defines the familiar distance in Euclidean space. Notice that the magnitude of d_{ij} depends upon the magnitudes of the x_{ij}'s used in the array. The magnitudes of x_{ij}, in turn, are dependent upon the units of measurement. For example, if attribute 1 is measured in feet rather than in inches, that fact implicitly carries with it a weight that will eventually alter the value of d_{12}. Furthermore, if attribute 1 is measured in feet and attribute

2 measured in kilograms, then how can one interpret the meaning of Eq. (2) with the first term in feet and the second term in kilograms? Therefore, there is a need to equalize the attribute scores in an appropriate manner. A number of ideas to handle this problem are currently in vogue. For instance,

1. Replace x_{ij} by $y_{ij} = x_{ij}/\bar{x}_j$, where $\bar{x}_j = (1/m)\sum_{i=1}^{m} x_{ij}$.
2. Replace x_{ij} by $y_{ij} = x_{ij}/r_j$, where r_j is the range of attribute j. Here, range is defined to mean the largest difference between any two x_{ij}'s in column j. With this operation $|y_{ij} - y_{kj}|$ always lies between 0 and 1.
3. Replace x_{ij} by $y_{ij} = x_{ij}/\sigma_j$, where σ_j is the standard deviation of the entries in column j. With this operation, the equalized variables have unit variance and the quantity $|y_{ij} - y_{kj}|$ is in terms of standard deviations.
4. Replace x_{ij} by $y_{ij} = (x_{ij} - \bar{x}_j)/\sigma_j$.

None of these measures is completely satisfactory under all types of problem conditions. There are other alternatives to the equalizing procedures described above. One of these, which utilizes subjective judgments, will be briefly described in a subsequent section.

A Hierarchical Clustering Procedure

The starting point of the hierarchical clustering procedure is the distance matrix **D** (or the similarity matrix **S**). After the attribute scores are equalized, the entries of **D** (or **S**) can be calculated using one of the metric or nonmetric measures of distance (or similarity). Starting from this matrix, one proceeds to construct a treelike structure showing the process of aggregation or clustering. This process is best understood in terms of a specific example.

Consider the problem of choosing one out of five possible plans. Assume that each plan is characterized by three attributes. The values of the attributes in each plan are measured using some suitable technique and the data thus obtained are displayed in Table 6.2.

Now the following steps are implemented.

TABLE 6.2

Raw Scores in a Plan–Attribute Matrix

Plans \ Attributes	a_1	a_2	a_3
Plan 1	2	6	6
Plan 2	3	4	7
Plan 3	5	4	4
Plan 4	6	3	2
Plan 5	8	1	3

TABLE 6.3

Equalized Scores in a Plan–Attribute Matrix

Plans \ Attributes	a_1	a_2	a_3
Plan 1	−1.173	1.321	0.772
Plan 2	−0.754	0.220	1.254
Plan 3	0.084	0.220	−0.193
Plan 4	0.503	−0.330	−1.157
Plan 5	1.340	−1.431	−0.675

Step 1 First, the values in the above table are equalized using method 4 described above and the equalized values are displayed in Table 6.3.

Step 2 Using a suitable distance or similarity measure, the distance between pairs of plans is calculated. In this example, the distance measure defined in Eq. (2) is chosen. For instance, the distance between plan 1 and itself is zero. The distance between plans 1 and 2 is

$$d_{12} = |-1.173 + 0.754| + |1.321 - 0.220| + |0.772 - 1.254| \approx 2.0$$

Similarly, all other plans are taken in pairs, and the resulting distances are arranged as a distance matrix:

$$\mathbf{D} = \begin{bmatrix} d_{21} & & \text{symmetric} & \\ d_{31} & d_{32} & & \\ d_{41} & d_{42} & d_{43} & \\ d_{51} & d_{52} & d_{53} & d_{54} \end{bmatrix} = \begin{bmatrix} 2.0 & & \text{symmetric} & \\ 3.31 & 2.27 & & \\ 5.17 & 4.20 & 1.94 & \\ 6.71 & 5.67 & 3.39 & 2.42 \end{bmatrix}$$

Step 3 Now the distance matrix is scanned to locate the pair of plans that are closest to each other. This is tantamount to searching for the smallest entry in the distance matrix **D**. In this example d_{43} is the smallest indicating that plans 3 and 4 are closest. That is, plans 3 and 4 are more alike than any other pair. If **D** is a large matrix, systematic search techniques, such as the sorting algorithms described in Chapter 3, would be required to locate the smallest entry. Now merge plans 3 and 4 and call the merged plan as plan 3′. This merger is pictorially shown in Fig. 6.4. Now there are only four plans, namely, 1, 2, 3′, and 5.

Step 4 Reconstruct the distance matrix, which now looks as shown

$$\mathbf{D} = \begin{bmatrix} d_{21} & & \\ d_{3'1} & d_{3'2} & \\ d_{51} & d_{52} & d_{53'} \end{bmatrix} = \begin{bmatrix} 2.0 & & \\ ?(3.31) & ?(2.27) & \\ 6.71 & 5.67 & ?(2.42) \end{bmatrix}$$

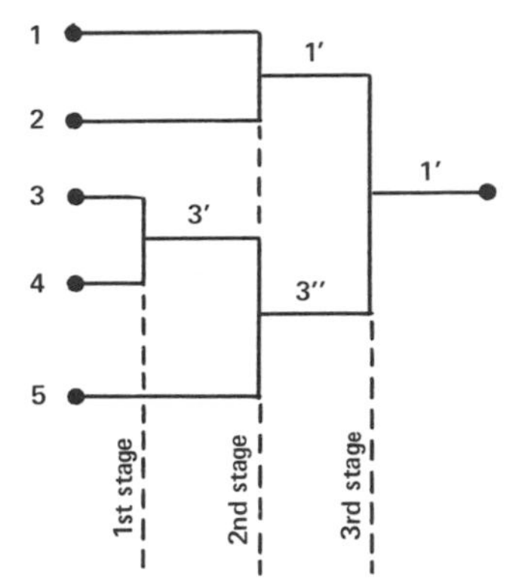

Fig. 6.4 Merging procedure in hierarchical clustering.

Note that there is no ambiguity in filling entries d_{21}, d_{51}, and d_{52}. One has to decide the meaning of entries like $d_{3'1}$. Such entries can be filled using the following rule:

$$d_{3'2} = \min(d_{32}, d_{42}) = \min(2.27, 4.20) = 2.27$$
$$d_{53'} = d_{3'5} = \min(d_{35}, d_{45}) = \min(3.39, 2.42) = 2.42$$

and so on. These modified distances are shown in the parentheses, next to the question mark, in the above matrix.

Step 5 Go back to Step 3. Scanning the new distance matrix reveals that now $d_{53'}$ is the smallest, and so plans 3′ and 5 are closest among the remaining four plans. Now plans 3′ and 5 are merged and the merged plan is now labeled as plan 3″ (the label for the merged plan is the smaller of the two merging plan numbers). Now we are left with three plans, namely 1, 2, and 3″. The new distance matrix now looks as follows:

$$\mathbf{D} = \begin{bmatrix} d_{21} & \\ d_{3''1} & d_{3''2} \end{bmatrix} = \begin{bmatrix} 2.0 & \\ 3.31 & 2.27 \end{bmatrix}$$

where, for instance,

$$d_{3''2} = \min(d_{3'2}, d_{52}) = 2.27$$

This matrix reveals that plans 1 and 2 are now close to each other. Therefore, plans 1 and 2 are merged and labeled as 1′, and finally plans 1′ and 3′ are merged.

The procedure above can be generalized and described as follows.

Step 1 Begin with m clusters each consisting of exactly one plan. Let the clusters be labeled by the numerals 1 through m.

Step 2 Choose a convenient distance (or similarity) measure and construct a distance matrix **D** (or similarity matrix **S**).

Step 3 For each row of the distance (similarity) matrix find the minimum (maximum) entry and record the column in which it occurs. As the ith row of the matrix involved contains $(i - 1)$ entries, the search in the ith row involves $(i - 2)$ comparisons. The entire search from row 2 to row m then involves $\sum_{i=2}^{m} (i - 2) = \frac{1}{2}(m - 1)(m - 2)$ comparisons.

Step 4 To find the pair of plans that are closest to each other search for the minimum (maximum) of the row minima (maxima) that were found in Step 3. Label the entry thus located as d_{pq} (or s_{pq}) with $p > q$. The total number of comparisons involved at this (i.e., the kth) stage of the search is $\sum_{k=1}^{m-1} (m - k - 1) = \frac{1}{2}m(m - 3) + 1$.

Step 5 Merge the clusters p and q and label the new cluster as q'. Update the entries in the distance (similarity) matrix to reflect the revised distances (proximities) between the cluster q' and all other clusters. Delete the row and column of the distance (similarity) matrix pertaining to plan p. The number of updatings required at this stage are $\sum_{k=1}^{m-1} (m - k) = \frac{1}{2}m(m - 1)$. Go back to Step 3.

Evidently, as the size of the plan–attribute matrix increases, the number of comparisons and updatings to be made grow very fast. A sound background in sorting and merging algorithms and a knowledge in handling lists in a computer come in handy at this stage. The procedure described above is only one of many possible ideas. A number of minor variations are also possible. One should consult the very extensive literature on pattern recognition and cluster analysis for further treatment of this topic.

4 NONHIERARCHICAL APPROACHES TO CLUSTERING

The hierarchical clustering procedure discussed in the previous section gave a nested classification of plans. In this section, attention is focused on the problem of clustering m plans into a single classification of k clusters. The central idea of the procedures to be described here is to choose some initial partition of plans and then gradually alter the cluster membership to get an improved partition. The various methods differ in defining what an improved partition is. Indeed there is an element of art involved in devising algorithms of this kind and minor variations lead to different algorithms. Some of the more popular algorithms are described below.

A Simple Cluster-Seeking Algorithm

In this procedure a seed point $\mathbf{z}_1$ is selected and the set of all plans that are at a specified distance (called the *threshold*) from the point $\mathbf{z}_1$ are assigned to the cluster labeled Z_1. If a plan lies beyond the threshold distance from $\mathbf{z}_1$, then a new seed point is selected and the procedure repeated. Specifically, consider a set of plans $\{\mathbf{p}_1, \mathbf{p}_2, \ldots, \mathbf{p}_m\}$. Let the first seed point be $\mathbf{p}_1$; that is, $\mathbf{z}_1 = \mathbf{p}_1$. Let the threshold be a nonnegative real number T. Now we compute the distance d_{21} from $\mathbf{p}_2$ to $\mathbf{z}_1$. If this distance exceeds T, a new cluster center $\mathbf{z}_2 = \mathbf{p}_2$ is started; otherwise, plan $\mathbf{p}_2$ belongs to the cluster Z_1. Now the distances d_{31} and d_{32}, respectively, from $\mathbf{p}_3$ to $\mathbf{z}_1$ and $\mathbf{p}_3$ to $\mathbf{z}_2$ are computed. If both d_{31} and d_{32} are greater than T, a third cluster center $\mathbf{z}_3$ is started. This procedure is repeated until all plans are assigned to one of the clusters.

The results of the foregoing procedure depend upon a number of factors: the first cluster center chosen, the order in which the plans are considered, the value of T, and the geometrical properties of the data. The influence of these factors on the clustering patterns is pictorially shown in Fig. 6.5.

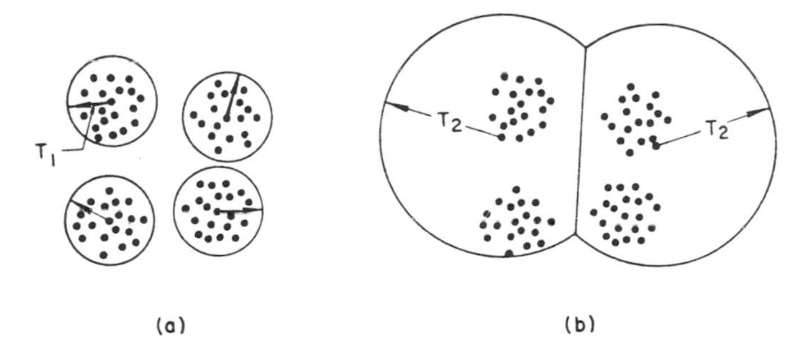

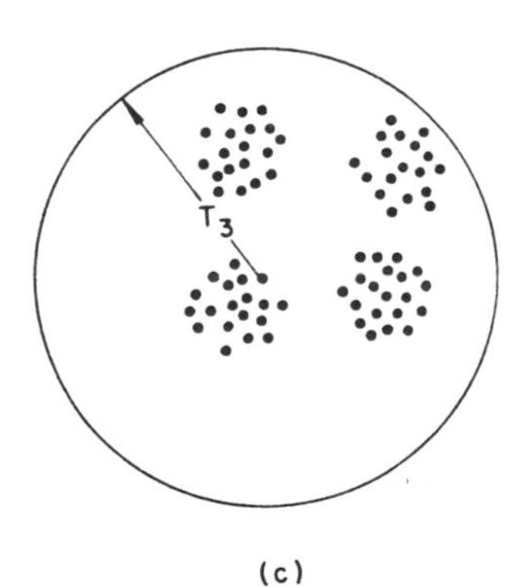

Fig. 6.5 Effects of the threshold and starting points in a simple cluster-seeking scheme. (a) Four starting points and one small value of threshold; (b) two starting points and a larger value of threshold; (c) only one starting point and one threshold. (Reproduced with permission from *Pattern Recognition Principles* by J. T. Tou and R. C. Gonzalez, Copyright © 1974, Addison-Wesley, Reading, Massachusetts.)

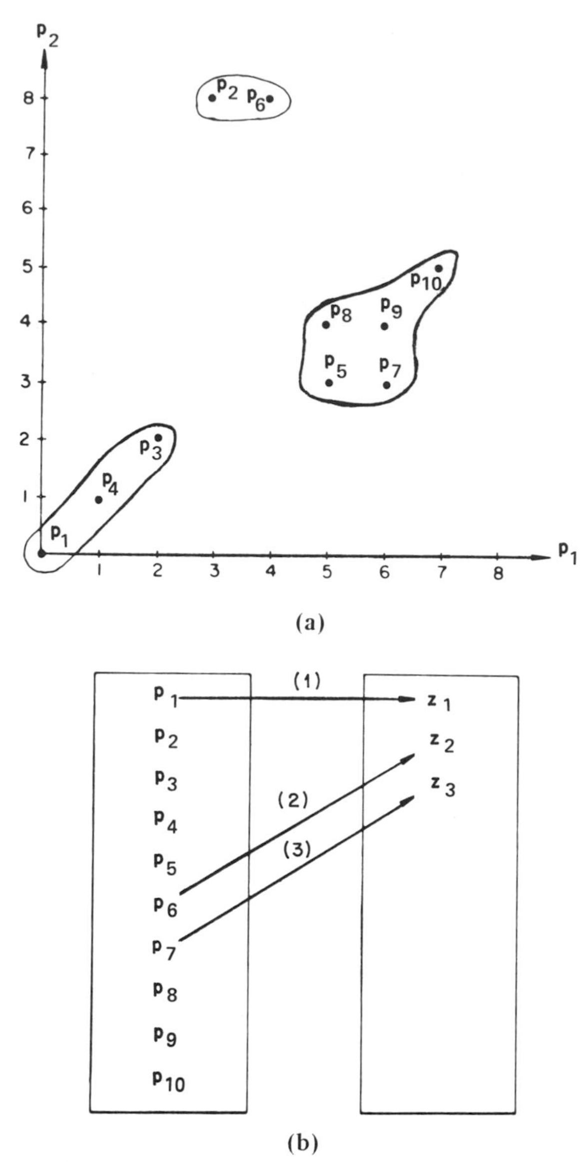

Fig. 6.6 (a) Plan patterns used in illustrating the maximum distance algorithm. (b) Plan and category tables. (Reproduced with permission from *Pattern Recognition Principles* by J. T. Tou and R. C. Gonzalez, Copyright © 1974, Addison-Wesley, Reading, Massachusetts.)

Evidently, this clustering procedure is heuristic and requires extensive experimentation with various values of parameters in order to gain useful insight into a problem.

Maximum Distance Algorithm

This is another heuristic procedure that utilizes the concept of Euclidean distance. This is similar to the preceding algorithm except that it first identifies the cluster regions that are farthest apart. This procedure is best explained by a simple example. For convenience in plotting on a graph paper, a two-dimensional example is chosen. Consider ten plans $\mathbf{p}_1, \mathbf{p}_2, \ldots, \mathbf{p}_{10}$, each characterized by two attributes. The location of these plans in the attribute space are plotted as shown in Fig. 6.6a. The first step of the maximum distance algorithm is to choose, as before, a seed point and designate it as $\mathbf{z}_1$. Let this seed point be $\mathbf{p}_1$; that is, $\mathbf{z}_1 = \mathbf{p}_1$. This step is indicated in Fig. 6.6b by an arrow from $\mathbf{p}_1$ to $\mathbf{z}_1$. The number on the arrow reminds us that this designation took place in Step 1. The second step is to calculate the distance between the cluster center $\mathbf{z}_1$ and $\mathbf{p}_k$ for all k. These distances are displayed in the first row of Table 6.4. Inspection of this row shows that the maximum distance occurs between $\mathbf{z}_1$ and $\mathbf{p}_6$. So $\mathbf{p}_6$ is designated by the label $\mathbf{z}_2$, and now this will serve as a new cluster center. This step is indicated in Fig 6.6b by drawing an arrow from $\mathbf{p}_6$ to $\mathbf{z}_2$ and labeling the arrow by the numeral (2). The third step is to calculate the distances from $\mathbf{z}_1$ and $\mathbf{z}_2$ to $\mathbf{p}_k$ for all k. The distances from $\mathbf{z}_1$ to $\mathbf{p}_k$ are already available in the first row of Table 6.4 and therefore need not be calculated. The distances from $\mathbf{z}_2$ to $\mathbf{p}_k$, for all k, are now calculated and displayed in the second row of the table. Next, each column of this table is scanned and the minimum of the two entries is recorded in the third row. Next, the maximum of the third row is located and circled as shown. This operation shows that plan $\mathbf{p}_7$ is farthest from both cluster centers $\mathbf{z}_1$ and $\mathbf{z}_2$. Now do we designate $\mathbf{p}_7$ as a new cluster center or not? Looking back we notice that the distance between $\mathbf{z}_1$ and $\mathbf{z}_2$ (which, at this stage, is really the distance between $\mathbf{p}_1$ and $\mathbf{p}_6$) is $\sqrt{80} = 8.94$

TABLE 6.4

First Stage of Implementing the Minimum Distance Algorithm

	p_1	p_2	p_3	p_4	p_5	p_6	p_7	p_8	p_9	p_{10}
z_1 to p_k	0	$\sqrt{75}$	$\sqrt{8}$	$\cdots$		$\sqrt{80}$	$\sqrt{45}$		$\cdots$	
z_2 to p_k	$\sqrt{80}$	1	$\sqrt{40}$	$\cdots$		0	$\sqrt{29}$		$\cdots$	
Column minima	0	1	$\sqrt{8}$	$\cdots$		0	$\boxed{\sqrt{29}}$		$\cdots$	

and the circled entry is $\sqrt{29} = 5.39$. Because 5.39 is more than half of 8.94 (this is a heuristic rule) $\mathbf{p}_7$ is considered a new cluster center. This is indicated in Fig. 6.6b by an arrow from $\mathbf{p}_7$ to $\mathbf{z}_3$, and the arrow is labeled with the numeral (3).

The next step is to calculate the distances from each of the three established cluster centers to plans $\mathbf{p}_k$. The results can be arranged in a new table in which the third row now would show the distances $\mathbf{z}_3$ to $\mathbf{p}_k$. Again the minimum of each column can be found and entered as an entry in the last row and the maximum of this row would be circled. Now do we designate the plan corresponding to the circled entry as a new cluster center or not? If this entry is an appreciable fraction of the previous maximum distances, then the newly circled plan would qualify as a new cluster center. The previous maximum distances at this stage are $\sqrt{29}$ and $\sqrt{80}$. The average of these two values is 7.16. So the newly circled entry would qualify as a new cluster center only if its value is at least one half of 7.16, that is, 3.58. If it does not qualify, the algorithm for locating the cluster centers is over.

The unfinished task is to assign the remaining plans to one of the cluster centers determined. This is done by simply assigning each of the remaining plans to the nearest cluster center. Figure 6.6a shows what the final answer should look like. The student is advised to carry out the computations and verify this answer.

K-Means Algorithm

The preceding two algorithms are based on heuristic and intuitive arguments. The *K*-means algorithm is based on the minimization of a performance index, which is defined as the sum of the squared distances from all points in a cluster domain to the cluster center. The key implication of this process is that the cluster center (or centroid) is computed on the basis of the membership of the cluster at a given stage of the computational cycle. This algorithm can be explained by the following steps.

Step 1 Choose the first K plans as the cluster centers. That is $\mathbf{z}_1 = \mathbf{p}_1$, $\mathbf{z}_2 = \mathbf{p}_2, \ldots, \mathbf{z}_K = \mathbf{p}_K$.

Step 2 Assign each of the remaining $m - K$ plans to the cluster with the nearest centroid.

That is, at the rth iterative step, assign the plans $\{\mathbf{p}\}$ to the K clusters using the rule

$$\mathbf{p} \in A_j(r) \qquad \text{if} \quad \|\mathbf{p} - \mathbf{z}_j(r)\| < \|\mathbf{p} - \mathbf{z}_i(r)\| \tag{1}$$

for all $i = 1, 2, \ldots, K$, $i \neq j$, where $A_j(r)$ denotes the cluster whose centroid is $z_j(r)$. Ties are resolved in an arbitrary manner.

Step 3 After all plans have been assigned to some cluster or other, take the existing centroids as seed points and make one more pass, assigning each plan to the nearest seed point.

That is, from the results of Step 2 compute new cluster centers $\mathbf{z}_j(r+1)$, $j = 1, 2, \ldots, K$ such that

$$J_j = \sum_{\mathbf{p} \in A_j(r)} \|\mathbf{p} - \mathbf{z}_j(r+1)\|^2; \qquad j = 1, 2, \ldots, K \tag{2}$$

is minimized. The $\mathbf{z}_j(r+1)$ that minimizes J_j is simply the centroid of $A_j(r)$. Therefore, the new cluster center is given by

$$\mathbf{z}_j(r+1) = (1/C_j) \sum_{\mathbf{p} \in A_j(r)} \mathbf{p}; \qquad j = 1, 2, \ldots, K \tag{3}$$

where C_j is the number of plans in cluster $A_j(r)$.

Step 4 If $\mathbf{z}_j(r+1) = \mathbf{z}_j(r)$, $j = 1, 2, \ldots, K$, then the algorithm has converged. If no convergence is evident, go back to Step 2.

The K-means algorithm is illustrated here by means of a simple example. Consider 20 plans each characterized by two attributes. The plan–attribute matrix is assumed to be given. This information is displayed in Fig. 6.7. For a simple two-dimensional case, such as the one under discussion, the answer

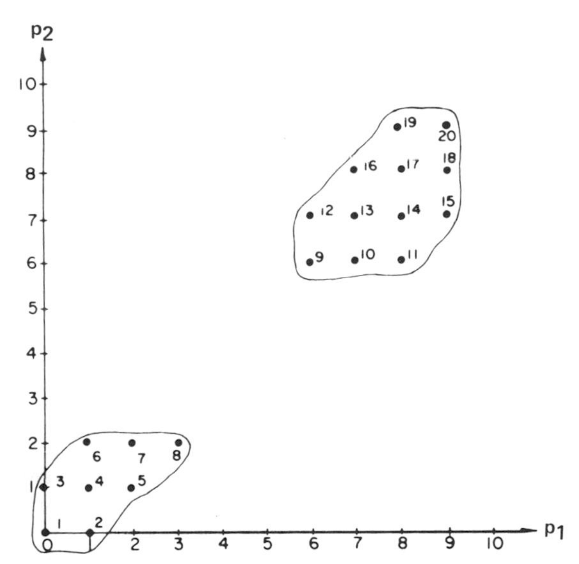

Fig. 6.7 Plan patterns used in the K-means algorithm. (Reproduced with permission from *Pattern Recognition Principles* by J. T. Tou and R. C. Gonzalez, Copyright © 1974, Addison-Wesley, Reading, Massachusetts.)

is obvious. However, the more general algorithmic procedure will be used here.

Step 1 Suppose that we are interested in clustering the plans into two clusters. Then $K = 2$. For the first iteration $r = 1$. Now, let us choose

$$\mathbf{z}_1(1) = \mathbf{p}_1 = (0, 0)^T, \qquad \mathbf{z}_2(1) = \mathbf{p}_2 = (1, 0)^T$$

as the two cluster centers respectively for clusters $A_1(1)$ and $A_2(1)$.

Step 2 Evidently, plan $\mathbf{p}_1$ belongs to $A_1(1)$ and $\mathbf{p}_2$ belongs to $A_2(1)$. To decide whether $\mathbf{p}_3$ belongs to $A_1(1)$ or $A_2(1)$, we compute $\|\mathbf{p}_3 - \mathbf{z}_1(1)\|$, which can be verified to be 1 and $\|\mathbf{p}_3 - \mathbf{z}_2(1)\|$, which when computed becomes 2. Because $\|\mathbf{p}_3 - \mathbf{z}_1(1)\| < \|\mathbf{p}_3 - \mathbf{z}_2(1)\|$, plan $\mathbf{p}_3$ belongs to cluster $A_1(1)$. Therefore, $A_1(1) = \{\mathbf{p}_1, \mathbf{p}_3\}$. Similar calculations show that $A_2(1) = \{\mathbf{p}_2, \mathbf{p}_4, \ldots, \mathbf{p}_{20}\}$.

Step 3 Now we have to update the cluster centers using Eq. (3):

$$\mathbf{z}_1(2) = (1/C_1) \sum_{\mathbf{p} \in A_1(1)} \mathbf{p} = \tfrac{1}{2}(\mathbf{p}_1 + \mathbf{p}_3) = (0.0, 0.5)^T$$

$$\mathbf{z}_2(2) = (1/C_2) \sum_{\mathbf{p} \in A_2(1)} \mathbf{p} = \tfrac{1}{18}(\mathbf{p}_2 + \mathbf{p}_4 + \cdots + \mathbf{p}_{20}) = (5.67, 5.33)^T$$

These are the new cluster centers.

Step 4 Since $\mathbf{z}_j(2) \neq \mathbf{z}_j(1)$, for $j = 1, 2$, there is no convergence. Go back to Step 2.

Step 2 With the new clusters this step is repeated. The computation shows (do this and verify!) that

$$A_1(2) = \{\mathbf{p}_1, \mathbf{p}_2, \ldots, \mathbf{p}_8\} \qquad \text{and} \qquad A_2(2) = \{\mathbf{p}_9, \mathbf{p}_{10}, \ldots, \mathbf{p}_{20}\}$$

Step 3 The updated cluster centers are

$$\mathbf{z}_1(3) = \tfrac{1}{8}(\mathbf{p}_1 + \mathbf{p}_2 + \cdots + \mathbf{p}_8) = (1.25, 1.13)^T$$
$$\mathbf{z}_2(3) = \tfrac{1}{12}(\mathbf{p}_9 + \mathbf{p}_{10} + \cdots + \mathbf{p}_{20}) = (7.67, 7.33)^T$$

Step 4 Since $\mathbf{z}_j(3) = \mathbf{z}_j(2)$, return to Step 2.

Step 2 Implementation of this step now shows convergence. The resulting clusters are shown in Fig. 6.7.

5 INDIVIDUAL AND GROUP PREFERENCE PATTERNS

The clustering procedures described in the previous two sections are essentially objective. Such procedures play a useful role in technological problems involving classification (via recognition of patterns) of objects such as electrocardiograms, fingerprints, chromosomes, cells, and ground cover (such as crops, vegetation), where it is possible to define well-understood objective criteria for classification. Pattern classification ideas with a different twist can also be found in public planning contexts. For example, an interesting application of the concepts developed so far arises in the context of using public goods, such as air and water, by a community. As public goods are not generally owned by individuals, or by a select group of people, their use depends on a collective decision. In other words, for an equitable use of public goods a society needs a collective decision rule. How does a society choose a collective decision rule? That is, what rule will one use to decide upon a collective decision rule?

Microeconomic theory has given us a very general mathematical model of the production and consumption of private goods. Private goods are characterized by the property that their consumption by one individual (or one sector of the society) excludes the consumption of the same goods by another individual. That is, private goods are divisible, or the services associated with private goods are divisible. (An apple is a private good; if one eats it, others do not have it.) Public goods, on the other hand, are indivisible commodities. They are jointly consumed by a large number of individuals who may or may not pay an equal price. (Natural resources, such as air and water, are examples of public goods. The use of air by one does not preclude, or should not preclude, its use by others.) The use of public goods is, therefore, not governed by individual decisions but by a *collective decision*. Most of the microeconomic models related to production and consumption are no longer valid in the context of collective utilization of public goods. The reason for this is that microeconomic models use decision rules of utility maximization for the individual consumer and profit maximization to the individual producer and are of little help in the context of collective decision making.

The study of collective decision making has two aspects. First, how is, or how should such a decision be chosen? In other words, as stated earlier, what rule shall we use to decide upon a collective decision rule? This is the classical *constitutional choice problem*. Second, how would a disinterested outside advisor or judge choose a collective decision rule on the basis of purely rational considerations? This is the problem of *individual preference aggregation* of classical economics.

To motivate further discussion, let us consider a group of three individuals who are required to decide and pick one plan collectively from among three possible alternative plans *a*, *b*, and *c*. While talking about collective preference, it is implicitly assumed that one has a rule for combining preferences. For concreteness, a majority voting rule has been chosen here for combining individual preferences. Let the individual preferences be described as follows:

$$a \mathsf{P}_1 b \mathsf{P}_1 c$$
$$b \mathsf{P}_2 c \mathsf{P}_2 a$$
$$c \mathsf{P}_3 a \mathsf{P}_3 b$$

where the notation $a \mathsf{P}_i b$ means "a is preferred to b by judge i." Inspection of the above relation reveals that

$$a \mathsf{P}_g b \quad \text{by } \tfrac{2}{3} \text{ majority}$$
$$b \mathsf{P}_g c \quad \text{by } \tfrac{2}{3} \text{ majority}$$
$$c \mathsf{P}_g a \quad \text{by } \tfrac{2}{3} \text{ majority}$$

where the notation $a \mathsf{P}_g b$ means "a is preferred to b by the group collectively." Obviously, this collective preference cannot be ordered. For instance, the above result forces one to conclude that because $a \mathsf{P}_g b$ and $b \mathsf{P}_g c$, then $a \mathsf{P}_g c$. However, this conclusion violates the last of the above three relations. This phenomenon is called the *voting paradox*. In a practical context, implication of the voting paradox is the following. If a group of individuals, such as a society, makes its social decisions using a majority rule, then the final winning alternative often depends upon the order of the voting process. That is, the common legislative procedure of considering one alternative at a time for voting and excluding each defeated alternative from further consideration is not an infallable procedure. This simple example illustrates the need for a practical procedure to aggregate individual preferences. Before embarking on the problem of preference aggregation and group decision making, it is instructive to understand how an individual decision maker (a judge) makes decisions.

The Individual as a Judge

Suppose an individual is presented with a plan–attribute matrix and is asked to render a judgment by picking a plan as the most desirable plan within the context of a global objective. The judgmental process can proceed through several different stages. For the purpose of the present discussion, it is sufficient to identify two of these stages.

The Subliminal Stage

This stage is explained with respect to a specific scenario. Consider a situation in which a judge is observing a viewscape in order to assess the construction of a dam, or the construction of a transmission tower. In this case, the information coming from the stimulus, the viewscape, is a visual signal corrupted by noise, where the word "noise" stands for all distractive influences other than the viewscape of interest. Modern research supports the view that transmission of information through receptors (visual, in this case) is inextricably associated with statistical uncertainties related to events at the molecular and atomic level. Both the signal and the noise may trigger activity in the same sensory receptors. Depending on the relative strengths of the signal and noise, different patterns of excitation will be perceived by the sensory apparatus of the judge.

In modern theories of human behavior, it is widely assumed that an observer stores in his memory system a representation of the signal for which he is looking. Any stimulus, such as a signal plus noise, will then be compared with the stored representation in the process of passing value judgments. In the case of the visual scene under discussion, the judge may have an idea of what the ideal viewscape should be. It is conceivable that this ideal viewscape can be characterized by several, say, n, attributes. If each attribute is assigned a dimension, then the sensation experienced in watching a viewscape can be represented by a point in an n-dimensional space. The coordinates of this point can be arranged as the elements of a vector. As the attributes can assume a range of values, each dimension may represent a continuum of values. As the measuring process is limited by a finite resolution, it is more realistic to assume that each attribute takes on a finite number of possible values. In the limit, it is convenient to assume, without loss of generality, that the attributes are binary values. Thus, a representation of the ideal viewscape in the memory system of the judge could be a vector such as $(1 \;\; 1 \;\; 1 \;\; -1 \;\; -1 \;\; -1)^T$. The point represented by this vector in the n-dimensional attribute space can be referred to as the *ideal point*.

Now when a judge observes a viewscape of a proposed plan, an input is received by the judge's sensory system and this input can also be characterized by a vector with binary entries. The judge's memory system then compares each input vector with the ideal vector and produces an output which is representative of the similarity or dissimilarity between the two vectors. The concept of similarity can be formalized in several different ways—each depending on the metric chosen in the n-space under consideration. From the standpoint of the neurophysiology of perceptual and memory systems, the appropriate metric appears to be the inner product $(\mathbf{I}, \mathbf{M})$ of the input vector $\mathbf{I}$ and the memory vector $\mathbf{M}$. That is,

$$(\mathbf{I}, \mathbf{M}) = i_1 m_1 + i_2 m_2 + \cdots + i_n m_n$$

where $\mathbf{I} = (i_1, i_2, \ldots, i_n)^{\mathrm{T}}$ and $\mathbf{M} = (m_1, m_2, \ldots, m_n)^{\mathrm{T}}$. Figure 6.8 illustrates a typical relation between the input, memory, and output of such a process. A large output value in this case is indicative of a higher degree of similarity between the input and memory vectors. That is, by this process, the judge is really ranking the plans subliminally, and the plan with the highest score is the preferred plan.

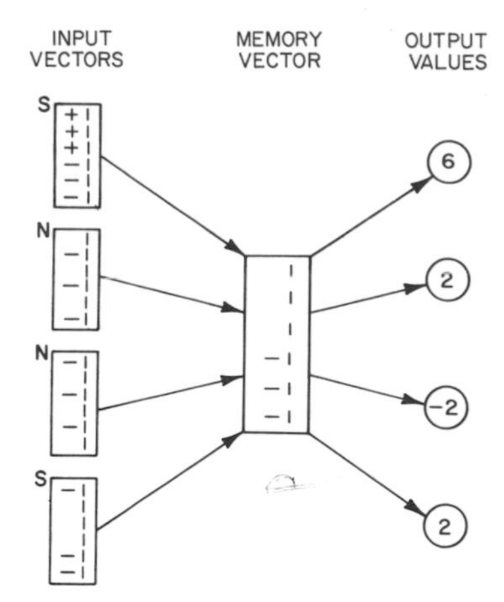

Fig. 6.8 Vector representation of a sequence of stimulus inputs from signal (*S*) and noise (*N*). Sources are compared in turn with a memory vector, which serves as a filter.

Policy-Capturing Stage

The judgmental process need not and in general does not stop after each plan is associated with a rank. One can proceed to construct a model to the judgmental process itself by asking the question: What importance or salience w_i does the judge attach to the ith attribute in arriving at the above rankings? The judge would not know the answer because the rankings were assigned in a subliminal manner. Denoting by J_k the rank of the kth plan, one can solve the regression problem

$$(J_j - J_k)^2 = \sum_{i=1}^{n} |w_i(x_{ji} - x_{ki})|^2 + e_{jk} \tag{1}$$

for the weights w_i. That is, once the judge's rankings are known, the weights w_i can be determined to "capture" or simulate the judge's policy of assigning importance to various attributes.

Group Judgments

When a number of judges, such as a committee, are involved in picking a plan there are several possible ways of aggregating individual preferences into a social choice; that is, there are a number of ways of defining the so-called social welfare function. Probably the most intuitive approach to amalgamating individual preferences into a group preference is a voting procedure. The method of using voting as a means of expressing social choice is not without its difficulties. The voting paradox, discussed earlier, pointed out one of the difficulties. To see other difficulties with voting, consider a group of 60 judges voting for one of three plans a, b, and c. Suppose 23 out of the 60 expressed a preference order $a \mathsf{P} c \mathsf{P} b$. That is, 23 out of 60 ranked plan a above plan c, and plan c above plan b. Similarly, 19 expressed a preference $b \mathsf{P} c \mathsf{P} a$, 16 expressed a preference $c \mathsf{P} b \mathsf{P} a$ and 2 expressed $c \mathsf{P} a \mathsf{P} b$. Therefore, if *plurality* is used to select a winner (i.e., if plurality is used as a social welfare function), then plan a has 23 first place votes, plan b has 19 and plan c has 18. Thus, plan a is the winner. However, if a majority is required to pick a winning plan, there must be a runoff selection between plans a and b. Keeping the preference orders unchanged, we see that $23 + 2 = 25$ judges express $a \mathsf{P} b$ either directly or by transitivity. Similarly, $19 + 16 = 35$ judges express $b \mathsf{P} a$. Thus, with a majority rule as a social welfare function, plan b is the winner.

There are other approaches to the problem of aggregating individual preferences into a social choice. For example, in the *intensity method* of voting, a weight of 2 is given to the first place vote and a weight of 1 to the second place vote. Votes in the third- and higher-order places are not counted. Under this scheme, plan a scores $(23 \times 2) + (2 \times 1) = 48$. Plan b scores $(19 \times 2) + (16 \times 1) = 54$. The score for plan c is $(23 \times 1) + (19 \times 1) + (16 \times 2) + (2 \times 2) = 78$. Thus, plan c is the winner.

Another way of defining a social welfare function is to use the so called *Borda count*. The Borda count $B(a)$ of plan a is defined as

$$B(a) = \sum_{i=1}^{n} B_i(a)$$

where $B_i(a)$ is the number of plans ranked below plan a by judge i. Suppose there are four judges and four plans and the preferences (or rankings) of the judge i are indicated in the column labeled P_i:

P_1	P_2	P_3	P_4
a	b	d	a
b	c	a	b
c	a	b	c–d
d	d	c	

That is, judge 1 ranked the plans as $a \mathsf{P}_1 b \mathsf{P}_1 c \mathsf{P}_1 d$. Judge 4 ranked the plans as $a \mathsf{P}_4 b \mathsf{P}_4 c \mathrm{I}_4 d$; that is, judge 4 is indifferent to plans c and d. Now, according to the above definition, $B_1(a) = 3$, $B_2(a) = 1$, $B_3(a) = 2$, and $B_4(a) = 3$. Therefore, the Borda count $B(a)$ for plan a is $3 + 1 + 2 + 3 = 9$. Similarly, $B(b) = 2 + 3 + 1 + 2 = 8$, $B(c) = 1 + 2 + 0 + 0 = 3$, and $B(d) = 0 + 0 + 3 + 0 = 3$. Thus, if the Borda count is used as an aggregation tool, the group preference is $a \; \mathsf{P}_g \; b \; \mathsf{P}_g \; c \; \mathrm{I}_g \; d$. That is, collectively, the group of judges ranked plan a over b, b over both c and d, and c and d are tied. The Borda count always leads to a ranking but it does not always agree with our intuitive notion of what is "fair" (See exercise 9).

Finally, one can use the ideas of cluster analysis and pattern recognition in aggregating individual preferences to a group preference. In this approach, each judge will be asked to rank the plans as was done earlier. These rankings are now plotted as points in the n-dimensional attribute space. Thus, the position of each judge with reference to a given plan becomes a point in the attribute space; with reference to m plans, one gets m points that can be regarded as the "preference pattern" of that judge. All such preference patterns, obtained by polling all the judges, can now be compared with each other to detect similarities that will allow one to partition the set of points into preference classes.

EXERCISES

1. If D satisfies the requirements of a distance measure as defined in Section 2, then show that $D' = D/(w + D)$, where w is any positive number, also satisfies the requirements of a distance measure.

2. Consider an array a_{ij} of real numbers. Denote the jth column of this array by $\mathbf{A}_j = (a_{1j}, a_{2j}, \ldots, a_{nj})^{\mathrm{T}}$. Then the distance between the jth and the kth columns can be defined by the so-called Minkowski metric as

$$D_p(\mathbf{A}_j, \mathbf{A}_k) = \left[\sum_{i=1}^{n} |a_{ij} - a_{ik}|^p \right]^{1/p}$$

where $p \geq 1$. The set of points for which $D_p(A_j, A_k) = 1$ is called the unit ball. For a two-dimensional problem sketch the unit balls for $p = 1$, 2, and ∞. The case for $p = 1$ is called "city block" or "taxicab", or L_1 metric. The case for $p = 2$ is called the Euclidean or L_2 metric. The case for $p = \infty$ is called the Chebyshev metric or L_∞ metric.

3. Two variables X and Y are sampled n times and their means $\bar{x}$ and $\bar{y}$ are calculated. Now two vectors $\mathbf{x}$ and $\mathbf{y}$ of centered scores are defined by

$$\mathbf{x} = [(x_1 - \bar{x}), (x_2 - \bar{x}), \ldots, (x_n - \bar{x})]^{\mathrm{T}}$$
$$\mathbf{y} = [(y_1 - \bar{y}), (y_2 - \bar{y}), \ldots, (y_n - \bar{y})]^{\mathrm{T}}$$

The inner product $\mathbf{x}^{\mathrm{T}}\mathbf{y}$ is called the scatter of X and Y, which when divided by n gives the covariance between X and Y. Similarly the scatter of X, namely, $\mathbf{x}^{\mathrm{T}}\mathbf{x}$, when divided by n gives the variance of X. Variance of Y is analogously defined. The so-called product moment correlation between

X and Y is defined as

$$r = r(X, Y) = \frac{\text{cov}(X, Y)}{\text{var}(X)\,\text{var}(Y)}$$

Show that r is equal to the cosine of the angle between $\mathbf{x}$ and $\mathbf{y}$.

4. Give a physical interpretation to the Tanimoto distance defined in Eq. (2.12).

5. Complete the unfinished calculations in the description of the hierarchical clustering procedure of Section 3.

6. Complete all the calculations in the example discussed with reference to Fig. 6.6. Extend the idea to the general case and write a computer program that can handle m plans and n attributes.

7. Complete all the calculations in the example discussed with reference to Fig. 6.7. Extend the idea to the general case and write a computer program that can handle m plans and n attributes in implementing the K-means algorithm.

8. Apply the cluster seeking algorithms discussed in Section 4 to the data set $\{(0, 0)^T, (0, 1)^T, (5, 4)^T, (5, 5)^T, (4, 5)^T, (1, 0)^T\}$ and compare the results.

9. Consider a problem with five judges and six plans. Each judge ranked the plans and these rankings are summarized:

P_1	P_2	P_3	P_4	P_5
a	a	a	a	b
b	b	b	b	c
c	c	c	c	d
d	d	d	d	e
e	e	e	e	f
f	f	f	f	a

Show that the Borda count gives the counterintuitive result if it is used as a social welfare function. Why is the result considered counterintuitive? (Hint: $B(b) = 21$, $B(a) = 20$.)

SUGGESTIONS FOR FURTHER READING

1. Excellent books on cluster analysis and pattern recognition are available. For best satisfaction, one has to pick a book that emphasizes the kinds of applications in which one is interested. For general reference, consult

M. R. Anderberg, *Cluster Analysis for Applications*. Academic Press, New York, 1973.

J. T. Tou and R. C. Gonzalez, *Pattern Recognition Principles*. Addison-Wesley, Reading, Massachusetts, 1974.

2. Several wide ranging applications of cluster analysis along with listings of computer programs can be found in

J. A. Hartigan, *Clustering Algorithms.* Wiley, New York, 1975.

3. For a discussion of group judgments, refer to

K. Arrow, *Social Choice and Individual Values*, 2nd ed. Wiley, New York, 1963.

F. S. Roberts, *Discrete Mathematical Models.* Prentice-Hall, Englewood Cliffs, New Jersey, 1976.

D. R. Farris and A. P. Sage, Introduction and survey of group decision making, *IEEE Trans. Systems*, Man, Cybernetics **5**, No. 3, 346–358 (1975).

R. G. Niemi and W. H. Riker, The choice of voting systems, *Sci. Am.* **234**, No. 6, 21–27, (June 1976).

I. J. Good and T. N. Tideman, From individual to collective ordering through multidimensional attribute space, *Proc. Roy. Soc.* (*London*) **A347**, 371–385 (1976).

A. K. Sen, *Collective Choice and Social Welfare.* Holden-Day, San Francisco, California, 1970.

7

Static Equilibrium Models

INTRODUCTION

The complexity of structure of many systems is best revealed by a graphical approach. The term "graph" used here is something quite different from the graphs that one may be familiar with from analytic geometry or function theory. The graphs that are about to be presented are simple geometric figures consisting of points (nodes) and lines (edges) that connect some of these points. These graphs are sometimes called "linear graphs." As graphs are pictorial in nature, they convey and carry information most suitable for human processing. As directed lines represent relations, collections of directed lines, or directed graphs, can be profitably used to represent interactions among various entities of a system. Thus, while modeling large systems, a graphical representation of the various activities is a useful first step.

The revealing nature of graphs is not only interesting from a mathematical point of view but also considered to be highly rewarding by sociologists and psychologists. For instance, the sociogram made it possible to reveal the structure of a group with its subgroups, cliques, leaders, isolates, and rejectees. Star patterns of psychology, simplexes of topology, circuit diagrams of electrical technology, network or arrow diagrams of operations research, work flow charts of management, Gantt or bar charts of industrial engineers, and flow charts of computer science are all, in essence, graphs of some kind. However, the nature of the process used by different investigators in different disciplines led to bewildering variations in what is otherwise a simple concept. There were several attempts to systematize and bring a semblance of orderliness to this choatic situation. Nevertheless, the subjectivity inherent in the drawing of a graph remained until matrix representations of the information contained in a graph was introduced. Matrix notation not only removes the element of subjectivity but renders the information contained in a graph suitable for processing by computers. In addition, matrix notation is compact and causes no additional difficulties when interest shifts from a static to a dynamic nature of relationships.

1 GRAPHICAL MODELS AND MATRIX MODELS

A *graph* $G(V, E)$, or simply a graph G, can be defined as a nonempty set $V = \{v_1, v_2, \ldots, v_n\}$ of points typically referred to as *vertices*, *nodes*, or *junctions* and a (possibly empty) set $E = \{e_1, e_2, \ldots, e_n\}$ of lines called *edges*, *branches*, or *arcs*. This statement is not complete. Each end point of an edge is necessarily a vertex and the converse is not necessarily true. Thus, a graph may be visualized as a collection of points some of which are joined together by lines. This definition gives rise to the possibility of having *isolated vertices* in a graph. This situation is quite uninteresting from a practical point of view. However, the idea is often helpful in connection with constructive mathematical proofs involving successive deletion or addition of edges from or to a graph.

In a pictorial representation, vertices are usually denoted by points or small circles and edges by straight or curved lines (see Fig. 7.1). The vertices are usually labeled by integers or subscripted alphabetical characters such as $v_1, v_2, \ldots,$. The edges are labeled by pairs of integers such as 1–7, 4–3, 2–1, ... or by a different set of subscripted alphabetical characters such as $e_1, e_2, \ldots$. Each edge that connects two and only two nodes can be thought of as expressing a relation between the two vertices in question.

Nodes and edges are indeed very general concepts. In an electrical circuit a node may actually represent a junction of two or more wires. A node may also stand for such diverse things as a city, an event in time, a person, or an animal species. Similarly an edge may stand for a street, a time-consuming job, a relation between two individuals, or a relation, such as the prey–predator relation, between two animal species. Thus, the graphical concept, aside from stripping the incidental geometric features away from the essential combinatorial aspects of a problem, enlarges the prospects for applications since many real-world entities have combinatorial characteristics that can be profitably viewed as graphs.

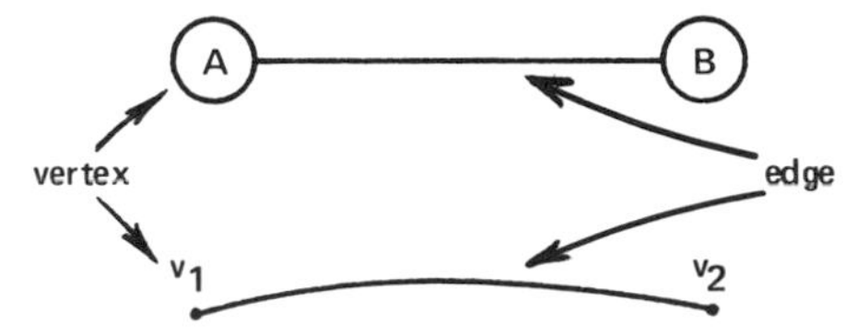

Fig. 7.1 Typical representation of vertices and edges.

Undirected Graphs

A *graph* $G(V, E)$, or simply G is a set of vertices between some or all pairs of which a certain symmetrical relation exists. The existence of such a relation is indicated by joining pairs of vertices by edges. If the set of edges in a linear graph is finite, it is called a *finite graph.* A subset of a graph is called a *subgraph* (see Fig. 7.2b). If the edges of a graph can be ordered such that each edge has one node in common with the preceding edge and one with the succeeding edge, then the ordered sequence of edges is called a *path.* That is, if one can go from vertex i to vertex j by passing along an ordered sequence of n edges, then there exists a path from i to j. If the said vertices i and j are distinct, the path is sometimes referred to as an *arc*. If i and j are identical, the path is called an *n circuit* or *n loop.* A special case of an n loop is a 1 *loop* or *self-loop* and is represented by an edge that curves back on itself so that both extremes of the edge lie on the same vertex.

If there exists a path (direct or through other vertices) between every pair of vertices in a graph, then the graph is said to be *connected.*

A graph is *complete* if every two distinct vertices are joined by an edge (i.e., every two distinct vertices are adjacent to each other).

The number of edges connected to a node is called the *degree of the node.* If the degree of every vertex of a graph is equal to k, the graph is said to be *regular of degree k or k regular.* For instance, a complete graph of k vertices is $(k - 1)$ regular.

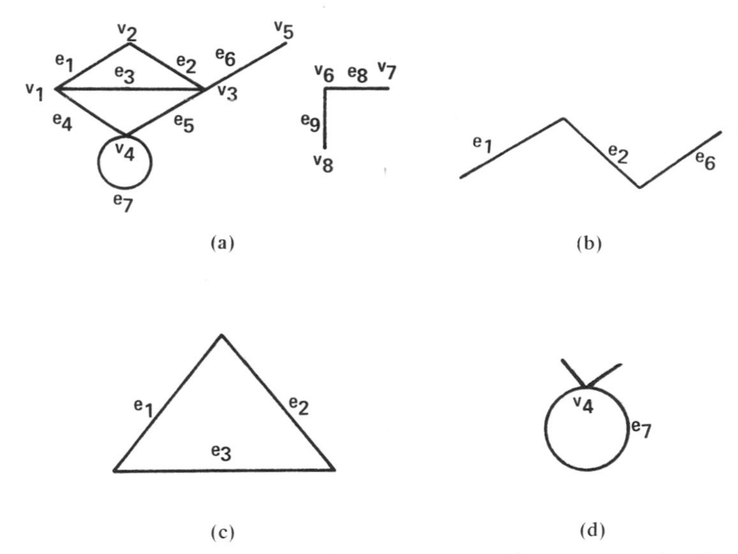

Fig. 7.2 (a) A finite linear graph G. This graph includes edges e_1–e_9 and vertices v_1–v_8. (b) A path in G that is also a subgraph of G. (c) A closed path (or a loop) in G. (d) A self-loop at vertex v_4 of G.

Another important concept in graph theory is that of a *forest*. A forest can be defined as a graph that has no circuits. A connected forest is a *tree*. These definitions lead to the picturesque result that a tree is always a forest but a forest is not necessarily a tree! As most of the treatment in this chapter deals with connected graphs, no further discussion of a forest will appear. A tree can also be defined as a graph in which there is one and only one path between every pair of nodes.

In many practical applications, it is necessary to associate each edge of a graph with a value, called the *edge measurement* or *weight*. For instance, an edge measurement may represent time of travel, flow rate, capacity of a pipeline, or merely the existence or nonexistence of a certain relation.

Directed Graphs

In many applications, it is necessary to associate with each edge of a graph an orientation or direction. In some applications, the orientation of the edges is a "true orientation" in the sense that the system represented by the graph exhibits some unilateral property. For example, the direction of one-way streets is a true orientation. In other situations, the orientation used is a "pseudoorientation" used in lieu of an elaborate reference system.

A *directed graph* $G_d(V, E)$ consists of a set V of *vertices* or *nodes* together with a set E of *directed edges* or *arcs*. (Some authors prefer to use the word *arc* in lieu of *directed edge*.) Pictorially, an arc is represented by placing an arrow on the edge. Similarly, a bidirected edge is represented by two arrows pointing in opposite directions. At times, it is convenient to replace an undirected edge by a pair of equal and oppositely directed edges. Similarly, it is sometimes useful to replace a bidirected edge by two directed edges.

In directed graphs, many of the preceding definitions take new meaning and depth. For instance, a path in a directed graph is defined as before, but there is an obvious added requirement that the arcs of the path must be directed in the same sense. Care is also required in defining a tree. In speaking of a tree "on" node k (or *rooted* on node k), it is to be understood that every arc is directed away from k. Similarly a directed graph is *strongly connected* if for any ordered pair of nodes, say, i and j, there exists a directed path: from i to 1; l to 2; . . . ; from $(j - 1)$ to j.

Directed graphs often exhibit such properties as symmetry and reflexivity. A directed graph is *symmetric* if there exists an arc from node i to node j as well as an arc from node j to node i. A directed graph is *reflexive* if every node has a self-loop. A graph is *complete* if every pair of nodes is connected in at least one direction.

Two arcs in a directed graph are said to be *parallel* if they connect the same two nodes and are oriented in the same direction.

A *net* is a directed graph with a finite and nonempty set of nodes and a finite (and possibly empty) set of arcs. Thus, an isolated node is a trivial case of a net.

A *relation* is a directed graph which satisfies the requirements of a net but does not have any parallel arcs.

A *digraph* is an *irreflexive* relation. That is, a digraph is a directed graph or net having no loops and parallel arcs.

Acyclic graphs are a special case of digraphs in which no two nodes are mutually reachable.

A *network* is a relation with *edge measurements.* The edge measurements can be used to represent capacities of flow such as signals, cars, people, oil, and other items. Some of these ideas are elaborated pictorially in Fig. 7.3.

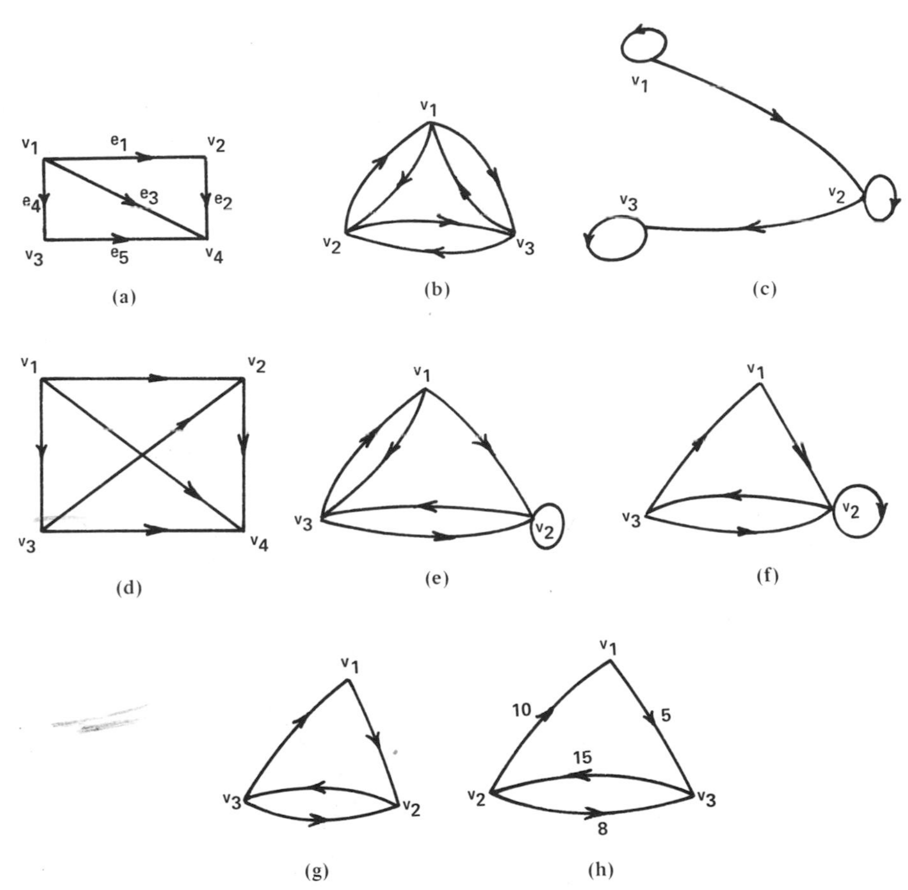

Fig. 7.3 Some special directed graphs. (a) Directed graph. (b) Symmetric graph. (c) Reflexive graph. (d) Complete graph. (e) Net. (f) Relation. (g) Digraph. (h) Network.

Distances and Deviations

The concept of distance developed in connection with sets can be adapted to define distance in a graph. Let $G = \{V, E\}$ be a connected graph. One way of defining distance between two vertices v and w is as follows. Let

$$d(v, w) = \{\text{minimum number of edges contained in any path joining } v \text{ and } w\}$$
$$d(v, v) = 0 \qquad \text{for all vertices}$$

Then, for a fixed vertex v, the integer

$$R(v) = \max_{w \in V} d(v, w) \tag{1}$$

measures the distance from v to the vertex (or vertices) most remote from v. It is intuitively clear that a vertex is relatively central if $R(v)$ is relatively small. Thus, it is natural to call

$$R_0 = \min_{v \in V} R(v) \tag{2}$$

the *radius* of the graph G and refer to any vertex v_0 as *center* if

$$R(v_0) = R_0 \tag{3}$$

The *diameter* of a connected graph is the maximum distance between pairs of vertices. Symbolically

$$D = \max_{u, w \in V} d(v, w) \tag{4}$$

For instance, the graph in Fig. 7.4a has a radius of unity and that of Fig. 7.4b has a radius of 3. In Fig. 7.4b, every vertex is a center.

At this point it is important to emphasize the difference between *distance* between two vertices v and w and *length of a path* between the vertices v and w. The *length of a path* in a directed graph is defined as the number of arcs that delineate the path. It is quite possible to find graphs in which there may be a multiplicity of paths connecting two vertices v and w. Among all such paths, there may be one path, or several paths, having a minimal length. The number of minimal paths connecting the two vertices v and w is called

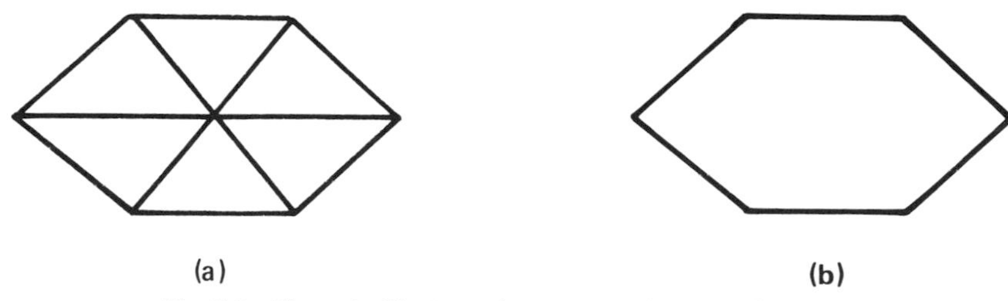

Fig. 7.4 Figure to illustrate the concept of radius of a graph.

the *deviation* of v to w. This concept is quite useful while examining the structure of a graph. Note that the deviation defined as above is not necessarily a measure of distance.

From Graphs to Matrices

To render the information contained in a graph suitable for quantitative manipulation it is often necessary to be able to represent a graph equivalently as a matrix and vice versa. Matrix representation is a convenient way of recording, storing, and manipulating the topological structure of a graph in a computer. Furthermore, there is a close relationship between various kinds of matrices charterizing graphs, and knowledge of these relations is often revealing. Matrix representation is valid for directed and undirected, as well as weighted, graphs. The selection of a particular matrix representation, of course, depends upon the problem at hand.

The Adjacency Matrix

If two nodes of a graph which are connected by an edge are defined as *adjacent nodes*, then a convenient way of specifying this adjacency relation is by an adjacency matrix. A typical element of an adjacency matrix is defined by

$$a_{ij} = \begin{cases} 1 & \text{if there is an edge connecting vertex } i \text{ to vertex } j \\ 0 & \text{otherwise} \end{cases}$$

The adjacency matrices of typical undirected and directed graphs are shown respectively in Figs. 7.5 and 7.6.

As an adjacency matrix essentially describes the topological connections, it is customary to restrict its use precisely for that purpose. Thus, the adjacency matrix of a graph with edge measurements is the same as that obtained for the graph with no edge measurements. Note also that in Fig. 7.6, all entries of the fifth row of the adjacency matrix are zero. However, the last row cannot be omitted from the matrix because it carries some significant information. The adjacency matrix in both the directed and undirected case

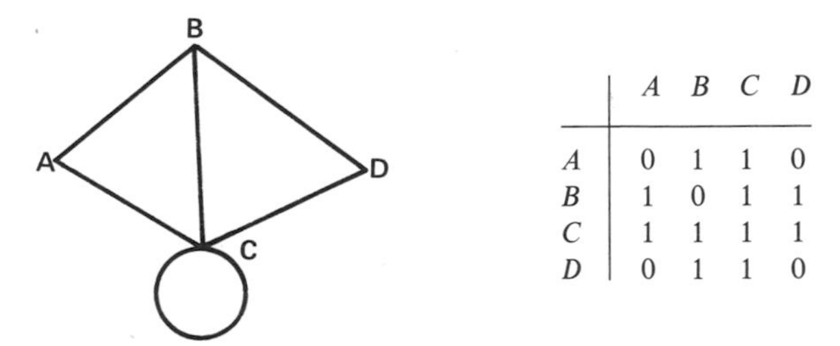

	A	B	C	D
A	0	1	1	0
B	1	0	1	1
C	1	1	1	1
D	0	1	1	0

Fig. 7.5 A graph and its adjacency matrix.

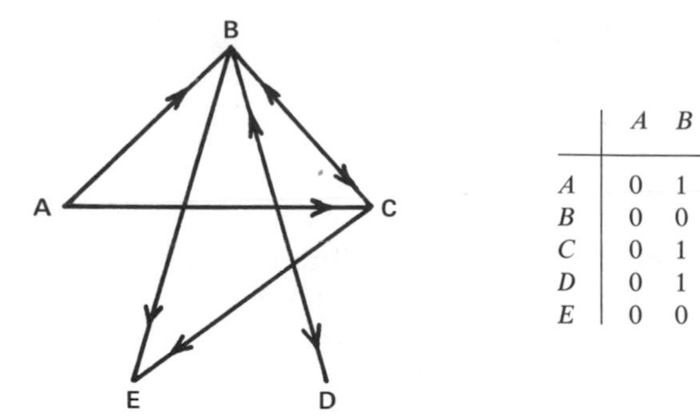

	A	B	C	D	E
A	0	1	1	0	0
B	0	0	1	1	1
C	0	1	0	0	1
D	0	1	0	0	0
E	0	0	0	0	0

Fig. 7.6 A directed graph and its adjacency matrix.

has only nonnegative entries. If the directed graph G_d is strongly connected, the adjacency matrix has an additional important property—it is said to be irreducible. The concept of irreducibility is extremely useful while solving large systems of algebraic equations.

Some interesting properties of a graph can be observed in its adjacency matrix. If the adjacency matrix is asymmetric, then the corresponding graph is a directed graph. A row and a column of zeros in the adjacency matrix is indicative of an isolated vertex. (For other properties, see the exercises.)

The Precedence Matrix

The adjacency matrix of an acyclic directed graph plays an important role and is referred to hereafter as a *precedence matrix*. That is, a precedence matrix carries precedence relations such as "*a* precedes *b*." Using this broad definition, four types of precedence matrices can be defined.

A vertex precedence matrix $\mathbf{P}_{00}$ is a square matrix with one column and one row for each vertex of the graph such that

$$p_{ij} = \begin{cases} 1 & \text{if vertex } i \text{ is the immediate predecessor of } j \\ \varnothing & \text{otherwise} \end{cases}$$

where $\varnothing$ stands for an undefined quantity and can be thought of as a blank. The double zero subscript on $\mathbf{P}$ is indicative of vertex precedence (because another name for a vertex is zero-order simplex and that for an arc is first-order simplex). It is easy to verify that $\mathbf{P}_{00}{}^{\mathrm{T}}$ defines a succedence matrix.

An *arc precedence matrix* $\mathbf{P}_{11}$ is a square matrix with as many rows and columns as there are arcs in a directed graph such that:

$$p_{ij} = \begin{cases} 1 & \text{if arc } i \text{ is the immediate predecessor of arc } j \\ \varnothing & \text{otherwise} \end{cases}$$

Mixed precedence relations $\mathbf{P}_{01}$ and $\mathbf{P}_{10}$ can likewise be defined. A directed graph and its associated precedence matrices are shown in Fig. 7.7.

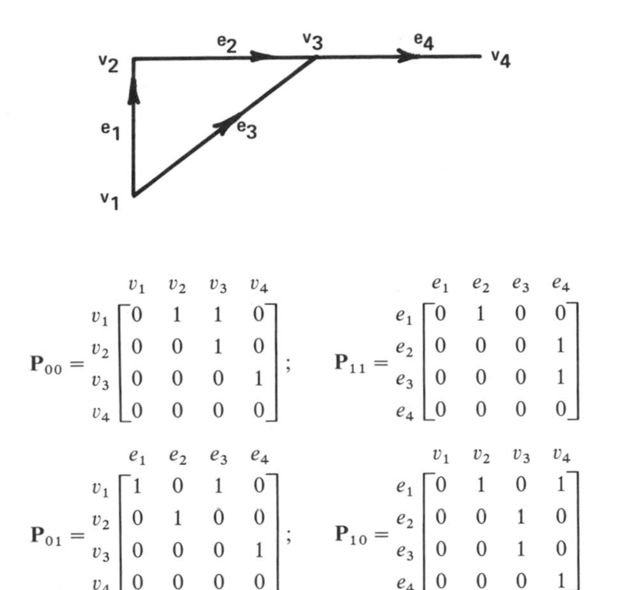

$$
\mathbf{P}_{00} = \begin{array}{c} \\ v_1 \\ v_2 \\ v_3 \\ v_4 \end{array} \begin{array}{c} \begin{array}{cccc} v_1 & v_2 & v_3 & v_4 \end{array} \\ \begin{bmatrix} 0 & 1 & 1 & 0 \\ 0 & 0 & 1 & 0 \\ 0 & 0 & 0 & 1 \\ 0 & 0 & 0 & 0 \end{bmatrix} \end{array}; \quad
\mathbf{P}_{11} = \begin{array}{c} \\ e_1 \\ e_2 \\ e_3 \\ e_4 \end{array} \begin{array}{c} \begin{array}{cccc} e_1 & e_2 & e_3 & e_4 \end{array} \\ \begin{bmatrix} 0 & 1 & 0 & 0 \\ 0 & 0 & 0 & 1 \\ 0 & 0 & 0 & 1 \\ 0 & 0 & 0 & 0 \end{bmatrix} \end{array}
$$

$$
\mathbf{P}_{01} = \begin{array}{c} \\ v_1 \\ v_2 \\ v_3 \\ v_4 \end{array} \begin{array}{c} \begin{array}{cccc} e_1 & e_2 & e_3 & e_4 \end{array} \\ \begin{bmatrix} 1 & 0 & 1 & 0 \\ 0 & 1 & 0 & 0 \\ 0 & 0 & 0 & 1 \\ 0 & 0 & 0 & 0 \end{bmatrix} \end{array}; \quad
\mathbf{P}_{10} = \begin{array}{c} \\ e_1 \\ e_2 \\ e_3 \\ e_4 \end{array} \begin{array}{c} \begin{array}{cccc} v_1 & v_2 & v_3 & v_4 \end{array} \\ \begin{bmatrix} 0 & 1 & 0 & 1 \\ 0 & 0 & 1 & 0 \\ 0 & 0 & 1 & 0 \\ 0 & 0 & 0 & 1 \end{bmatrix} \end{array}
$$

Fig. 7.7 A directed graph and its associated precedence matrices.

The Value Matrix

An adjacency matrix cannot completely describe all the information in a network as the latter carries edge measurements—a significant information. To distinguish the case of a network, the adjacency matrix of a network, with a slight modification, is called the *value matrix* **V**. If the values on the arcs are associated with probabilities or cost, then the value matrix is called a *probability matrix* or *cost matrix*. A probability network and its associated probability matrix are shown in Fig. 7.8. A cost network and its associated cost matrix are shown in Fig. 7.9.

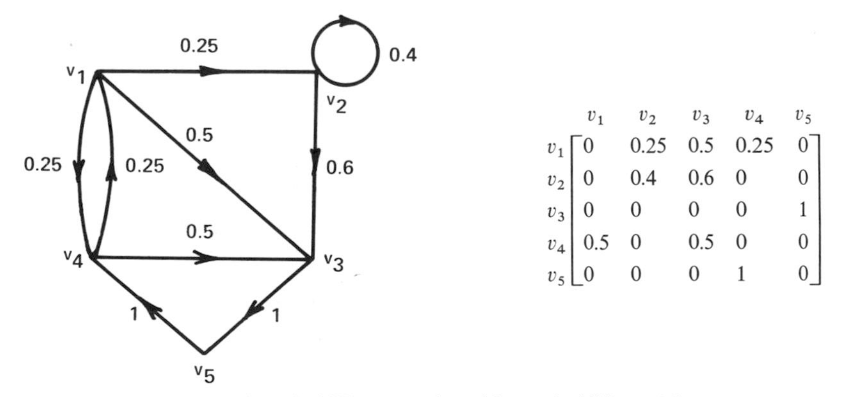

$$
\begin{array}{c} \\ v_1 \\ v_2 \\ v_3 \\ v_4 \\ v_5 \end{array} \begin{array}{c} \begin{array}{ccccc} v_1 & v_2 & v_3 & v_4 & v_5 \end{array} \\ \begin{bmatrix} 0 & 0.25 & 0.5 & 0.25 & 0 \\ 0 & 0.4 & 0.6 & 0 & 0 \\ 0 & 0 & 0 & 0 & 1 \\ 0.5 & 0 & 0.5 & 0 & 0 \\ 0 & 0 & 0 & 1 & 0 \end{bmatrix} \end{array}
$$

Fig. 7.8 A probability network and its probability matrix.

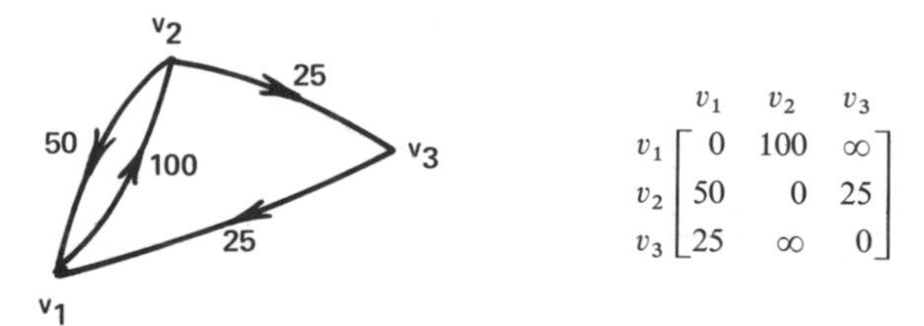

Fig. 7.9 A cost network and its cost matrix.

Notice that the row sums of the value matrix in Fig. 7.8 are all equal to unity. If, in addition, the probabilities associated with the arcs are time dependent, then the graph is called a *Markov chain.* Notice also that not all nonzero entries of the cost matrix in Fig. 7.9 are zero.

The Incidence Matrix

The incidence matrix $\mathbf{E}_{10}$ associated with a directed graph has as many rows as there are arcs and as many columns as there are vertices and is defined as

$$e_{ij} = \begin{cases} +1 & \text{if an arc is directed away from vertex } i \text{ and toward } j \\ -1 & \text{if an arc is directed toward vertex } i \text{ from vertex } j \\ 0 & \text{otherwise} \end{cases}$$

A directed graph and its incidence matrix are shown in Fig. 7.10.

The matrices defined thus far can be thought of as fundamental matrices. There are several matrices that are derivable from the fundamental matrices by performing simple algebraic operations on them.

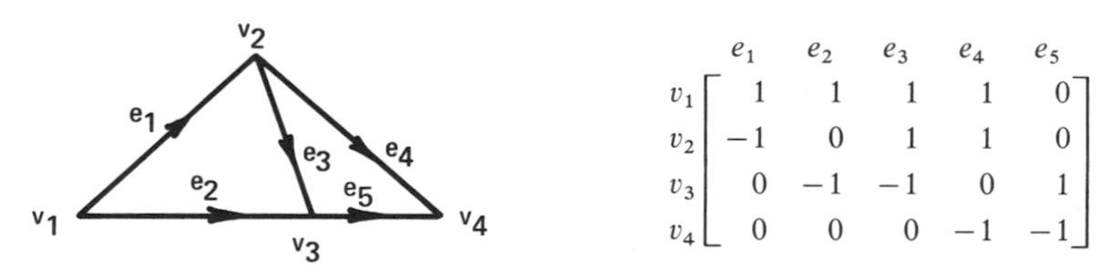

Fig. 7.10 A directed graph and its incidence matrix.

The Reachability Matrix

The *reachability matrix* **R** gives useful information about a graph and can be defined as

$$r_{ij} = \begin{cases} 1 & \text{if one can travel from node } i \text{ to node } j \text{ over any path} \\ & \text{regardless of its length} \\ 0 & \text{otherwise} \end{cases}$$

The reachability matrix **R** can be derived from the adjacency matrix **A** by using the equation

$$\mathbf{R} = f(\mathbf{I} + \mathbf{A} + \mathbf{A}^2 + \cdots + \mathbf{A}^n) \tag{5}$$

That is, **R** is a function of the identity matrix and the powers of **A**, up to the nth power where n is the dimension of **A**. This function is denoted by $f(a)$ and is defined as

$$f(a) = \begin{cases} 1 & \text{for any nonzero number } a \\ 0 & \text{if } a = 0 \end{cases} \tag{6}$$

The meaning of the above two equations can be given as follows. The nonzero entries of the adjacency matrix **A** show the number of paths of length one, that is, the number of arcs, between the corresponding nodes.

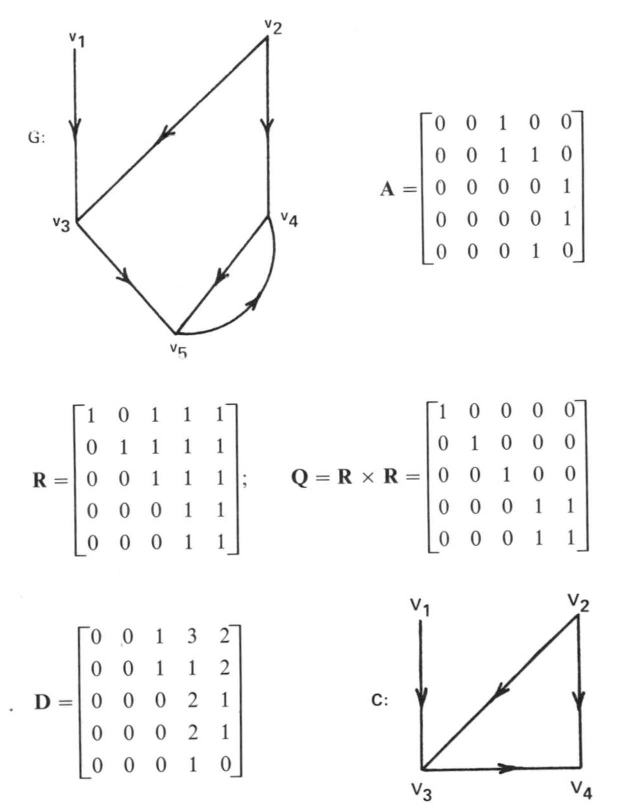

$$\mathbf{A} = \begin{bmatrix} 0 & 0 & 1 & 0 & 0 \\ 0 & 0 & 1 & 1 & 0 \\ 0 & 0 & 0 & 0 & 1 \\ 0 & 0 & 0 & 0 & 1 \\ 0 & 0 & 0 & 1 & 0 \end{bmatrix}$$

$$\mathbf{R} = \begin{bmatrix} 1 & 0 & 1 & 1 & 1 \\ 0 & 1 & 1 & 1 & 1 \\ 0 & 0 & 1 & 1 & 1 \\ 0 & 0 & 0 & 1 & 1 \\ 0 & 0 & 0 & 1 & 1 \end{bmatrix}; \qquad \mathbf{Q} = \mathbf{R} \times \mathbf{R} = \begin{bmatrix} 1 & 0 & 0 & 0 & 0 \\ 0 & 1 & 0 & 0 & 0 \\ 0 & 0 & 1 & 0 & 0 \\ 0 & 0 & 0 & 1 & 1 \\ 0 & 0 & 0 & 1 & 1 \end{bmatrix}$$

$$\mathbf{D} = \begin{bmatrix} 0 & 0 & 1 & 3 & 2 \\ 0 & 0 & 1 & 1 & 2 \\ 0 & 0 & 0 & 2 & 1 \\ 0 & 0 & 0 & 2 & 1 \\ 0 & 0 & 0 & 1 & 0 \end{bmatrix}$$

Fig. 7.11 A directed graph *G*, its adjacency, reachability, and deviation matrices and the condensation *C*.

Similarly, the entries of $\mathbf{A}^2$, where $\mathbf{A}^2$ is obtained by regular matrix multiplication of $\mathbf{A}$ by itself, indicates the number of possible paths of length two. This idea can be extended. The identity matrix $\mathbf{I}$ is added to render the meaning of reachability consistent.

A graph G_d, its adjacency matrix $\mathbf{A}$, and its reachability matrix $\mathbf{R}$ are shown in Fig. 7.11. The reader can readily verify Eq. (5).

The elementwise product $\mathbf{Q} = \mathbf{R} \times \mathbf{R}^{\mathrm{T}}$ is obviously a symmetric matrix. The nonzero entries q_{ij} of $\mathbf{Q}$ indicate that vertices i and j are mutually reachable. The set of nodes that are mutually reachable in this fashion is called a *strong component*. (In view of this, a *strongly connected* graph can also be defined as one all of whose vertices constitute a strong component.) For instance, the graph in Fig. 7.11 has four strong components, $V_1 = \{v_1\}$, $V_2 = \{v_2\}$, $V_3 = \{v_3\}$, and $V_4 = \{v_4, v_5\}$. Now using V_1, V_2, V_3, and V_4, one can construct a new graph. Node V_i is connected, in this new graph, to node V_j by an arc if there exists a path from any node in the set V_i to any node in the set V_j. The new graph is called the *condensation* of the graph G_d and is denoted by $\mathbf{C}$ in Fig. 7.11.

The Deviation Matrix

The deviation matrix $\mathbf{D}$ of a graph is defined as:

$$d_{ij} = \alpha \qquad \text{where } \alpha \text{ is the deviation from } i \text{ to } j$$

In Fig. 7.11, $\mathbf{D}$ is the deviation matrix of the graph G.

Deviations can be interpreted in a number of ways. In a manufacturing process, the vertices may represent various stages of production, product development, or assembly of parts. Then deviations can be used to locate efficient paths along which a product can be developed from stage i to stage j.

2 PATTERNS OF GROUP STRUCTURE

Building models to human interactions in groups is an important part in the overall task of building models to large scale systems. Essentially, this involves a study of the structure of relations among individuals and groups of individuals. Toward this end, one can start with a *dyad* or two-person relation where the individuals are represented by points and relations by directed edges. Then, the results can be progressively generalized to build models to small groups and on to entire societal systems. The final level of complexity could be, for example, a city, a nation, or an international organization.

The sociologist Moreno was one of the first who attempted to reveal the structure of human interactions using directed graphs. Then the concept of using a connection matrix to represent such binary relations as "a friend of" or "is in communication with" was introduced.

Consider the statement "an individual A is the paternal grandmother of B if, for some C, A is the mother of C and C is the father of B." Here the statement "paternal grandmother of" is a compound relation whose structure can be simplified by expressing it as a *relative product* of two dyads or simple binary relations, namely, M, "the mother of" and F, "the father of." The elements of these relational matrices can be considered Boolean, that is,

$$\mathbf{M} = [m_{ij}], \qquad \mathbf{F} = [f_{ij}]$$

where

$$m_{ij} = \begin{cases} 1 & \text{if} \quad iMj \text{ is true (i.e., } i \text{ is the mother of } j) \\ 0 & \text{if} \quad iMj \text{ is false} \end{cases}$$

$$f_{ij} = \begin{cases} 1 & \text{if} \quad iFj \text{ is true (i.e., } i \text{ is the father of } j) \\ 0 & \text{if} \quad iFj \text{ is false} \end{cases}$$

Then, the relational matrix for the relative product is simply the Boolean product **MF**. The Boolean product **MF** is obtained by multiplying **M** and **F** according to the usual multiplication rules and observing the rules of Boolean algebra (see Appendix 3).

These elementary ideas can be extended to studies aimed at revealing the structures of more complicated relations. Initial attempts at extending the above approach for an analysis and description of the patterns of relations among various members of a group or various groups took the form of drawing complicated diagrams called sociograms. A *sociogram* is nothing but a collection of vertices and edges where each vertex represents an individual, an agency, or a subgroup and an edge represents some kind of relation or communication link. Indeed, a sociogram is an equivalent representation of a directed graph.

Consider a sociogram such as the one shown in Fig. 7.12. Here the vertices stand for individuals in a group and an arc is drawn from vertex x to y if the influence of x on y is sufficiently great. Many interesting problems can be studied using sociograms of this type: Who is the most powerful or influential or dominant member of a group? Who are the people whose loss can destroy the unity and cohesion of an organization? Which communication center(s) should be destroyed to isolate one or more regions of an enemy territory?

Certainly, the concept of a center can be invoked to identify a leader. During a qualitative inspection of a graph, one is tempted to select the most dominant individual as the leader. For instance, in Fig. 7.12 vertex y is the person who has the most direct influence, since y dominates five persons;

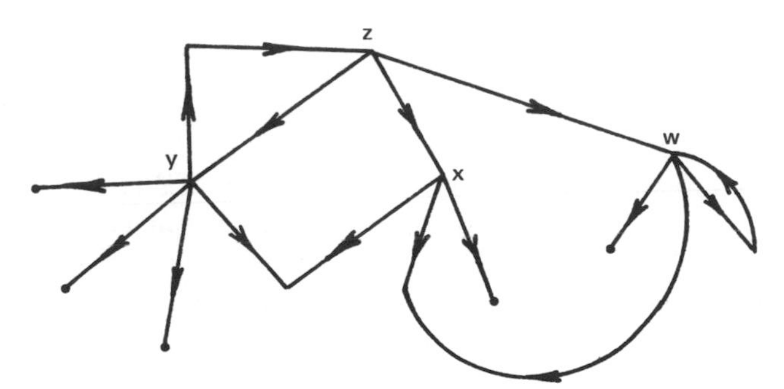

Fig. 7.12 A sociogram.

however, these five persons are not themselves influential. The individual z is the most powerful; z is chosen directly by only three people, but these three are themselves very influential. The connections of w, x, and y are certainly important; such points are called *articulation points.*

Analysis of some of the aspects of group structures becomes relatively easy if sociograms are represented by matrices. A *group matrix* **G** associated with a sociogram S is nothing but an adjacency matrix of order n, where n is the number of interacting individuals or groups. The entries of S are binary; a 1 in the g_{ij} position is indicative of the existence of some kind of relation between group i and group j and a 0 is indicative of the absence of such a relation. For instance, a directed edge from vertex i to vertex j implies that group i reports or sends information to group j. This is reflected by the entry 1 in row i and column j of the matrix **G**. A sociogram and its group matrix are shown in Fig. 7.13.

One of the immediate benefits that accrue by studying social structure using graphical models is the ability to detect the formation of cliques. A *clique* may be defined as a subset consisting of two or more individuals (or groups) all of whom choose each other mutually for communication purposes. In other words, direct one-step connections exist between every possible pair of members of such a clique. Cliques may be detected by raising the group matrix **G** to a power. For example, it can be easily shown that if a group matrix is raised to the kth power, each nondiagonal matrix entry represents the number of k-step connections that exist between any two groups and the diagonal entry refers to the number of k-step connections back to itself. Similarly, by adding the original group matrix to the one raised to the kth power, it is possible to obtain such information as: Who influences the greatest number of people in less than a specified number of steps? Which people are influenced by the greatest number of people? Which of the individuals are most remotely connected to each other?

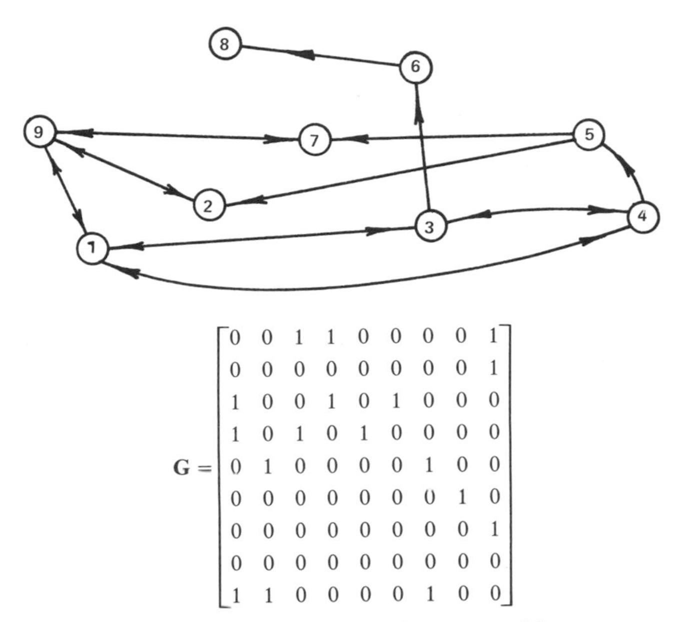

$$\mathbf{G} = \begin{bmatrix} 0 & 0 & 1 & 1 & 0 & 0 & 0 & 0 & 1 \\ 0 & 0 & 0 & 0 & 0 & 0 & 0 & 0 & 1 \\ 1 & 0 & 0 & 1 & 0 & 1 & 0 & 0 & 0 \\ 1 & 0 & 1 & 0 & 1 & 0 & 0 & 0 & 0 \\ 0 & 1 & 0 & 0 & 0 & 0 & 1 & 0 & 0 \\ 0 & 0 & 0 & 0 & 0 & 0 & 0 & 1 & 0 \\ 0 & 0 & 0 & 0 & 0 & 0 & 0 & 0 & 1 \\ 0 & 0 & 0 & 0 & 0 & 0 & 0 & 0 & 0 \\ 1 & 1 & 0 & 0 & 0 & 0 & 1 & 0 & 0 \end{bmatrix}$$

Fig. 7.13 A sociogram and its group matrix.

Example 1. Consider a large interdisciplinary research project consisting of, say, nine research groups. Given the interactions among the groups, the interest is to get an insight into the properties of this group structure.

Let the group interactions be described by the sociogram S in Fig. 7.13. The matrix obtained by squaring the group matrix **G** is

$$\mathbf{G}^2 = \begin{bmatrix} 3 & 1 & 1 & 1 & 1 & 1 & 1 & 0 & 0 \\ 1 & 1 & 0 & 0 & 0 & 0 & 1 & 0 & 0 \\ 1 & 0 & 2 & 1 & 1 & 0 & 0 & 1 & 1 \\ 1 & 1 & 1 & 2 & 0 & 1 & 1 & 0 & 1 \\ 0 & 0 & 0 & 0 & 0 & 0 & 0 & 0 & 2 \\ 0 & 0 & 0 & 0 & 0 & 0 & 0 & 0 & 0 \\ 1 & 1 & 0 & 0 & 0 & 0 & 1 & 0 & 0 \\ 0 & 0 & 0 & 0 & 0 & 0 & 0 & 0 & 0 \\ 0 & 0 & 1 & 1 & 0 & 0 & 0 & 0 & 3 \end{bmatrix}$$

The diagonal entry $g_{11}{}^2$ of matrix $\mathbf{G}^2$ says, for instance, that there are three two-step connections from group 1 back to itself. This is the same as saying that there are three closed communication loops originating from group 1 back to itself (namely, 1 to 3 to 1, 1 to 4 to 1, and 1 to 9 to 1). An off-diagonal entry, say, ${g_{12}}^2 = 1$, says that there is only one two-step connection between groups 1 and 2 (namely, 1 to 9 to 2). Similarly, ${g_{59}}^2 = 2$ says that there are two two-step connections between groups 5 and 9 (namely, 5 to 2 to 9 and 5 to 7 to 9).

Similarly, the number of three-step connections can be detected by cubing the matrix **G**. For instance,

$$\mathbf{G}^3 = \begin{bmatrix} 2 & 1 & 4 & 4 & 1 & 1 & 1 & 1 & 5 \\ 0 & 0 & 1 & 1 & 0 & 0 & 0 & 0 & 3 \\ 3 & 2 & 2 & 3 & 1 & 2 & 2 & 0 & 1 \\ 4 & 1 & 3 & 2 & 2 & 1 & 1 & 1 & 3 \\ 2 & 2 & 0 & 0 & 0 & 0 & 2 & 0 & 0 \\ 0 & 0 & 0 & 0 & 0 & 0 & 0 & 0 & 0 \\ 0 & 0 & 1 & 1 & 0 & 0 & 0 & 0 & 3 \\ 0 & 0 & 0 & 0 & 0 & 0 & 0 & 0 & 0 \\ 5 & 3 & 1 & 1 & 1 & 1 & 3 & 0 & 0 \end{bmatrix}$$

Inspection of this matrix reveals that there are only two three-step connections from group 1 back to itself. (Inspection of Fig. 7.13 reveals that these are 1 to 3 to 4 to 1 and 1 to 4 to 3 to 1.) These two, indeed, involve the same groups and therefore are symmetrical. Thus groups 1, 3 and 4 forms a clique. In fact, this observation can be formalized. The nonzero diagonal entries indicate the groups that form a clique. ■

Thus, an important property of the adjacency matrix is that the kth power of this matrix gives all the k-step paths between vertices. Each nonzero element of this matrix $\mathbf{A}^k$ (or $\mathbf{G}^k$) indicates there is a path going through k edges (or a k-step path) from vertex j to vertex i.

It is important to emphasize that raising the adjacency matrix to the kth power only reveals the *number* of k-step connections, not their location. In the preceding example, these cliques were located by inspection of the graph. This is not always easy if the graph is large. An algorithm for this purpose can be developed in terms of the *reachability matrix*, which was defined earlier.

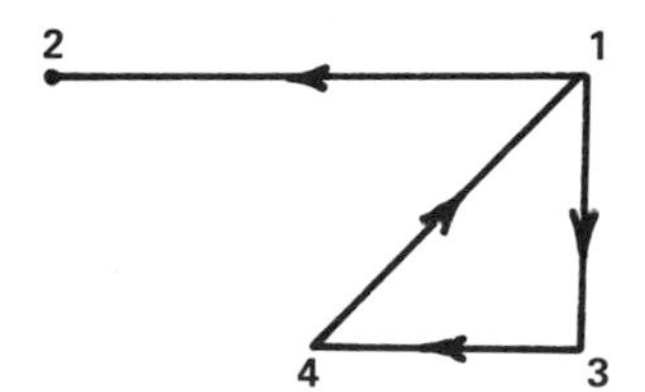

Fig. 7.14 Directed graph used in Example 2.

Example 2. Consider the directed graph shown in Fig. 7.14. The adjacency matrix and its various powers are

$$\mathbf{A} = \begin{bmatrix} 0 & 1 & 1 & 0 \\ 0 & 0 & 0 & 0 \\ 0 & 0 & 0 & 1 \\ 1 & 0 & 0 & 0 \end{bmatrix} \qquad \mathbf{A}^2 = \begin{bmatrix} 0 & 0 & 0 & 1 \\ 0 & 0 & 0 & 0 \\ 1 & 0 & 0 & 0 \\ 0 & 1 & 1 & 0 \end{bmatrix}$$

$$\mathbf{A}^3 = \begin{bmatrix} 1 & 0 & 0 & 0 \\ 0 & 0 & 0 & 0 \\ 0 & 1 & 1 & 0 \\ 0 & 0 & 0 & 1 \end{bmatrix} \qquad \mathbf{A}^4 = \begin{bmatrix} 0 & 1 & 1 & 0 \\ 0 & 0 & 0 & 0 \\ 0 & 0 & 0 & 1 \\ 1 & 0 & 0 & 0 \end{bmatrix}$$

$$\mathbf{I} + \mathbf{A} + \mathbf{A}^2 + \mathbf{A}^3 + \mathbf{A}^4 = \begin{bmatrix} 2 & 2 & 2 & 1 \\ 0 & 1 & 0 & 0 \\ 1 & 1 & 2 & 2 \\ 2 & 1 & 1 & 2 \end{bmatrix}$$

The reachability matrix $\mathbf{R}$ and the elementwise product $\mathbf{R} \times \mathbf{R}^{\mathrm{T}}$ are

$$\mathbf{R} = \begin{bmatrix} 1 & 1 & 1 & 1 \\ 0 & 1 & 0 & 0 \\ 1 & 1 & 1 & 1 \\ 1 & 1 & 1 & 1 \end{bmatrix} \qquad \mathbf{Q} = \mathbf{R} \times \mathbf{R}^{\mathrm{T}} = \begin{bmatrix} 1 & 0 & 1 & 1 \\ 0 & 1 & 0 & 0 \\ 1 & 0 & 1 & 1 \\ 1 & 0 & 1 & 1 \end{bmatrix}$$

Inspection of $\mathbf{Q}$ reveals that for $i, j = 1, 3,$ and 4 the condition $q_{ij} = q_{ji}$ is satisfied. Therefore, the vertices 1, 3 and 4 lie on a clique which can be verified by inspection of the graph. ■

A Simple Computational Rule

Notice that some of the computational steps above can be avoided by computing various powers of the adjacency matrix using Boolean algebra and standard matrix multiplication rules. For large matrices these operations could be cumbersome. Suppose, it is required to perform $\mathbf{C} = \mathbf{A} \cdot \mathbf{B}$ using Boolean arithmetic. Evidently

$$c_{ij} = (a_{i1} \cdot b_{1j}) \cup (a_{i2} \cdot b_{2j}) \cup \cdots \cup (a_{in} \cdot b_{nj}) \tag{1}$$

where all a_{ik}'s and b_{kj}'s are either 0 and 1 and

$$\begin{aligned} 0 \cdot 0 &= 0 \qquad & 0 \cup 1 &= 1 \\ 1 \cdot 1 &= 1 \qquad & 1 \cup 1 &= 1 \\ 1 \cdot 0 &= 0 \qquad & 1 \cup 0 &= 1 \end{aligned}$$

If $a_{im} \equiv 0$, then all terms with $k = m$ in the above equation do not contribute to c_{ij} for all j. That is, if $a_{im} = 0$, all terms with $k = m$ do not make a contribution to the ith row of **C**. However, if $a_{im} = 1$, every 1 in row m of **B** appears in c_{ij}. Therefore, row i of **C** is just the Boolean union of the rows of **B** corresponding to the nonzero entries in row i of **A**. For instance

$$\begin{bmatrix} 0 & 1 & 1 \\ 1 & 0 & 0 \\ 0 & 0 & 1 \end{bmatrix} \cdot \begin{bmatrix} 1 & 1 & 0 \\ 0 & 1 & 1 \\ 1 & 0 & 1 \end{bmatrix} = \begin{bmatrix} 1 & 1 & 1 \\ 1 & 1 & 0 \\ 1 & 0 & 1 \end{bmatrix}$$

$$\mathbf{A} \quad \cdot \quad \mathbf{B} \quad = \quad \mathbf{C}$$

Here, the first row of **C** is obtained by the Boolean union of the *second* and *third* rows of **B** because the *second* and *third* entries of the first row of **A** are nonzero. Thus, time-consuming matrix multiplication is avoided.

Similarly, the arithmetic involved in computing the entries of the reachability matrix can be simplified by noting that

$$(\mathbf{A} \cup \mathbf{I})^n = \mathbf{A}^n \cup \mathbf{A}^{n-1} \cup \cdots \cup \mathbf{A} \cup \mathbf{I}$$

3 INPUT–OUTPUT TYPE MODELS

Many important and interesting models of interacting entities can be developed using the so-called input–output approach. Leontief developed such *input–output models* to describe the equilibrium condition of economic systems. Biologists and ecologists use analogous methods in their *compartmental models* to study interactions between organs, or between animal species. These concepts can be applied to several other fields.

The fundamental step in building this type of model is to conceptually separate the system into a number of distinct and interconnected entities, called *compartments*, between which material, energy, or some such quantity is transported. The compartments need not be spatial regions, but they must be distinguishable on some basis (e.g., different plant, animal, or chemical species, different industries, chemical phases). Ideally the compartments are assumed to be homogeneous. The resolution level required in setting up compartmental models must take into account the accuracy of simulation desired and the purpose of the study.

Examples of compartmental (or input–output) models abound. In studying translocation of water in a plant, one might have soil, roots, stem, leaves, and atmosphere as compartments. The exchange of mineral elements in the blood stream would probably include bones, blood, kidneys, and urine as compartments. For energy transfer in a forest food chain, one might choose vegetation, herbivores, carnivores, decomposers, litter, etc. In building models for economic and industrial systems, one chooses an industry producing a single good using one process as constituting a compartment. The manner in which the various components of a system are lumped or aggregated into a compartmental model is also partially dictated by the kinds of observations one can make and the quality of data that is available.

Another important characteristic of these models is the nature of transfer between compartments. Although nonlinear and time-varying transfer processes are the rule in realistic cases and have been the subject of some study, analysis in this chapter is confined exclusively to linear, time-invariant transfer characteristics. Also the various compartments are assumed to be at steady state, that is, transients have subsided and the system is at equilibrium.

The Equilibrium Equation

Consider a simple model of a system consisting of n interacting compartments. The compartments are labelled by integers from 1 to n. The amount of material in compartment i is denoted by x_i. In a biological system this x_i could represent the quantity, or concentration, of a chemical substance and in an industrial system, this x_i could represent the production of industry i. In any time period (say, a unit time period), the quantity of material (or energy, for instance) that is transported from compartment j to compartment i is denoted by f_{ij} and is called the *flux* from j to i. (Noticc thc slight diffcrcncc between the sense of direction implied by this definition and that implied in the definition of an adjacency matrix.) In an industrial system f_{ij} could represent the fraction of the output of the jth industry required for use in industry i. The ratio f_{ij}/x_j is therefore the fraction of flux transfer from j to i and is denoted by a_{ij} and is called the *fractional transfer coefficient.*

With these definitions, consider the flows taking place with respect to the ith compartment as shown in Fig. 7.15. In the said unit time period, let I_i be the material "imported" (or "ingested") into the ith compartment from the environment of the system. Then $-f_{0i}x_i$ is the amount of material "exported" (or "excreted") from i into the environment. The quantity $f_{ji}x_i$

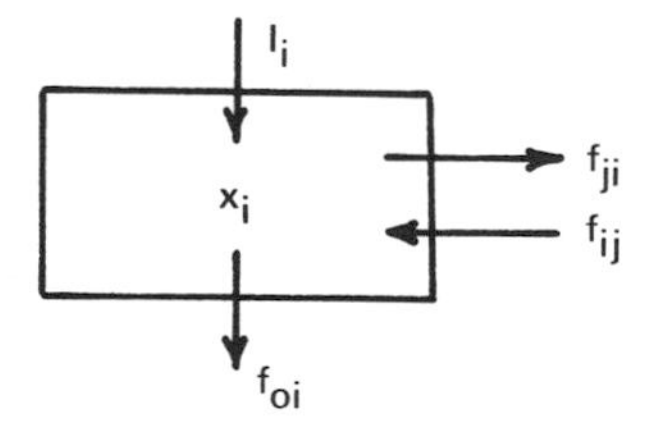

Fig. 7.15 An isolated compartment in a compartmental model.

represents material transfer from the ith to the jth compartment. All f's are positive as per the definition. Then, under equilibrium conditions,

$$\text{quantity going in} \quad - \text{quantity coming out} = 0$$

$$\sum_{j=1, j\neq i}^{n} a_{ij}x_j + I_i - a_{0i}x_i - \sum_{j=1, j\neq i}^{n} a_{ji}x_i = 0 \tag{1}$$

Example 3. A three-compartmental model with concentrations and fluxes at some point in time is shown in Fig. 7.16. The concentration vector, the flux matrix, and the fractional transfer coefficient matrix for this system are

$$\mathbf{x} = \begin{bmatrix} 10 \\ 5 \\ 1 \end{bmatrix}, \quad \mathbf{F} = \begin{bmatrix} -8 & 1 & 0 \\ 5 & -3 & 0 \\ 3 & 2 & 0 \end{bmatrix}, \quad \mathbf{A} = \begin{bmatrix} -0.8 & 0.2 & 0 \\ 0.5 & -0.6 & 0 \\ 0.3 & 0.4 & 0 \end{bmatrix} \quad \blacksquare$$

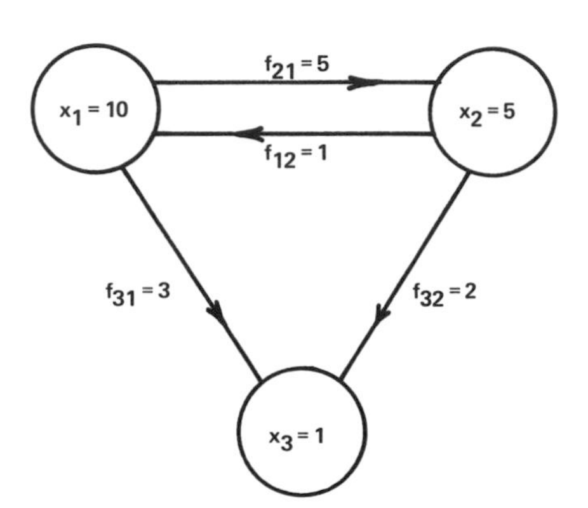

Fig. 7.16 A three-compartment model.

The Leontief Closed Model

In *Leontief models*, an economy is assumed to consist of a number of interacting industries. Each industry is assumed to produce a single good and uses only one process of production to make this good. To produce its good, a given industry needs as input goods made by other industries, labor, and perhaps other inputs from outside the industrial complex. Each industry must produce enough to supply the needs of other industries as well as to meet external consumer demand. If the set of commodities that appear as inputs at least once in the system is identical with the set of commodities that appear as outputs, and if there is no source of inputs other than current production and no use for outputs except as inputs, then the model is referred to as a Leontief *closed model*. Otherwise, it is a *Leontief open model*. Therefore, the condition for a Leontief closed model can be achieved if inputs coming from and outputs going to the environment external to the compartmental system are eliminated. This can be achieved in two different ways. One way is to set both I_i and a_{0i} for all i equal to zero. The other alternative is to assume that I_i is coming from another compartment and a_{0i} is going to still another compartment. Setting $I_i = 0 = a_{0i}x_i$, a Leontief model for a closed economic system can be written as

$$\sum_{j=1, j\neq i}^{n} a_{ij}x_j = \sum_{j=1, j\neq i}^{n} a_{ji}x_i \tag{2}$$

For the first industry, that is, for $i = 1$, the above equation becomes

$$a_{12}x_2 + a_{13}x_3 + \cdots + a_{1n}x_n = (a_{21} + a_{31} + \cdots + a_{n1})x_1 \tag{3}$$

The right side of the above equation represents the net quantity of material going from industry 1 to all other industries. Notice that the product manufactured by industry 1 is not used in the production process of industry 1 itself, which of course is not true. Generation of electricity requires some electricity to excite the field coils of generators. During the production of coke, it is common practice to use coke oven gas to heat the ovens. To avoid confusion, it is conventional to measure the net production x_i and set $a_{ii} = 0$. With this assumption it is easy to see that the quantity inside the parentheses on the right side of Eq. (3) is unity and one can write, for $i = 1$,

$$a_{12}x_2 + a_{13}x_3 + \cdots + a_{1n}x_n = x_1 \tag{4}$$

A similar equation holds good for $i = 2, 3, \ldots, n$ and the entire set can be written as

$$\mathbf{Ax} = \mathbf{x} \tag{5}$$

where the matrix $\mathbf{A}$ is sometimes referred to as the *technology matrix* and the vector $\mathbf{x}$ as output vector. From physical considerations, it is obvious that all inputs and outputs are nonnegative and that every output requires at least one input and every input gives rise to at least one output. Therefore, the entries of $\mathbf{A}$ are necessarily nonnegative, and every row and every column of $\mathbf{A}$ must necessarily contain at least one nonzero entry. Therefore, $\mathbf{A}$ is a *semipositive* matrix.

The Leontief closed system is said to be in *interior equilibrium* if

$$\mathbf{Ax} = \mathbf{x}; \qquad \mathbf{x} > \mathbf{0} \tag{6}$$

That is, every commodity is produced and the production just equals the requirement of every other commodity. According to matrix theory, a necessary condition for the existence of $\mathbf{x}$ which satisfies Eq. (6) is that one of the eigenvalues of $\mathbf{A}$ be unity.

The Leontief Open Model

A Leontief open model is essentially a Leontief closed model plus at least one extra input (which is not the output of a production process) and an external demand for some or all outputs over and above their use as inputs. The extra input is most often identified as labor, but this identification is not essential. Neglecting this extra input for the time being, let x_j be the total quantity industry j produces to meet its demands. Then

$$\mathbf{x} = \mathbf{A}\mathbf{x} + \mathbf{b} \tag{7}$$

where the first term on the right side is the total internal demand and b_i is the external demand on industry i. Notice that b_i is being used here instead of a_{0i} merely for notational convenience.

To determine the gross demand, Eq. (7) is solved for $\mathbf{x}$ as

$$\mathbf{x} - \mathbf{A}\mathbf{x} = (\mathbf{I}\mathbf{x} - \mathbf{A}\mathbf{x}) = (\mathbf{I} - \mathbf{A})\mathbf{x} = \mathbf{b}$$

or

$$\mathbf{x} = (\mathbf{I} - \mathbf{A})^{-1}\mathbf{b}; \qquad \mathbf{b} \geq \mathbf{0} \tag{8}$$

As it stands, Eq. (8) does not guarantee that $\mathbf{x} \geq \mathbf{0}$. However, if all the entries of $(\mathbf{I} - \mathbf{A})^{-1}$ are positive, then $(\mathbf{I} - \mathbf{A})^{-1}\mathbf{b}$ is always nonnegative if $\mathbf{b}$ is nonnegative. Thus, the central mathematical problem here is to find the conditions under which $(\mathbf{I} - \mathbf{A})^{-1}$ is guaranteed to be positive (i.e., to have positive entries). An important result in matrix theory states that a sufficient condition for $(\mathbf{I} - \mathbf{A})^{-1}$ to be positive is that the absolute value of the largest eigenvalue of $\mathbf{A}$ be less than unity (see Appendix 3). Assuming that this condition is satisfied, the right side of Eq. (8) can be written as a converging infinite series:

$$\mathbf{x} = (\mathbf{I} + \mathbf{A} + \mathbf{A}^2 + \cdots)\mathbf{b} = \mathbf{b} + \mathbf{A}\mathbf{b} + \mathbf{A}(\mathbf{A}\mathbf{b}) + \cdots \tag{9}$$

The above procedure may not be the best way to compute $(\mathbf{I} - \mathbf{A})^{-1}$. However, an interesting physical interpretation can be given to Eq. (9). The gross production $\mathbf{x}$ must first consist of the bill of goods or external demand $\mathbf{b}$. In order to meet this demand, each industry i—in addition to its obligations to the demands of other industries—must produce an additional quantity $\sum_j a_{ij}b_j = \mathbf{A}\mathbf{b}$. However, to produce $\mathbf{A}\mathbf{b}$, an additional quantity $\mathbf{A}(\mathbf{A}\mathbf{b}) = \mathbf{A}^2\mathbf{b}$ must be produced, and so forth. In other words the entries of $(\mathbf{I} - \mathbf{A})^{-1}$, denoted here as $\tilde{a}_{ij}$, give the *total requirement* of input i for unit operation of industry j. Remember that the entries of $\mathbf{A}$, denoted by a_{ij}, represent the *direct requirement* of input i in industry j.

The Gozinto Problem

As an illustration of the application of Leontief type models, consider a simple problem in production scheduling.

Imagine a hypothetical company wishing to manufacture three models $M1$, $M2$, and $M3$ of fishing equipment. Assume that each model contains essentially two basic parts: rods, $R1$, and reels, $R2$. The models differ in details caused by assembling rods and reels to form intermediate subassemblies $S1$, $S2$, $S3$, and $S4$. In addition to selling finished products, the company wishes to supply its dealers with an adequate supply of spare parts. The company wishes to ascertain the number of rods and reels it has to order from its suppliers.

The true magnitude of a problem like this can be easily visualized if the hypothetical company is replaced by a real one such as an automobile manufacturing company. A typical company manufactures several models of automobiles and may have as many as 40,000 suppliers.

A useful first step in a problem like this is to pictorially represent the assembly process by means of a weighted directed graph such as the one in Fig. 7.17. The nodes are numbered in technological order. Such a graph is called gozinto (goes + into) graph as it pictorially represents what part goes into what subassembly and what assembly goes into what model.

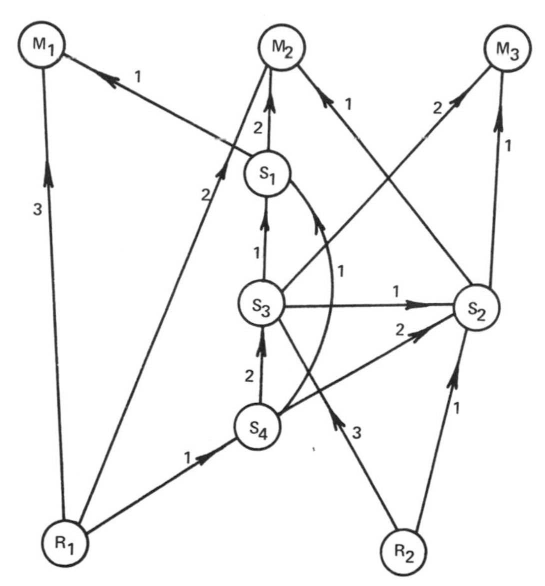

Fig. 7.17 A gozinto graph for the parts requirement problem.

The first step of this problem is to construct what is called the *next assembly quantity matrix*. This is a square matrix with as many rows and columns as there are nodes in the gozinto graph. The entry q_{ij} of $\mathbf{Q}$ is the number of ith items *directly needed* to assemble one jth item. For this problem, the matrix $\mathbf{Q}$ is displayed in Table 7.1.

TABLE 7.1
The **Q** Matrix

		Old label	M_1	M_2	M_3	S_1	S_2	S_3	S_4	R_1	R_2
Old label	New label	New label	A_1	A_2	A_3	A_4	A_5	A_6	A_7	A_8	A_9
M_1	A_1		0	0	0	0	0	0	0	0	0
M_2	A_2		0	0	0	0	0	0	0	0	0
M_3	A_3		0	0	0	0	0	0	0	0	0
S_1	A_4		1	2	0	0	0	0	0	0	0
S_2	A_5		0	1	1	0	0	0	0	0	0
S_3	A_6		0	0	2	1	1	0	0	0	0
S_4	A_7		0	0	0	1	2	2	0	0	0
R_1	A_8		3	2	0	0	0	0	1	0	0
R_2	A_9		0	0	0	0	1	3	0	0	0

Note that the labels on the basic parts, subassemblies, and finished products have been changed for uniformity and notational convenience. In the above matrix, $q_{82} = 2$ means that two items of R_1 (i.e., A_8) *directly* go to the assembly of one M_2 (i.e., A_2). Note that one A_8 goes into one A_7 and one A_7 in turn goes into one A_4, two of which in turn go into one A_2. However, this is an *indirect* use. Evidently $q_{ii} = 0$ for all i as no product can be a part of itself. The upper triangular portion of $\mathbf{Q}$ is also zero because no item can be used as a part of a system before it itself is assembled. Stated differently, *technological order* is a prerequisite for the assembly of any system. It is also of interest to note that a row of zeros in the matrix indicates a final product and a column of zeros indicates a basic component.

For simplicity and concreteness, consider the problem of computing the total number of units of A_7 required to assemble one unit of A_2. This quantity, defined as n_{72}, can be readily ascertained from the gozinto graph as

$$n_{72} = q_{71}n_{12} + q_{72}n_{22} + \cdots + q_{78}n_{82} + q_{79}n_{92}$$

or, in general

$$n_{ij} = \sum_{\text{all } k} q_{ik}n_{kj} \tag{10}$$

where n_{ij} is the total number of A_i's going into the assembly of one A_j. If the above is summed over all j, one gets the total number of A_i's required to manufacture one unit of A_j for all j, that is, to meet an internal demand of one unit of each. Therefore,

$$\sum_{\text{all } j} n_{ij} = \sum_{\text{all } k} \sum_{\text{all } j} q_{ik} n_{kj}$$

or

$$n_i = \sum_{\text{all } k} q_{ik} n_k \tag{11}$$

If there is an additional external demand of b_i units of A_i (such as for spare parts), then

$$n_i = \sum_{\text{all } k} q_{ik} n_{kj} + b_i \tag{12}$$

In vector–matrix notation

$$\mathbf{n} = \mathbf{Q}\mathbf{n} + \mathbf{b}$$

or

$$\mathbf{n} = (\mathbf{I} - \mathbf{Q})^{-1}\mathbf{b}$$

The actual computation of **n** is left as an exercise.

The Aggregation Problem

Application of Leontief models to real problems is not without its difficulties. In any viable economy, there are thousands of different activities which, in a Leontief model, could be considered to be industries. First, the task of collecting data is monumental. Even if the data could be collected, none of the available computers could be made to solve the resulting system of thousands of equations. To get a practical grip on the problem, it is necessary to aggregate a number of activities into a single industry.

The problem of aggregation is very difficult, and the questions raised by it are not yet resolved completely. Only an attempt is made here to get a flavor of the problem. Three of the more frequent ideas used for aggregation are (a) substitutability, (b) complementarity, and (c) similarity of production functions. If the first concept is adopted, one would aggregate products which are close substitutes. According to the second idea, one would aggregate items that complement each other and are used in roughly fixed proportions. According to the final criterion, one would aggregate items requiring essentially the same type of production process. The above methods are not exhaustive and are also not foolproof. In an ecological problem, for instance, one may want to aggregate animals in an ecological niche, say by the speed of the hunter or the size of the prey. Another possibility is to aggregate on the basis of the eating habits of a predator, such as carnivorous and herbivorous.

What happens to the set of equations $\mathbf{x} = \mathbf{A}\mathbf{x} + \mathbf{b}$ after aggregation? Suppose that it is decided to aggregate the industries $m, m+1, \ldots, n$ into a single industry. If $\hat{b}_m$ is the net demand for this single industry, then

$$\hat{b}_m = \sum_{i=m}^{n} b_i \tag{13}$$

The sales from the aggregated industry m to any other industry j will be denoted by $\hat{y}_{mj}$, and therefore,

$$\hat{y}_{mj} = \sum_{i=m}^{n} y_{ij} \tag{14}$$

Similarly, if $\hat{y}_{im}$ represents the sales from industry i to the aggregated industry j, then

$$\hat{y}_{im} = \sum_{j=m}^{n} y_{ij} \tag{15}$$

If, for the period under study, the gross sales of industry j are denoted by x_j and if x_m denotes the gross sales of the aggregated industry m, then

$$\hat{x}_m = \sum_{j=m}^{n} x_j \tag{16}$$

The new technological coefficients a_{mj} of the aggregated model can be computed as follows:

$$a_{mj} = \hat{y}_{mj}/x_j = (1/x_j) \sum_{i=m}^{n} y_{ij} = \sum_{i=m}^{n} a_{ij}; \qquad j = 1, 2, \ldots, m-1 \tag{17}$$

$$a_{im} = \hat{y}_{im}/\hat{x}_m = \sum_{j=m}^{n} y_{ij} \Big/ \sum_{j=m}^{n} x_j = \sum_{j=m}^{n} a_{ij}x_j \Big/ \sum_{j=m}^{n} x_j; \qquad i = 0, \ldots, m \tag{18}$$

The above procedure ignores several details. For instance, $\hat{a}_{im}$ is not independent of the bill of goods. Whenever the bill of goods changes, one has to recompute the $\hat{a}_{im}$'s again. For further details on the aggregation procedure when the aggregated industries are exactly complementary (i.e., exact substitutes of each other) and when the production functions of the aggregated industries are not identical, one has to consult advanced books on economic theory.

4 DECOMPOSITION OF LARGE SYSTEMS

An alternative to aggregation, briefly discussed in the preceding section, is to seek simplification in the original system by means of decomposition. The decomposition approach is motivated by two major factors. First, decomposition permits a better understanding of the structural relations of a system, which would in turn allow one to reorganize a system into subsystems that may be more meaningful. Second, decomposition allows one to consider the possibility of solving a group of small systems of equations than one large system of equations. This fact itself is of considerable significance as the computational effort required roughly goes up as the cube of the number of equations. Also, calculating the eigenvalues of a small system is considerably easy. Solving systems of equations and determining eigenvalues are fundamental mathematical operations in the study of static equilibrium structure. Therefore, this section will be devoted to a brief discussion of these concepts.

Decomposition by Partitioning

Consider the problem of solving

$$\mathbf{Cx} = \mathbf{b} \tag{1}$$

which is really a rearranged version of Eq. (3.7). In the absence of any knowledge about decomposition principles, one can proceed to simplify the problem of solving Eq. (1) by partitioning $\mathbf{C}$ and rewriting Eq. (1) as

$$\begin{bmatrix} \mathbf{E} & \mathbf{F} \\ \mathbf{G} & \mathbf{H} \end{bmatrix} \begin{bmatrix} \mathbf{x}_1 \\ \mathbf{x}_2 \end{bmatrix} = \begin{bmatrix} \mathbf{b}_1 \\ \mathbf{b}_2 \end{bmatrix} \tag{2}$$

where $\mathbf{E}$ is an $r \times r$ matrix, $\mathbf{H}$ is an $(n - r) \times (n - r)$ matrix, $\mathbf{F}$ is a $r \times (n - r)$ rectangular matrix, and $\mathbf{G}$ is $(n - r) \times r$ matrix, with $\mathbf{x}_1, \mathbf{x}_2, \mathbf{b}_1$, and $\mathbf{b}_2$ having appropriate dimensions. It can be easily verified that Eq. (2) can also be written as

$$\begin{bmatrix} \mathbf{I} & \mathbf{0} \\ \mathbf{GE}^{-1} & \mathbf{I} \end{bmatrix} \begin{bmatrix} \mathbf{E} & \mathbf{0} \\ \mathbf{0} & \mathbf{G} - \mathbf{GE}^{-1}\mathbf{F} \end{bmatrix} \begin{bmatrix} \mathbf{I} & \mathbf{E}^{-1}\mathbf{F} \\ \mathbf{0} & \mathbf{I} \end{bmatrix} \begin{bmatrix} \mathbf{x}_1 \\ \mathbf{x}_2 \end{bmatrix} = \begin{bmatrix} \mathbf{b}_1 \\ \mathbf{b}_2 \end{bmatrix} \tag{3}$$

Using the fact that $(\mathbf{AB})^{-1} = \mathbf{B}^{-1}\mathbf{A}^{-1}$, Eq. (3) can be solved for $\mathbf{x}_1$ and $\mathbf{x}_2$ as

$$\begin{bmatrix} \mathbf{x}_1 \\ \mathbf{x}_2 \end{bmatrix} = \begin{bmatrix} \mathbf{I} & -\mathbf{E}^{-1}\mathbf{F} \\ \mathbf{0} & \mathbf{I} \end{bmatrix} \begin{bmatrix} \mathbf{E}^{-1} & \mathbf{0} \\ \mathbf{0} & (\mathbf{H} - \mathbf{GE}^{-1}\mathbf{F})^{-1} \end{bmatrix} \begin{bmatrix} \mathbf{I} \\ -\mathbf{GE}^{-1} \end{bmatrix} \begin{bmatrix} \mathbf{0} & \mathbf{b}_1 \\ \mathbf{I} & \mathbf{b}_2 \end{bmatrix} \tag{4}$$

Thus, the inversion of a large square matrix $\mathbf{C}$ has been reduced to the inversion of two matrices of lower order. In the case discussed above, inversion of $\mathbf{C}$ is accomplished by inverting $\mathbf{E}$ and $(\mathbf{H} - \mathbf{G}\mathbf{E}^{-1}\mathbf{F})$.

As this method is only valid when $\mathbf{C}$ is partitioned into four submatrices, the utility of this method quickly decreases as the dimensions of the submatrices increase with increasing size of $\mathbf{C}$. This approach loses its simplicity if $\mathbf{C}$ is partitioned into submatrices of size 3×3, 4×4, etc. Furthermore, the brute-force nature of the method sheds little insight into the nature of the structural relations of the system under study.

Decomposition by Permutations

Under some special conditions, the rows and columns of the matrix $\mathbf{A}$, associated with a system can be rearranged by permuting rows and columns, such that the nonzero entries appear only on the main diagonal. Then $\mathbf{A}$ is said to be diagonal and the system is said to be *completely decomposable* into its constituent elements. This case, which implies no interconnections among the elements, is not very interesting. However, if $\mathbf{A}$ can be rearranged and then partitioned such that all of the submatrices other than those on the main diagonal are filled with zeros, then $\mathbf{A}$ is said to be *block diagonal*, and the system it represents is said to be completely decomposable into its constituent subsystems. Therefore, if one is dealing with a block diagonalizable matrix $\mathbf{A}$, considerable simplification can be achieved by partitioning $\mathbf{A}$ into suitable constituent blocks and solving the equations represented by each block separately.

A more complicated situation arises if permuting the rows and columns of $\mathbf{A}$ leads to a set of equations that, after a suitable partition, can be displayed as

$$\begin{bmatrix} \mathbf{C}_{1,1} & \mathbf{C}_{1,2} \\ \mathbf{0} & \mathbf{C}_{2,2} \end{bmatrix} \begin{bmatrix} \mathbf{x}_1 \\ \mathbf{x}_2 \end{bmatrix} = \begin{bmatrix} \mathbf{b}_1 \\ \mathbf{b}_2 \end{bmatrix} \tag{5}$$

where $\mathbf{C}_{1,1}$ is a $r \times r$ submatrix, $\mathbf{C}_{2,2}$ is a $(n - r) \times (n - r)$ submatrix, $1 \leq r \leq n$, of $\mathbf{C}$ and $\mathbf{x}_1$, $\mathbf{x}_2$, $\mathbf{b}_1$, and $\mathbf{b}_2$ are suitably permuted and partitioned versions of $\mathbf{x}$ and $\mathbf{b}$. Now, the second of the above equations can be solved first for $\mathbf{x}_2$, and substituting this in the first, the value of $\mathbf{x}_1$ is obtained. Thus, the solution of the original matrix equation is *reduced* to the solution of two lower-order matrix equation.

Therefore, an $n \times n$ square matrix $\mathbf{C}$ is said to be *reducible* if there exists a permutation matrix $\mathbf{P}$ such that

$$\mathbf{P}\mathbf{C}\mathbf{P}^{\mathrm{T}} = \begin{bmatrix} \mathbf{C}_{1,1} & \mathbf{C}_{1,2} \\ \mathbf{0} & \mathbf{C}_{2,2} \end{bmatrix} \tag{6}$$

where the $\mathbf{C}_{i,j}$'s are submatrices of $\mathbf{C}$ as defined earlier. Otherwise, $\mathbf{C}$ is said to be *irreducible*.

A graphical interpretation of the concept of irreducibility would be an extremely useful aid in detecting reducibility or the lack of it. For instance, if the original $(n \times n)$ matrix $\mathbf{A}$ can be regarded as a connection matrix of a directed graph $G_d(\mathbf{A})$, then a nonzero entry in the (i, j)th position of $\mathbf{A}$ can be regarded as an edge connecting nodes i to j of the graph. For example, two matrices $\mathbf{A}$ and $\mathbf{B}$ and their directed graphs $G_d(\mathbf{A})$ and $G_d(\mathbf{B})$ are shown in Fig. 7.18. By inspection, it is clear that $G_d(\mathbf{A})$ is strongly connected, whereas $G_d(\mathbf{B})$ is not. It can be verified that matrix $\mathbf{A}$ is irreducible and $\mathbf{B}$ is evidently reducible. This observation leads to a well known theorem which states that *a matrix* $\mathbf{A}$ *is irreducible if and only if its directed graph is strongly connected.* Using this theorem, it is at least possible to observe the graph of interconnections of a system and find out whether the system equations can be decomposed by merely rearranging their order. As an illustration of what is involved, consider an economy whose technology matrix is given by

$$\mathbf{C} = \begin{bmatrix} 0 & 0 & 0 & 0 & 0 \\ 0 & 0 & x & 0 & x \\ 0 & x & x & x & x \\ x & 0 & 0 & 0 & x \\ 0 & 0 & 0 & x & 0 \end{bmatrix}$$

where the nonzero entries are indicated by the x's.

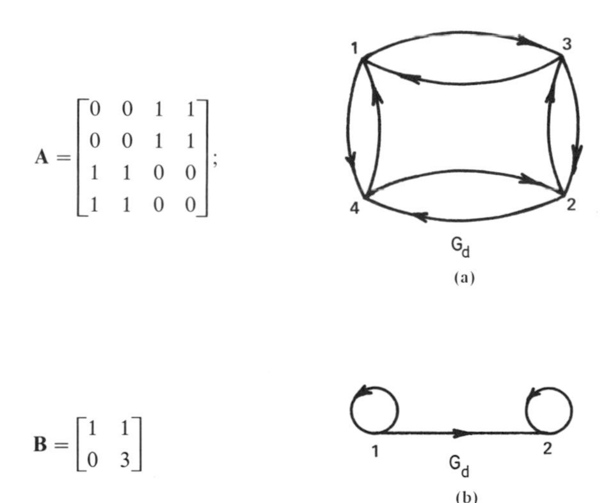

Fig. 7.18 A figure to illustrate the concept of reducibility.

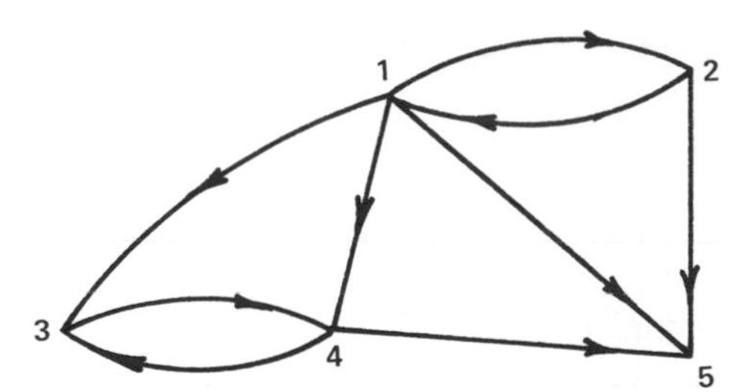

Fig. 7.19 A graph of an economy.

A graph of this economy is shown in Fig. 7.19. Visual inspection reveals that the directed graph in Fig. 7.19 is not strongly connected. Therefore, matrix **C** is reducible. Indeed, **C** can be rearranged into the form

$$\mathbf{C} = \left[\begin{array}{cc|cc|c} 0 & a_{12} & a_{13} & a_{14} & a_{15} \\ a_{21} & 0 & 0 & 0 & a_{25} \\ \hline 0 & 0 & 0 & a_{34} & 0 \\ 0 & 0 & a_{43} & 0 & a_{45} \\ \hline 0 & 0 & 0 & 0 & 0 \end{array}\right] = \begin{bmatrix} \mathbf{C}_{11} & \mathbf{C}_{12} & \mathbf{C}_{13} \\ \mathbf{0} & \mathbf{C}_{22} & \mathbf{C}_{23} \\ \mathbf{0} & \mathbf{0} & \mathbf{C}_{33} \end{bmatrix}$$

by simply interchanging rows according to 1 to 5, 3 to 1, 5 to 3, 2 to 4, and 4 to 2.

In a complex problem, visual inspection is not always an efficient way to determine reducibility or the lack of it. Given a matrix, systematic procedures are available to determine the permutation matrix **P**, if it exists, that will perform the required interchanges.

An Algorithm to Determine the Permutation Matrix

Consider a system whose directed graph D is given as shown in Fig. 7.20. If the system is reducible, a permutation matrix **P**, which performs the necessary row and column operations in order to bring the system equations into the form of Eq. (6), can be found if it exists by following the computations indicated in the steps below.

Step 1 Write down the adjacency matrix $\mathbf{A}(D)$ of the graph D as shown:

$$\mathbf{A} = \begin{bmatrix} 0 & 0 & 1 & 1 & 1 & 0 \\ 0 & 0 & 1 & 0 & 0 & 0 \\ 0 & 0 & 0 & 0 & 0 & 1 \\ 1 & 1 & 1 & 0 & 1 & 1 \\ 1 & 1 & 0 & 1 & 0 & 1 \\ 0 & 0 & 1 & 0 & 0 & 0 \end{bmatrix} \tag{7}$$

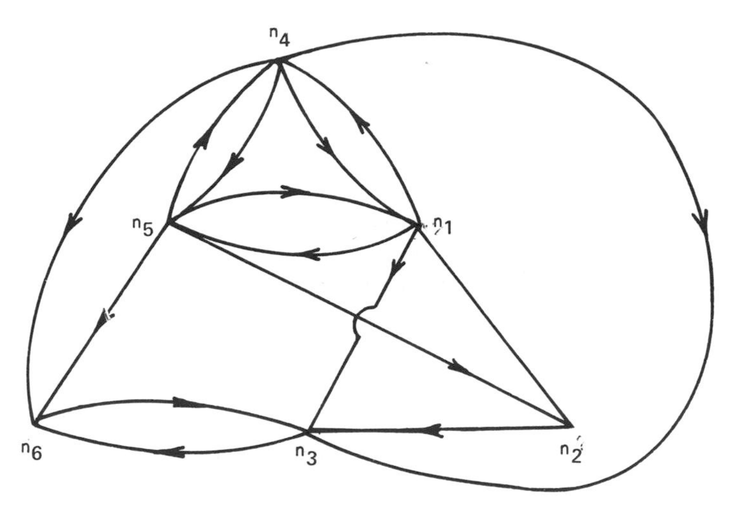

Figure 7.20 Directed graph of a system.

Step 2 Determine the reachability matrix **R** using the procedure described earlier. This results in

$$\mathbf{R} = \begin{bmatrix} 1 & 1 & 1 & 1 & 1 & 1 \\ 0 & 1 & 1 & 0 & 0 & 1 \\ 0 & 0 & 1 & 0 & 0 & 1 \\ 1 & 1 & 1 & 1 & 1 & 1 \\ 1 & 1 & 1 & 1 & 1 & 1 \\ 0 & 0 & 1 & 0 & 0 & 1 \end{bmatrix} \tag{8}$$

Step 3 Find the strong components of D by inspecting the entries of $\mathbf{R} \times \mathbf{R}$. The strong components are found to be

$$\begin{aligned} N_1 &= \{n_1, n_4, n_5\} \\ N_2 &= \{n_2\} \\ N_3 &= \{n_3, n_6\} \end{aligned} \tag{9}$$

Step 4 Using N_1, N_2, and N_3 as vertices, construct the condensation D^* of D. This is shown in Fig. 7.21.

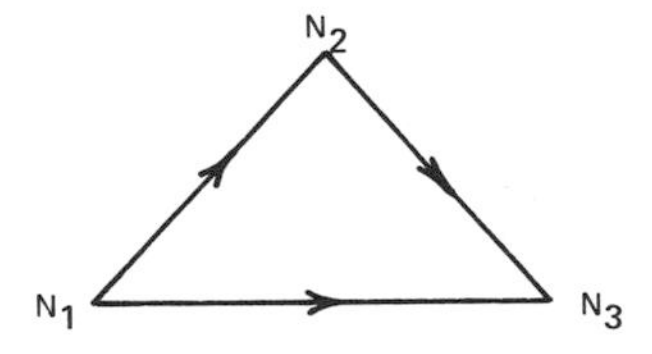

Fig. 7.21 Condensation of the graph in Fig. 7.20.

Step 5 Construct the adjacency matrix $\mathbf{A}^*$ of D^*

$$\mathbf{A}^* = \begin{bmatrix} 0 & 1 & 1 \\ 0 & 0 & 1 \\ 0 & 0 & 0 \end{bmatrix} \tag{10}$$

Step 6 Now inspect $\mathbf{A}^*$ and locate a column of zeros and label the strong component corresponding to this column as N_{i1}. Delete this row and column and repeat the procedure with the remaining matrix and label N_{i2}, etc. Now reorder the rows and columns of $\mathbf{A}^*$ as $N_{i1}, N_{i2}, \ldots,$ and call the resulting matrix $\mathbf{B}^*$. That is, $\mathbf{B}^*$ is defined as $\mathbf{QA}^*\mathbf{Q}^\mathrm{T}$, where $\mathbf{Q}$ is a permutation matrix. In the present example, it is easily observed that $\mathbf{B}^* = \mathbf{A}$.

Step 7 Now relabel the vertices of the original directed graph D as follows. Let $v_1, v_2, \ldots, v_{n1}$ be any ordering of vertices in the strong component N_{i1}. Let $v_{n1+1}, v_{n1+2}, \ldots, v_{n1+n2}$ be the ordering of vertices in the strong component N_{i2} and so on. Referring back to Eq. (9), this step yields

$$\begin{aligned} N_1 &= \{n_1, n_4, n_5\} \to \{v_1, v_2, v_3\} \\ N_2 &= \{n_2\} \qquad\quad\;\; \to \{v_4\} \\ N_3 &= \{n_3, n_6\} \qquad \to \{v_5, v_6\} \end{aligned} \tag{11}$$

That is, the relation between the old and new vertex labels is

$$\begin{array}{rcccccc} \text{old labels:} & n_1 & n_2 & n_3 & n_4 & n_5 & n_6 \\ & \downarrow & \downarrow & \downarrow & \downarrow & \downarrow & \downarrow \\ \text{new labels:} & v_1 & v_4 & v_5 & v_2 & v_3 & v_6 \end{array}$$

By looking at the subscripts of the old and new labels, the required permutation matrix is written as

$$\mathbf{P} = \begin{bmatrix} 1 & 0 & 0 & 0 & 0 & 0 \\ 0 & 0 & 0 & 1 & 0 & 0 \\ 0 & 0 & 0 & 0 & 1 & 0 \\ 0 & 1 & 0 & 0 & 0 & 0 \\ 0 & 0 & 1 & 0 & 0 & 0 \\ 0 & 0 & 0 & 0 & 0 & 1 \end{bmatrix} \tag{12}$$

That is a relabeling from n_1 to v_1 leads to unity in the (1, 1) position of $\mathbf{P}$. A transition from n_3 to v_5 corresponds to a unity in the (3, 5) entry of $\mathbf{P}$.

Step 8 Verify now that $\mathbf{PAP}^\mathrm{T}$ indeed gives a matrix of the desired structure.

$$\mathbf{PAP}^{\mathrm{T}} = \left[\begin{array}{ccc|c|cc} 0 & 1 & 1 & 0 & 1 & 0 \\ 1 & 0 & 1 & 1 & 1 & 1 \\ 1 & 1 & 0 & 1 & 0 & 1 \\ \hline 0 & 0 & 0 & 0 & 1 & 0 \\ \hline 0 & 0 & 0 & 0 & 0 & 1 \\ 0 & 0 & 0 & 0 & 1 & 0 \end{array}\right] \tag{13}$$

Notice that the matrix multiplication implied by $\mathbf{PAP}^{\mathrm{T}}$ need not be performed using the conventional rules. It is sufficient to remember that premultiplication of any matrix, say, **A**, by a permutation matrix **P** corresponds to an interchange of rows in **A** and post multiplication by **P** (or $\mathbf{P}^{\mathrm{T}}$ or $\mathbf{P}^{-1}$) corresponds to an interchange of the columns of **A**. While premultiplying, $p_{ij} = 1$ represents a replacement of the ith row of **A** by the jth row of **A**. While post multiplying, $p_{ij} = 1$ represents a replacement of the ith column by the jth column.

5 ROUTING PROBLEMS

The operational problems involved in the planning, evaluation, and execution of large complex systems have specific features and require a somewhat specialized approach for their solution. One of the important features of operational problems is that the control is usually a team of executives, with appropriate technical means and equipment required to achieve a certain goal. The system to be controlled is usually a project such as constructing a building, launching a space vehicle, or developing a new research program.

To control an operation effectively, it is usually necessary to solve two types of problems: (1) to work out the optimal plan for carrying out the stated operation, and (2) to ensure the realization is near to the optimal plan under changing conditions.

A scientific approach to the solution of these problems is possible only if a mathematical model is available that reflects sufficiently clearly the properties and characteristics of the object and is suitable for investigation by formalized methods. One of the most convenient and most extensively used models for solving problems of operational control is the so-called *network model.*

A useful application of network models arises when one wishes to determine the so-called *shortest route* or *quickest path* between two points. Applications of this general concept abound in many physical systems. For instance, during the preliminary stages of developing a master plan for a city, an engineer might be interested in the possible traffic patterns in the city under various circumstances. He might be interested in determining the quickest path from a residential neighborhood to an industrial neighborhood, or the quickest path to evacuate the city in an emergency.

In the first case the interest is to find the shortest path from a given point, say, s, to another given point, say, t. The problem of evacuating a city however, is slightly different; here the interest is to find the shortest routes from several points s_i to one particular point t. Sometimes the shortest path may not be available due to several reasons such as a traffic accident, or road repair. Then one would be interested in finding not only the shortest route but also the next best route. This problem can be stated, in general terms, as finding the k shortest routes between two points s and t. Finally, a traffic engineer may want to know the shortest path between every possible trip origin to every possible trip destination.

To appreciate the magnitude of the problem involved let us consider, as a first step, a single isolated passenger wishing to go from home to work by the quickest possible way. To begin with, a street map of the city is required. Then, one way of attacking the problem is to list *all* possible routes from home to work. Then one can conduct a sequence of experiments by actually traveling each possible route, measure the elapsed time for each and choose the route of shortest time. This method of *exhaustive enumeration* of all possibilities is obviously impractical in large scale problems.

The routing problem can be solved using more systematic methods based on graph theoretic ideas. First, one can define a *directed path* from a vertex v_1 to v_n as a collection of distinct vertices $v_1, v_2, \ldots, v_n$, together with the arcs $v_1v_2, v_2v_3, \ldots, v_{n-1}v_n$ considered in the following order: v_1, v_1v_2, $v_2v_3, \ldots, v_{n-1}v_n, v_n$. Then the length of a path, cost associated with a path, etc. can be defined (see Section 7.1) and the resulting information can be arranged in matrix form.

Consider a partial street map of a city as shown in Fig. 7.22. This figure can be considered as a crude model of the city streets because it contains a

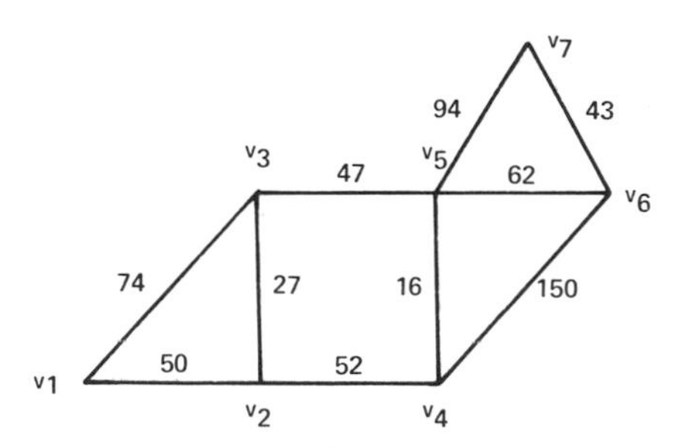

Fig. 7.22 A portion of a road map.

minimal amount of information. That is, such details as the number of traffic lanes, location of traffic signals, gradiants of streets, etc. were omitted. Only the general layout is shown. Now it is necessary to determine the time required to travel along each segment of the street map. The data so obtained are arranged in a matrix form:

From \ To	v_1	v_2	v_3	v_4	v_5	v_6	v_7
v_1	0	50	74				
v_2		0	27	52			
v_3		27	0		47		
v_4				0	16	150	
v_5				26	0	62	94
v_6					62	0	43
v_7							0

$= \mathbf{C}$ (1)

It should be constantly kept in mind that the numbers presented in the above array are obtained through hard field work and constitute the basic data for the problem. The number in a given row and column in the table indicates the time required (in some convenient unit) to go *from* the intersection bearing the number at the left end of the row to the intersection bearing the number at the top of the column. The blank spaces, therefore, indicate either (1) that there is no direct connection between the given intersections or (2) travel in that direction is obviously disadvantageous. For example, the table shows that one can go from v_1 to v_1 in 0 sec, from v_1 to v_2 in 50 sec, but since there are no other entries to the right, no other intersection can be reached *directly* from intersection v_1. Second, even though there is one edge connecting nodes v_2 and v_1, no entry appears at the intersection of the row marked v_2 and column marked v_1 because traveling from v_2 to v_1 is ruled out as it is obviously disadvantageous. Often these blank spaces are all filled with the symbol ∞ and the resulting square matrix $\mathbf{C}$ is called the "cost" matrix; here the "cost" involved is the time of travel.

As the present interest is to determine a minimum time path, the goal of the computational process should be to arrive at a matrix $\mathbf{F}$, where f_{ij} is the time of travel from vertex v_i to vertex v_j. This matrix $\mathbf{F}$ can be calculated from the cost matrix $\mathbf{C}$ using the following rules of modified arithmetic to calculate the "powers" $\mathbf{C}^{[2]}$, $\mathbf{C}^{[3]}$:

$$\text{modified multiplication } a \dot{\times} b = a + b \tag{2}$$

$$\text{modified addition } a \dot{+} b = \min(a, b) \tag{3}$$

For instance, $c_{25}^{[2]}$ of $\mathbf{C}^{[2]}$ is calculated using the modified arithmetic as

$$\begin{aligned} c_{25}^{[2]} &= (c_{21} \dot{\times} c_{15}) \dot{+} (c_{22} \dot{\times} c_{25}) \dot{+} \cdots \dot{+} (c_{27} \dot{\times} c_{75}) \\ &= (\infty \dot{\times} \infty) \dot{+} (0 \dot{\times} \infty) \dot{+} (27 \dot{\times} 47) \dot{+} (52 \dot{\times} 16) \dot{+} (\infty \dot{\times} 0) \dot{+} (\infty \dot{\times} 62) \dot{+} (\infty \dot{\times} \infty) \\ &= \min(\infty, \infty, 74, 68, \infty, \infty, \infty) = 68 \end{aligned}$$

Each term of the above equation, shown in parentheses, gives the time of travel from v_2 to v_5 along an edge sequence of length 2 and the "modified product" of these is the minimum travel time among the sequences. This result can be generalized. In general, $c_{ij}^{[n]}$ is the minimum cost path (quickest path, here) from v_i to v_j whose length is at most n. If $\mathbf{C}^{[n+1]} = \mathbf{C}^{[n]}$, then the entry $c_{ij}^{[n]} = f_{ij}$ gives the quickest time of travel from vertex v_i to v_j and $\mathbf{F} = \mathbf{C}^{[n]}$. This minimum cost (or time) path is called the *cost* (*or time*) *geodesic* because a geodesic is a path of minimum length. Carrying out the calculations, one gets

$$\mathbf{C}^{[2]} = \begin{bmatrix} 0 & 50 & 74 & 102 & 121 & \infty & \infty \\ 27 & 0 & 27 & 52 & 68 & 202 & \infty \\ 0 & 27 & 0 & 73 & 47 & 109 & 141 \\ \infty & \infty & \infty & 0 & 16 & 78 & 110 \\ \infty & \infty & \infty & 26 & 0 & 62 & 94 \\ \infty & \infty & \infty & 88 & 62 & 0 & 43 \\ \infty & \infty & \infty & \infty & \infty & \infty & 0 \end{bmatrix}$$

$$\mathbf{C}^{[3]} = \begin{bmatrix} 0 & 50 & 74 & 102 & 118 & 183 & 215 \\ 27 & 0 & 27 & 52 & 68 & 202 & 162 \\ 0 & 27 & 0 & 73 & 47 & 109 & 141 \\ \infty & \infty & \infty & 0 & 16 & 78 & 110 \\ \infty & \infty & \infty & 26 & 0 & 62 & 94 \\ \infty & \infty & \infty & 88 & 62 & 0 & 43 \\ \infty & \infty & \infty & \infty & \infty & \infty & 0 \end{bmatrix}$$

and

$$\mathbf{C}^{[4]} = \mathbf{C}^{[5]} = \mathbf{F} = \begin{bmatrix} 0 & 50 & 74 & 102 & 118 & 180 & 212 \\ 27 & 0 & 27 & 52 & 68 & 130 & 162 \\ 0 & 27 & 0 & 73 & 47 & 109 & 141 \\ \infty & \infty & \infty & 0 & 16 & 78 & 110 \\ \infty & \infty & \infty & 26 & 0 & 62 & 94 \\ \infty & \infty & \infty & 88 & 62 & 0 & 43 \\ \infty & \infty & \infty & \infty & \infty & \infty & 0 \end{bmatrix}$$

The matrix **F** derived from the above procedure gives the cost of going from v_i to v_j along *some* cost geodesic; it does not, however, identify the cost geodesics themselves. Further computation is required to determine the minimum time path(s). This calculation requires some additional concepts that are stated here without proof. Given a network N, a *cost geodetic network from v_i to v_j* is defined as a subnetwork N_g of N such that N_g contains all the vertices and edges of N that lie on at least one cost geodesic from v_i to v_j. Now it is easy to verify that (1) every subpath of a cost geodesic is also a cost geodesic, (2) a vertex v_k is on a cost geodesic from v_i to v_j if and only if

$$f_{ik} + f_{kj} = f_{ij} \tag{4}$$

where f_{ij} is the (i, j)th entry of **F**, and (3) the directed edge $v_r v_s$ is on a cost geodesic from v_i to v_j if and only if

$$f_{ir} + c_{rs} = f_{is} \tag{5}$$

The meaning of the above two equations is trivially simple. Equation (4) states that the minimum time path, say, from v_1 to v_7, is the same as the concatenation of the minimum time paths from v_1 to an intermediate vertex v_k and the minimum time path from this v_k to the destination v_j. Applying this rule to $i = 1$ and $j = 7$, the vertices that lie on the minimum time path from v_1 to v_7 are those v_k that satisfy $f_{1k} + f_{k7} = f_{17} = 217$, where f_{17} is obtained from **F** calculated above. Inspection of **F** reveals that $f_{1k} + f_{k7} = 217$ is satisfied for $k = 2$, 4, and 5. Therefore, the vertices 1, 2, 4, 5, and 7 lie on the quickest path of interest. In the present problem, there is no ambiguity, and the quickest path is seen to be $v_1 v_2 v_4 v_5 v_7$. If an ambiguity arises, it is possible to determine the edges that lie on the minimum time path using Eq. (5). This equation says that if one has already travelled from v_i to v_r along a minimum time path, then to reach an adjacent vertex s in minimum time, it is only necessary to choose that edge $v_r v_s$ that still keeps the travel time from v_i to v_s a minimum. As all entries of **F** have already been calculated, selection of c_{rs} in Eq. (5) is simple as demonstrated below. Starting from vertex v_1 the equation $f_{1r} + c_{rs} = f_{1s}$ was found to be satisfied only by the following sequence of equations:

$$\begin{aligned} f_{1,1} + c_{1,2} &= f_{12} = 50 \\ f_{1,2} + c_{2,4} &= f_{14} = 102 \\ f_{1,4} + c_{4,5} &= f_{15} = 118 \\ f_{1,5} + c_{5,7} &= f_{16} = 212 \end{aligned} \tag{6}$$

Any other combination of the subscripts r and s is not admissible. For instance, $f_{11} + c_{13} = 74$, which is an admissible value because there is an entry in matrix **F** whose value is 74, namely, f_{13}. This option brings us to vertex 3 and the next step is to look for an entry c_{3s}, which when added to f_{13}, gives a number f_{1s}, which can be found in the **F** matrix. Simple calculation shows that no such c_{3s} exists, and going from v_1 to v_3 is fruitless. This type of argument can be repeated at every step. Inspection of Eq. (6) clearly shows that the minimum-time path is $v_1v_2v_4v_5v_7$.

An Iterative Algorithm

Some of the ideas presented so far can be consolidated and presented from a different viewpoint—a viewpoint reminiscent of dynamic programming techniques. The central idea of this method is presented in conjunction with a trivially simple road map shown in Fig. 7.23, where t_{ij} indicates the travel time from vertex i to j. For concreteness, let vertex 1 be the starting point and vertex 4 be the destination.

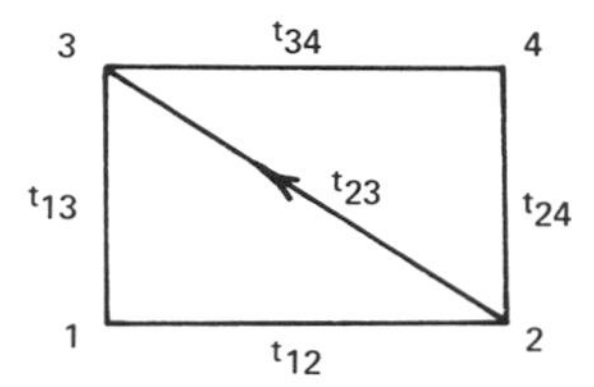

Fig. 7.23 A road map with a one-way street.

Let

$$f_1 = \text{least time to travel from node 1 to the destination}$$

or, in general,

$$f_i = \text{least time to travel from node } i \text{ to the destination}$$

For a very simple map, such as that shown in Fig. 7.23, it is straightforward to write down expressions for f_i as

$$\begin{aligned} f_1 &= \min[(t_{12} + t_{24}), (t_{12} + t_{23} + t_{34}), (t_{13} + t_{34})] \\ f_2 &= \min[t_{24}, (t_{21} + t_{13} + t_{34}), (t_{23} + t_{34})] \\ f_3 &= \min[t_{34}, (t_{31} + t_{12} + t_{24})] \\ f_4 &= 0 \end{aligned} \tag{7}$$

The above set of equations for f_i are obtained by inspection. By using a little more subtlety, it is possible to write a little more sophisticated set of equations. For instance, let P be the path of least time to go from node 1 to node 4. The starting point on this path is obviously 1. If the next node on P is 2, then the route thereafter must follow the quickest route from 2 to 4. (This is f_2 according to the definition of f_i.) This statement is based on pure common sense. Similarly, if the next point on P is node 3, then the path thereafter must follow the quickest route from 3 to 4. (This is f_3 according to the definition of f_i.) Thus, f_1 is either $(t_{12} + f_2)$ or $(t_{13} + f_3)$. That is,

$$f_1 = \min[(t_{12} + f_2), (t_{13} + f_3)] \tag{8a}$$

Similarly,

$$\begin{aligned} f_2 &= \min[t_{24}, (t_{23} + f_3), (t_{22} + f_2)] \\ f_3 &= \min[t_{34}, (t_{31} + f_1)] \\ f_4 &= 0 \end{aligned} \tag{8b}$$

Equation (8a), for instance, can be interpreted in terms of the "common-sense" principle: "No matter what junction we go to next, the continuation must be a path of quickest time from this new starting point to the fixed destination."

The next step is to solve the three simultaneous equations in Eq. (8) for the unknown quantities f_1, f_2, and f_3. An inspection of Eq. (8) reveals that the unknown quantities appear on both sides of the equations. Therefore, in general, the right side of the equations cannot be evaluated until the equations themselves are solved. Equations defined in this fashion are said to be defined *implicitly*. As implicit equations occur quite frequently in the mathematical analysis of systems, it is worthwhile to look for a general method of solving implicit equations.

A general and widely used technique of solving a set of implicitly defined simultaneous equations is the so-called *iterative method*. The plural, *iterative methods*, is more apt because these are really a class of methods with many variations. One particular variation of this method is *the method of successive approximations*, which is described below.

As both sides of Eq. (8) contain unknown quantities, elimination of some unknowns and evaluation of the remaining unknowns is not feasible. A possible alternative is to "shoot in the dark." That is, some arbitrarily selected values are first assigned to the unknown quantities, namely, f_1, f_2, and f_3. To emphasize the point that these values are only a guess, the values assigned are called *initial guess* and are denoted by $f_1^{(0)}$, $f_2^{(0)}$, and $f_3^{(0)}$. This initial guess is substituted in the right side of Eq. (2), and the right side is evaluated to yield a new set of values. To emphasize that these values are obtained after one substitution of an initial guess, these values are denoted by $f_1^{(1)}$, $f_2^{(1)}$, and $f_3^{(1)}$, where

$$\begin{aligned} f_1^{(1)} &= \min[(t_{12} + f_2^{(0)}), (t_{13} + f_3^{(0)})] \\ f_2^{(2)} &= \min[t_{24}, (t_{23} + f_3^{(0)}), (t_{21} + f_1^{(0)})] \\ f_3^{(1)} &= \min[t_{34}, (t_{31} + f_1^{(0)})] \end{aligned} \tag{9}$$

Had the initial guess been a correct guess of the solution (this rarely happens!), both sides of Eq. (9) would tally! That is, the three equations

$$f_1^{(1)} = f_1^{(0)}, \qquad f_2^{(1)} = f_2^{(0)}, \qquad f_3^{(1)} = f_3^{(0)} \tag{10}$$

would have been satisfied. If the initial guess happens to be an incorrect guess (as is usually the case), the errors defined by

$$\epsilon_1^{(0)} = f_1^{(1)} - f_1^{(0)}, \qquad \epsilon_2^{(0)} = f_2^{(1)} - f_2^{(0)}, \qquad \epsilon_3^{(0)} = f_3^{(1)} - f_3^{(0)} \tag{11}$$

would, in many cases, provide a clue as to how far the initial guess is away from a correct solution. If the errors $\epsilon_1^{(0)}$, $\epsilon_2^{(0)}$, and $\epsilon_3^{(0)}$ are too large, the initial guesses are replaced by $f_1^{(1)}$, $f_2^{(1)}$, and $f_3^{(1)}$. Using these values as a *first approximation*, the entire procedure is repeated and a second approximation, namely, $f_1^{(2)}$, $f_2^{(2)}$, and $f_3^{(2)}$ is obtained. This procedure is successively repeated until the errors

$$\begin{aligned} \epsilon_1^{(n)} &= f_1^{(n+1)} - f_1^{(n)} \\ \epsilon_2^{(n)} &= f_2^{(n+1)} - f_2^{(n)} \\ \epsilon_3^{(n)} &= f_3^{(n+1)} - f_3^{(n)} \end{aligned} \tag{12}$$

or, in vector notation,

$$\boldsymbol{\epsilon}^{(n)} = \mathbf{f}^{(n+1)}) - \mathbf{f}^{(n)} \tag{13}$$

fall below an acceptable level. The process of going from an nth approximation to $(n + 1)$th approximation is termed an *iteration*. In Eq. (13), ϵ is called the *error vector*, and ϵ_1, ϵ_2, ϵ_3 are called the *components of the error vector*.

An algorithm of this kind can terminate in three possibly different ways. If the absolute values $|\boldsymbol{\epsilon}^{(0)}|$, $|\boldsymbol{\epsilon}^{(1)}|, \ldots,$ $|\boldsymbol{\epsilon}^{(n)}|$ keep on decreasing, the iterative method is said to be *convergent* and the process can be terminated after $|\boldsymbol{\epsilon}^{(n)}|$ falls below an acceptable level. If $|\boldsymbol{\epsilon}^{(0)}|$ $|\boldsymbol{\epsilon}^{(1)}|, \ldots,$ $|\boldsymbol{\epsilon}^{(n)}|$ keep on increasing, the process is said to be *divergent*, and in such cases no useful solutions can be obtained. Alternatively, the values $|\boldsymbol{\epsilon}^{(0)}|$, $|\boldsymbol{\epsilon}^{(1)}|, \ldots,$ $|\boldsymbol{\epsilon}^{(n)}|$ may oscillate about some value instead of falling below a prescribed level. When either of the latter two things happen while one is solving the problem on a computer, either the results become meaningless or the computation goes into a never-ending loop unless one is careful to provide proper safeguards in the program.

Another problem of theoretical and computational interest arises in this context. This problem is concerned with techniques of *measuring* convergence: Ideally, in order to solve the problem exactly, it is necessary to make $\epsilon_1 = 0$, $\epsilon_2 = 0$, and $\epsilon_3 = 0$. Stated in another way, the problem is solved if set $E =$ set Z, where $Z = \{0, 0, 0\}$ and $E = \{\epsilon_1, \epsilon_2, \epsilon_3\}$, or equivalently if the distance $d(E, Z)$ between the sets E and Z is made zero. If it is not possible to make this distance zero, at least it is minimized. Thus, to test for convergence a yardstick to measure the distance between two sets is required. For instance, $d(E, Z)$ can be defined as

$$d(E, Z) = [(\epsilon_1)^2 + (\epsilon_2)^2 + (\epsilon_3)^2]^{1/2} \tag{14}$$

such a distance is called *root mean square distance* or in more familiar language as *root mean square error.*

Example 4. The above procedure is illustrated by solving the following example. Consider a road map shown in Fig. 7.23; once again, let $t_{12} = 2$, $t_{24} = 5$, $t_{34} = 1$, $t_{13} = 4$, and $t_{23} = 3$. With these values Eq. (5) becomes

$$\begin{aligned} f_1 &= \min[(2 + f_2), (4 + f_3)] \\ f_2 &= \min[5, (3 + f_3), (2 + f_1)] \\ f_3 &= \min[1, (4 + f_1)] \end{aligned}$$

It is more convenient to rewrite this equation as a recurrence relation using the superscripts as

$$\begin{aligned} f_1^{(n+1)} &= \min[2 + f_2^{(n)}, (4 + f_3^{(n)})] \\ f_2^{(n+1)} &= \min[5, (3 + f_3^{(n)}), (2 + f_1^{(n)})] \\ f_3^{(n+1)} &= \min[1, (4 + f_1^{(n)})] \end{aligned}$$

The rest of the procedure is illustrated in the following steps.

Step 1 Choose an initial guess by assigning arbitrary values to $f_1^{(0)}$, $f_2^{(0)}$, and $f_3^{(0)}$ as

$$f_1^{(0)} = 1, f_2^{(0)} = 2, f_3^{(0)} = 1$$

Step 2 Substitute these values on the right side of the above equation and get

$$\begin{aligned} f_1^{(1)} &= \min[(2 + 2), (4 + 1)] = 4 \\ f_2^{(1)} &= \min[5, (3 + 1), (2 + 1)] = 3 \\ f_3^{(1)} &= \min[1, (4 + 1)] = 1 \end{aligned}$$

Step 3 Calculate the components of error

$$\begin{aligned} \epsilon_1^{(0)} &= f_1^{(1)} - f_1^{(0)} = 4 - 1 = 3 \\ \epsilon_2^{(0)} &= f_2^{(1)} - f_2^{(0)} = 3 - 2 = 1 \\ \epsilon_3^{(0)} &= f_3^{(1)} - f_3^{(0)} = 1 - 1 = 0 \end{aligned}$$

and compute

$$d = [\epsilon_1^{(0)2} + \epsilon_2^{(0)2} + \epsilon_3^{(0)2}]^{1/2} = [(3)^2 + (1)^2 + 0]^{1/2} = 10^{1/2}$$

This is the end of the first iteration.

Step 4 As d is not small enough, $f_i^{(1)}$, $i = 1, 2, 3$, from Step 2 are treated as the first approximation to get a second improvement:

$$f_1^{(2)} = \min[(2 + 3), (4 + 1)] = 5$$
$$f_2^{(2)} = \min[5, (3 + 1), (2 + 4)] = 4$$
$$f_3^{(2)} = \min[1, (4 + 4)] = 1$$

Step 5 Calculate the components of error

$$\epsilon_1^{(1)} = f_1^{(2)} - f_1^{(1)} = 5 - 4 = 1$$
$$\epsilon_2^{(1)} = f_2^{(2)} - f_2^{(1)} = 4 - 3 = 1$$
$$\epsilon_3^{(1)} = f_3^{(2)} - f_3^{(1)} = 1 - 1 = 0$$

and compute

$$d = [(1)^2 + (1)^2 + 0]^{1/2} = 2^{1/2}$$

This is the end of the second iteration.

Step 6 As d is not small enough, $f_1^{(2)}$, $i = 1, 2, 3$, from Step 4 are treated as the second approximation to get a third approximation:

$$f_1^{(3)} = \min[(2 + 4), (4 + 1)] = 5$$
$$f_2^{(3)} = \min[5, (3 + 1), (2 + 5)] = 4$$
$$f_3^{(3)} = \min[1, (4 + 5)] = 1$$

Step 7 Calculate the components of error

$$\epsilon_1^{(2)} = f_1^{(3)} - f_1^{(2)} = 5 - 5 = 0$$
$$\epsilon_2^{(2)} = f_2^{(3)} - f_2^{(2)} = 4 - 4 = 0$$
$$\epsilon_3^{(2)} = f_3^{(3)} - f_3^{(2)} = 1 - 1 = 0$$

and $d = 0$.

The error is now zero. So the algorithm has converged and the values of f_1, f_2, and f_3 are taken as $f_1 = 5$, $f_2 = 4$, $f_3 = 1$. This can be verified to be correct by inspection of Fig. 7.23. ■

Programming the Routing Problem

Iterative algorithms, though laborious for hand calculation, are highly efficient when used in conjunction with digital computers. As methods and notations suitable for hand calculation are seldom useful for machine calculation, Eq. (8a) is rewritten in a format suitable for computer use as

$$f_1 = \min[(t_{12} + f_2), (t_{13} + f_3)] = \min_{j=2,3} [(t_{1j} + f_j)] \quad (15a)$$

The second equation above is an abbreviated form of the first and means that one has to set $j = 2$ and $j = 3$ in succession in the expression enclosed in the square brackets and choose the smaller of the two resulting values. Similarly,

$$f_2 = \min_{j=2,3,4} [(t_{2j} + f_j)] \tag{15b}$$

$$f_3 = \min_{j=1,4} [(t_{3j} + f_j)] \tag{15c}$$

This strategy resulted in some simplification. Still one has to write as many equations as there are nodes. Further compactness can be obtained by bunching all the above three equations as

$$f_i = \min_{j=?}[t_{ij} + f_j] \qquad \text{for all} \quad i \tag{16}$$

which can be rewritten once again in a format suitable for iterative operation as

$$f_i^{(k)} = \min_{j=?}[t_{ij} + f_j^{(k-1)}] \qquad \text{for all} \quad i \text{ and } k = 2, 3, \ldots \tag{17}$$

The problem is not finished yet. In the above equation the allowable values of j really correspond to nodes that are directly connected to node i. Hence, the values j assume depend on i. Since this dependence changes from map to map, it is not possible to indicate in Eq. (17) in a generic manner the values of j. This difficulty can be bypassed in several different ways.

A Method Using Fictitious Links

An obvious way to overcome the above difficulty is to assign an extremely large value, say, ∞, to the travel time t_{ij} whenever vertices i and j are not connected by an edge.

A Method Using the Adjacency Matrix

Let a_{ij} represent the adjacency matrix of the graph of the road map. In addition to reading in the time matrix $\mathbf{T} = [t_{ij}]$, another matrix $\mathbf{W} = [w_{ij}]$, defined below, will also be read into the computer:

$$w_{ij} = \begin{cases} t_{ij} & \text{if} \quad a_{ij} = 1 \\ 0 & \text{if} \quad a_{ij} = 0 \end{cases} \tag{18}$$

and in particular, $w_{ii} = 0$. Thus, the entry w_{ij} is always zero whenever t_{ij} is not defined.

A Method Using a Set of Accessible Points

Alternatively, one can determine the set of points $S(i)$ accessible from each vertex i by an edge. If, for each i, the elements of $S(i)$ are stored in the computer, then one can rewrite Eq. (17) as

$$f_i = \min_{j \in S(i)} [t_{ij} + f_j] \tag{19}$$

Of course, Eq. (15)–(19) can be further modified by recognizing that $f_N = 0$.

The iterative algorithm presented in this section is just one of many methods of finding the shortest route. Some of the so-called "labeling algorithms" are often more suitable if the networks are small and hand computation is contemplated.

EXERCISES

1. Consider the adjacency matrix $\mathbf{A}$ of a directed graph. Define $\mathbf{AA}^{\mathrm{T}} = [s_{ij}]$. Show, by using several illustrative examples, that s_{ij} gives the number of vertices to which both v_i and v_j are adjacent.

2. Show that $\mathbf{A}^{\mathrm{T}}\mathbf{A}$ also gives a result analogous to that of the previous exercise.

3. Let $G_{\mathrm{d}1}$ and $G_{\mathrm{d}2}$ be two directed graphs having the same number of vertices and the vertices are numbered in the same order as shown in Fig. E7.3. Let $\mathbf{A}_1$ and $\mathbf{A}_2$ be the respective adjacency matrices. Then show that the adjacency matrices of the intersection, union, and symmetric difference directed graphs are

$$\mathbf{A}(G_{\mathrm{d}1} \cap G_{\mathrm{d}2}) = \mathbf{A}_1 \times \mathbf{A}_2$$
$$\mathbf{A}(G_{\mathrm{d}1} \cup G_{\mathrm{d}2}) = (\mathbf{A}_1 + \mathbf{A}_2)_{\mathrm{b}}$$
$$\mathbf{A}(G_{\mathrm{d}1} \oplus G_{\mathrm{d}2}) = (\mathbf{A}_1 + \mathbf{A}_2)_{\mathrm{b}} - (\mathbf{A}_1 \times \mathbf{A}_2)$$

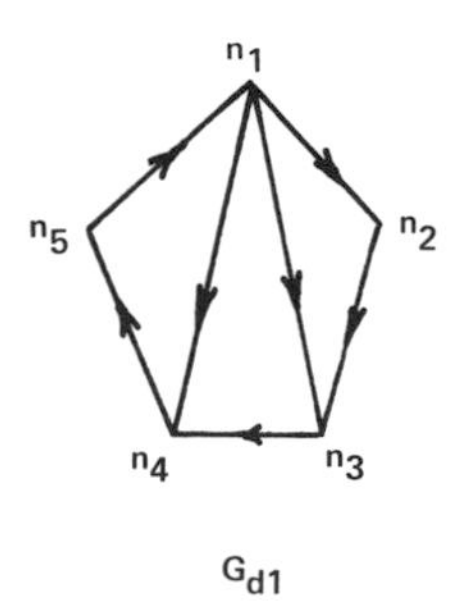

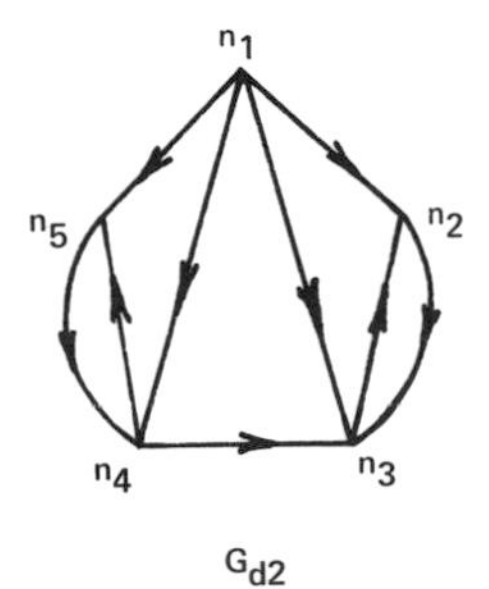

Fig. E7.3

where $\mathbf{A}_1 \times \mathbf{A}_2$ indicates the elementwise product and $(\mathbf{A}_1 + \mathbf{A}_2)_b$ indicates Boolean addition.

4. A directed graph is said to be *transitive* if whenever there is a path of length 2 from one vertex to another, then there is also an arc from the first vertex to the second vertex. A *transitive closure* $G_d{}^t$ of a directed graph G_d is the minimal transitive directed graph containing G_d and has the same set of vertices as G_d. What is the transitive closure $G_d{}^t$ of G_d shown in Fig. E7.4?

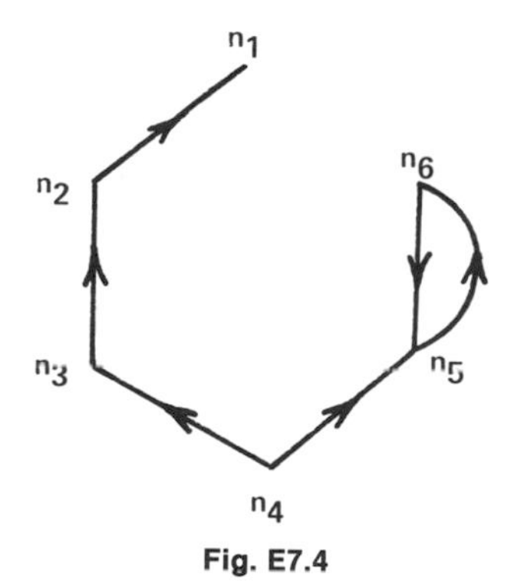

Fig. E7.4

5. Show that for any two distinct vertices v_i and v_j of G_d, the arc v_iv_j is in $G_d{}^t$ if and only if v_j is reachable from v_i in G_d.

6. Considering the statement in Exercise 5 as a theorem, show that the following corollaries hold:

(a) A directed graph is transitive if and only if it is its own transitive closure.

(b) For any directed graph $\mathbf{R}(G_d) = \mathbf{A}(G_d{}^t) + \mathbf{I}$, where $\mathbf{R}$ is the reachability matrix of G_d and $\mathbf{A}(G_d{}^t)$ is the adjacency matrix of $G_d{}^t$.

7. For the directed graph shown in Fig. E7.7, find $\mathbf{R}$, $\mathbf{R} \times \mathbf{R}^T$ and determine the strong components and the condensation C.

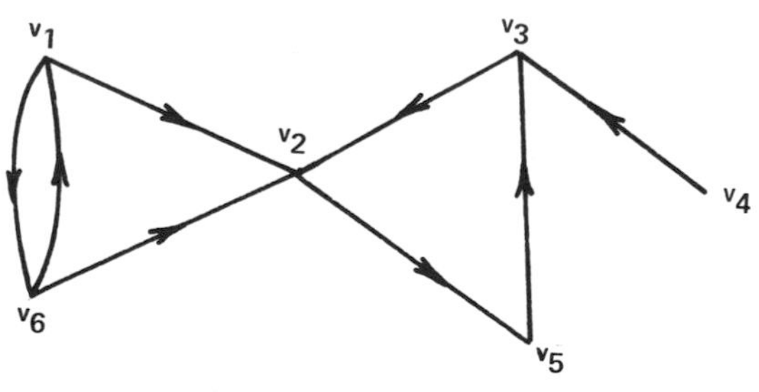

Fig. E7.7

8. Let $\mathbf{R}$ be the reachability matrix of G_d. Then show that $r_{ii}^{(2)}$, which is the (i, i)th element of $\mathbf{R}^2$, is the number of vertices in the strong component of G_d.

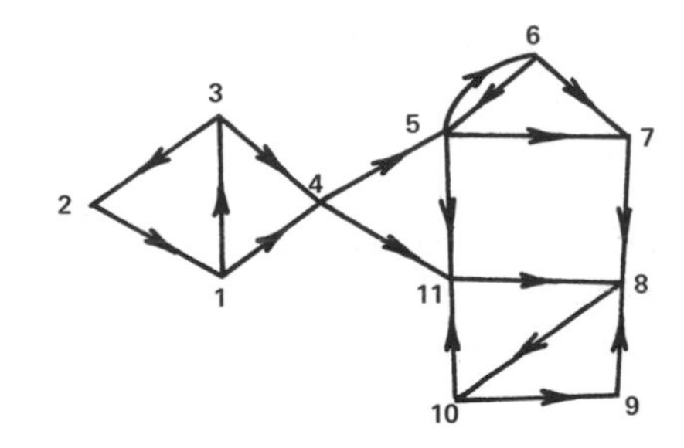

Fig. E7.9

9. Find the condensation C of the graph shown in Fig. E7.9.

10. For the directed graph shown in Fig. E7.10 develop a labeling algorithm so that it will relabel the nodes in such a way that the number on the initial node is smaller than the number on the terminal node in every arc. Your procedure should be general enough to handle any other graph. (Hint: if all the nodes are numbered arbitrarily first, then the algorithm relabels the nodes in a topological order by specifying that no arcs be allowed to start from a node until all branches merging into that node have been accounted for in your sorting procedure.)

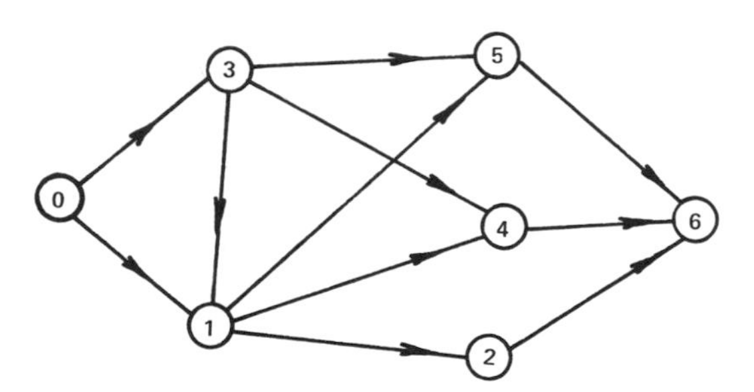

Fig. E7.10

* **11.** Given

$$\mathbf{A} = \begin{bmatrix} 0 & 0 & 0 & 2 & 1 & 0 & 0 & 0 & 0 & 3 \\ 5 & 0 & 1 & 3 & 0 & 0 & 0 & 0 & 0 & 0 \\ 0 & 0 & 0 & -1 & 0 & 0 & -1 & 1 & 0 & 0 \\ 1 & 0 & 0 & 1 & 1 & 0 & -2 & 0 & 0 & 0 \\ 1 & 0 & 0 & 1 & -1 & 4 & 1 & 0 & 0 & 0 \\ 0 & 0 & 0 & 0 & 0 & 0 & -1 & 0 & 0 & 0 \\ 0 & 0 & 0 & 0 & 0 & 0 & 1 & 0 & 0 & 1 \\ 0 & 0 & 0 & -1 & 0 & 0 & 0 & 0 & 1 & 1 \\ 0 & 1 & 0 & 0 & 0 & 0 & 0 & 0 & 0 & 2 \\ 0 & 0 & 0 & 0 & 0 & 2 & 0 & 0 & 0 & 0 \end{bmatrix}$$

(a) Find the reachability matrix $\mathbf{R}$.

(b) Using $\mathbf{R}$, identify the strong components of the directed graph of $\mathbf{A}$.

(c) Partition **A** according to the rule specified by the strong components. Call the partitioned submatrices by $\mathbf{M}_i$, $i = 1, 2, \ldots$.

(d) Find the eigenvalues of $\mathbf{M}_i$, $i = 1, 2, \ldots$. Let the eigenvalues of $\mathbf{M}_i$ be $\lambda_{i1}, \lambda_{i2}, \ldots$.

(e) Verify that the collection of all the eigenvalues for all admissible i constitute a set of eigenvalues of **A**.

(f) Is **A** reducible or irreducible?

12. For the graph shown in Fig. E7.12 find the cost-geodesic from v_5 to v_4 using two different methods.

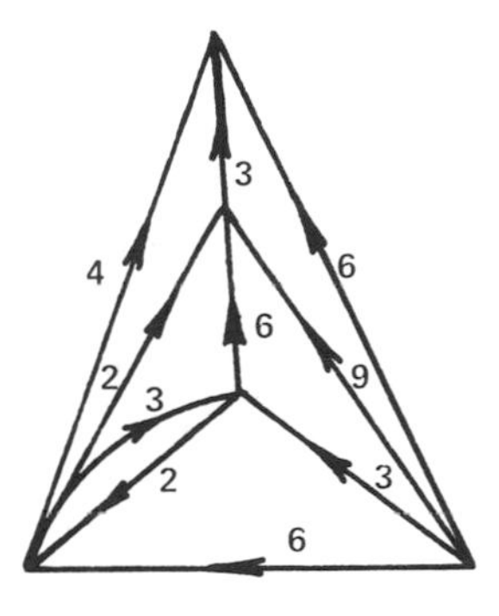

Fig. E7.12

13. Write a general computer program to implement the iterative algorithm of Section 5 and demonstrate the program by actually finding the shortest route from every possible starting point to every possible destination in Fig. 7.22.

14. Consider the problem of finding the minimum time route from vertex v_1 to vertex v_7 in Fig. 7.22. The travel times are same as those given in Eq. (5.1). Now, assume that the traveler incurs a loss of 2 min every time he makes a left turn. Is the minimum time route the same as before? Make suitable modifications in the algorithm of the previous exercise.

15. Consider the problem of going from v_1 to v_7 once again. Suppose that the time required to traverse a road (i.e., an edge in the linear graph) is a function of the time of the day. Formulate the routing problem by introducing minimum functions that are functions of time. Discuss how the successive approximation method can be applied here.

SUGGESTIONS FOR FURTHER READING

1. Directed graphs are widely used in a variety of application areas. Two of the more important references in this area are

   F. Harary, R. Z. Norman, and D. Cartwright, *Structural Models: An Introduction to Directed Graphs*. Wiley, New York, 1965.

R. G. Busacker and T. L. Saaty, *Finite Graphs and Networks: An Introduction with Applications*. McGraw-Hill, New York, 1965.

2. For an in-depth discussion on the patterns of group structure, consult

H. C. White, *An Anatomy of Kinship*. Prentice-Hall, Englewood Cliffs, New Jersey, 1963.

R. Stone, *Mathematics in the Social Sciences and Other Essays*. M.I.T. Press, Cambridge, Massachusetts, 1966.

3. For a discussion of graph theoretic computational algorithms at an elementary level with a wide variety of applications, see

R. Bellman and K. L. Cooke, *Algorithms, Graphs, and Computers*. Academic Press, New York, 1970.

4. For a comparison of a number of routing problems, see

S. Dreyfus, *An Appraisal of Some Shortest Path Algorithms*, The Rand Corporation, RM-5433-PR, October 1967, Santa Monica, California.

5. If the shortest path is not available for some reason, one can use the next best or, in general, the kth best route through a network. For a discussion of such problems, see

M. Pollack, Solutions of the kth best route through a network—a review, *J. Math. Anal. Appl.* **3**, 547–559 (1961).

6. For a number of ideas concerning decomposition of large scale systems, refer to

D. M. Himmelblau (ed.), *Decomposition of Large Scale Problems*. American Elsevier, New York, 1973.

8

Analysis of Competitive Situations

INTRODUCTION

Leonhard Euler (1707–1783) once wrote a philosophical comment: "Since the fabric of the world is the most perfect and established by the wisest creator, nothing happens in this world in which some reason of maximum or minimum would not come to light." Such sweeping generalizations were found to be true in many branches of natural sciences. Heron of Alexandria asserted that light travels between two points by the shortest path. Fermat's more general principle states that light travels between two points in the least time rather than least distance. (Implicit in this statement is the suggestion that distance can be measured in terms of time.) Hamilton's single minimum principle brought together light and mechanics. In thermodynamics, a system in equilibrium has minimum "free energy." Adam Smith attempted this idea by defining an "economic man" who always acts to maximize his profits. Pontryagin's minimum principle and Bellman's principle of optimality, both developed in the context of modern control theory, are other examples of Euler's belief.

One large and increasingly important class of problems in the domain of modern mathematics and system theory involves selecting a combination of variables that maximizes or minimizes some predefined variable or function of the problem. This problem, called the *optimization problem*, becomes formidable when the number of combinations in which the variables can be combined are large. With the development of modern computers, the computational difficulties are no longer insurmountable.

Many of these problems involve finding an optimum combination of a large, but finite, number of elements. For instance, one might seek the least expensive communication network linking a group of cities, or the least expensive means of distributing goods from warehouses to retail outlets. Modern optimization techniques have not only provided a means of solving these problems but also profoundly influenced the way of thinking itself: the art of *network planning* has been sharpened into the more exact Program Evaluation and Review Technique (PERT); critical-path scheduling technique now allows complex projects to be done on time, for the minimum possible cost. The most important of the optimization techniques that are widely used to tackle problems of the above nature is the *linear programming* (or LP) method. The LP method can be profitably used in conjunction with graph theoretic methods in order to derive efficient computational algorithms.

There are many practical situations in which one is forced to make decisions without complete knowledge of the consequences of possible alternative actions. For instance, a nation has to decide on a defense budget without a complete knowledge of the enemy's intentions. More paradoxically, a nation may have to plan for progress without a complete knowledge about the deleterious effects of progress. Similarly a scientist may have to decide on an experiment without knowing the actual experimental conditions. Many problems of this kind can be handled with the use of probability theory. Indeed, many problems of physics and genetics are of this kind, and they are being handled using probability. But what about innumerable kinds of situations in which probabilities cannot be computed? What was the probability of Columbus finding land before the crew ran out of food and water? An answer to this question can be found in what is now widely called *rational decision making*. The theory behind this rational decision making can be called *game theory*. This chapter, however, is concerned only with a description of linear programming; not with game theory.

1 ELEMENTARY IDEAS ABOUT OPTIMIZATION

A function can be defined, in very general terms, as a rule of correspondence between the elements of two sets. If for every $x \in A$, there exists a $y \in B$, then y is said to be a function of x and this relation is written as $y = f(x)$. In a majority of cases, the set A to which x belongs is a segment of the real line. Then the function f is called a function of a real variable. A function $y = f(x)$ may be termed *continuous* at a point x if

$$y(x) = \lim_{h \to 0} y(x + h)$$

This is not a rigorous definition. For intuitive guidance, the above equation implies that the function may be represented by an unbroken curve in the region of interest and the function is not attaining arbitrarily large values by going through steep gradients. A function is said to be *discrete* if the rule of correspondence is valid only at discrete values of the independent variable. Discrete functions are quite common in many large scale systems problems.

If the value of the independent variable x is changed in one direction (say, increased from an initial value x_0 to a final value x_1) and if $f(x)$ increases (or decreases) up to a maximum (or minimum) and then decreases (or increases), then $f(x)$ is said to be a *unimodal* function in the interval (x_0, x_1). Examples of some unimodal functions are shown in Fig. 8.1. *Concave functions* and *convex functions* are special cases of unimodal functions. Intuitively, a surface (or line) is concave (or, concave from below) if a line segment joining any two points on the surface (or line) lies on or below the surface (or line). A similar definition holds for convex functions. The general problem of finding the maximum or minimum value of a function of one or many varables, when the function may be linear or nonlinear, discrete valued or continuous, constrained or unconstrained, is a difficult problem to solve. Only some of the more elementary ideas will be described here. Most of the optimization methods can be broadly divided into two categories: analytical methods and search techniques. The analytical methods are indirect in the sense they ultimately involve solving an equation rather than searching for an optimum.

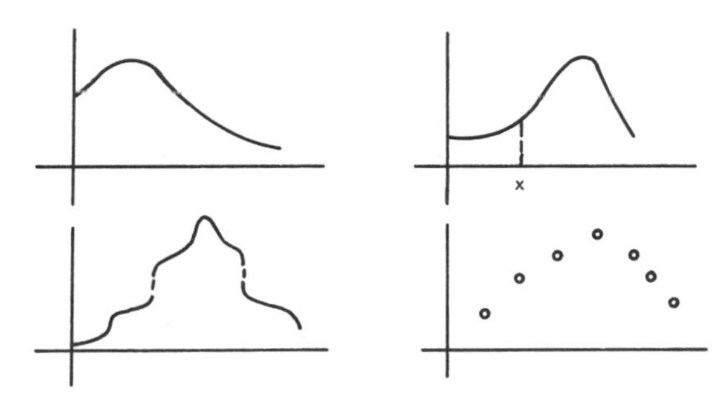

Fig. 8.1 Typical examples of unimodal functions.

Exhaustive Enumeration

This is an example of a search technique and this method is suitable, in general, for locating the maximum or minimum of a discrete valued function. If one is interested in selecting the best from a finite number of policies, then it is conceivable that one go about evaluating the merits and demerits of each and every policy. Theoretically simple, but practically tedious, this method is suitable when the number of alternative policies to be inspected is small compared to the speed of the computational facilities available.

Search Methods

There are better ways of finding out the maxima and minima of functions if something about the function is known. In many practical problems this is really not too much to ask. For instance, in economics, the Law of Diminishing Returns gives a profit function which is concave down with respect to the amount invested. In thermodynamics the *free energy* function is known to be concave up. Convexity or concavity of a function implies unimodality. Therefore, it is not unreasonable to assume a unimodal surface for the "profit" or criterion function, at least in the range of interest.

If a function, whose maximum or minimum value is to be determined, is known to be unimodal, the exhaustive enumeration method can be profitably replaced by a systematic search procedure. For instance, if a function is unimodal in an interval (a, b), then by evaluating the function at *two* points, say, at x_1 and x_2, with $x_1 < x_2$, it is possible to say that the maximum value of the function lies in either (a, x_1) or (x_1, x_2) or (x_2, b)—all being smaller than the original interval. Thus, the domain of search is considerably reduced.

Consider a unimodal function which is defined on an interval (a, b). The values y_1 and y_2 of the function are measured at two arbitrarily chosen points x_1 and x_2 both lying in the interval (a, b). In Fig. 8.2 three possible ways of choosing the locations of x_1 and x_2 are shown. Inspection of the

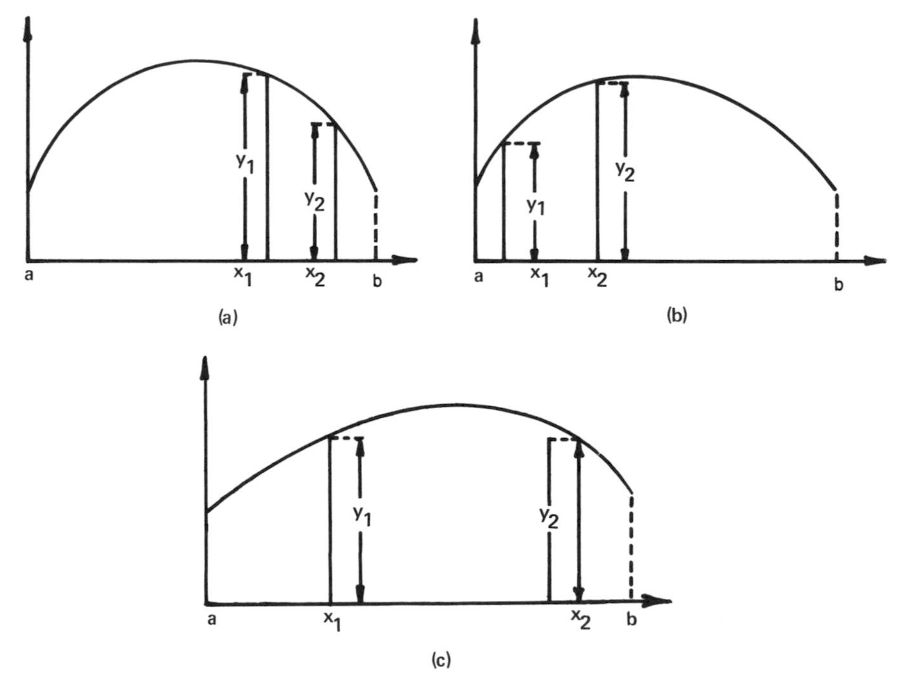

Fig. 8.2 Illustration of three search strategies.

figure reveals that for any choice of x_1 and x_2 there are only three possible outcomes in the measured values of y_1 and y_2:

$$y_1 > y_2, \qquad y_1 < y_2, \qquad \text{or} \qquad y_1 = y_2$$

If $y_1 > y_2$, as in Fig. 8.2a, the maximum cannot possibly lie to the right of x_2 without contradicting the unimodality assumption. If $y_1 < y_2$, the optimum cannot lie to the left of x_1, and if $y_1 = y_2$, the optimum must lie in (x_1, x_2). Thus, it is possible to reduce the search interval substantially by conducting a pair of simple experiments. Once the interval of interest is identified, the above procedure can be repeated a number of times to pinpoint the location of the maximum to any desired degree of accuracy.

Example 1. Determine the minimum of

$$y = (x - 9)^2$$

using a direct search technique.

The given function is unimodal, and so there is only one minimum. The first stage of the search assumes that this minimum occurs at an arbitrarily selected *base point* denoted by $x = x_0^{(1)} = 0$, where the superscript indicates the stage of the search. The value of y at this point is evaluated to be $y(x_0^{(1)}) = y_0^{(1)} = 81$. Now change the location of the base point by making a change in its value. For concreteness, let the new location of the base point be $x_1^{(1)} = -2$. The value of y at this point is $y(x_1^{(1)}) = y_1^{(1)} = 121$. Since $y(x_1^{(1)}) > y(x_0^{(1)})$ and the function is known to be unimodal, a minimum cannot possibly lie in the region described by $x < -2$. That is, the initial move from $x_0^{(1)}$ to $x_1^{(1)}$ has been taken in the wrong direction.

The location of the base point is now moved in the other direction with $x_2^{(1)} = 2$. Now $y(x_2^{(1)}) = 49$. As the value of y is decreasing, the search is continued in this direction and the results summarized in the table:

k:	0	1	2	3	4	5	6
$x_k^{(1)}$:	0	-2	2	4	6	8	10
$y(x_k^{(1)})$:	81	121	49	25	9	1	1

Inspection of this table reveals that the value of y keeps on decreasing until $x_k = 8$ and tends to reverse itself and increase somewhere in the interval $8 < x_k < 10$. This concludes the first stage of the search.

To pinpoint the location of the minimum, the search is now confined to the interval (8, 10) and the step size is reduced to, say, 0.3. The results of the second stage calculation are shown in the table:

k:	0	1	2	3	4
$x_k^{(2)}$:	8.0	8.3	8.6	8.9	9.2
$y(x_k^{(2)})$:	1.00	0.49	0.16	0.01	0.04

The optimum now lies in the range $8.9 < x < 9.2$. A further reduction in the step size could be used to determine the location of the minimum with better accuracy. ■

A systems analyst is generally not content with stopping at this point. To make the computational process efficient one should address himself to the question: Is there any "good" strategy in selecting the locations of x_1 and x_2 in order to make the search efficient? Many sophisticated search techniques such as the Fibonacci search, the golden section search, etc. are available for this purpose. A detailed discussion of these methods is beyond the scope of this book.

The Classical Calculus Method

The systematical search procedure described in the preceding paragraphs can be further modified and improved by imposing additional restrictions on the profit or criterion function. If a function is known to be continuous in (a, b), it will have a maximum or a minimum value either within the interval (a, b) or at its boundaries. This is guaranteed by a well-known theorem of Weierstrass, which states that every function which is continuous in a closed domain possesses a largest and smallest value either in the interior or on the boundary of that domain.

Thus, if the function, whose maximum or minimum is required to be located, has a nice analytical structure then some of the drudgery of the search techniques can be avoided by deriving the necessary conditions under which an extremum exists. This approach allows the location of those extrema that lie in the *interior* of a region of search. Such locations must then, in the general problem, be compared with the results obtained from other feasible optima locations, namely, discontinuities and boundaries.

For the purpose of illustration a function y of a single variable x is considered. Let $y(x)$ be defined in the interval $a \leq x \leq b$. A Taylor series expansion of y about any point α in the interval would look like

$$y(x) = y(\alpha) + (x - \alpha)y'(\alpha) + \cdots + [(x - \alpha)^{n-1}/(n - 1)!]\,y^{(n-1)}(\alpha) + \cdots \quad (1)$$

If $y(x)$ has a local minimum at $x = \alpha$, then $y'(\alpha)$ necessarily vanishes. This statement can be proved from the above equation as follows. Suppose that $y(x)$ has a local minimum at $x = \alpha$ but suppose $y'(\alpha) \neq 0$. If $h = (x - \alpha)$ is very small, then $hy'(\alpha)$ dominates the right side of Eq. (1) and

$$y(\alpha + h) - y(\alpha) = hy'(\alpha) \quad (2)$$

If $y'(\alpha) \neq 0$, then it is certainly possible to choose the sign of h so that $hy'(\alpha) > 0$. This implies $y(\alpha + h) < y(\alpha)$, and therefore $y(\alpha)$ cannot be a minimum. Therefore, if $y(\alpha)$ is to be a minimum, it is necessary that $y'(\alpha)$ vanish. The point $x = \alpha$ at which $y'(\alpha)$ vanishes is called a *stationary point*. By a similar argument it can be proved that a maximum or minimum is also possible only at a stationary point. Note that the vanishing of $y'(\alpha)$ is a necessary condition; it is not a sufficient condition.

Consider once again a stationary point $x = \alpha$. At this point if $y''(\alpha) = 0$, Eq. (1) can be written as

$$y(\alpha + h) - y(\alpha) = \tfrac{1}{2}h^2 y''(\alpha) + \text{higher-order terms} \tag{3}$$

If $y''(\alpha) > 0$, then the right side of Eq. (3) is definitely positive, implying that $y(\alpha + h) > y(\alpha)$. Thus, the requirement $y''(\alpha) > 0$, is a *sufficient* condition for a minimum to exist. A similar analysis reveals that $y''(\alpha) < 0$ is a sufficient condition for a maximum to exist. Table 8.1 shows a summary of various tests one can perform on a stationary point in order to understand its true nature.

TABLE 8.1

Summary of Tests at Stationary Values of a Curve $y(x)$[a]

$y'(a)$	$y''(a)$	$y'''(a)$	$y''''(a)$	*Nature of point*, $x = a$
0	−	Exists	Anything	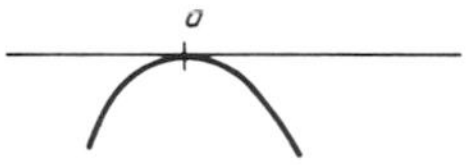
0	+	Exists	Anything	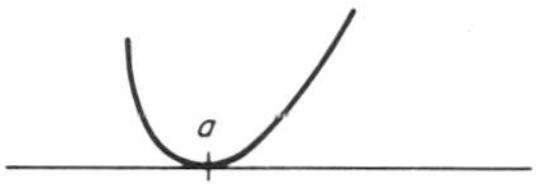
0	0	+	Exists	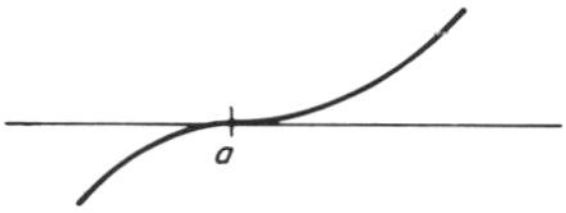
0	0	−	Exists	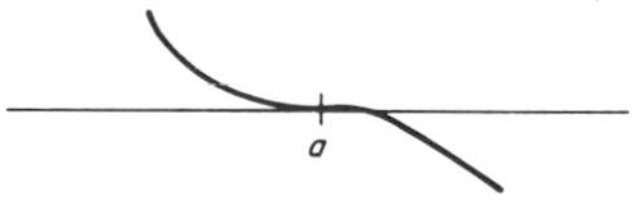
0	0	0	+	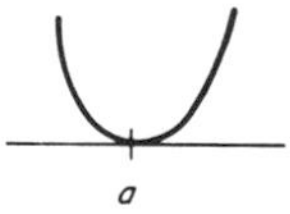 $y^v(a)$ exists
0	0	0	−	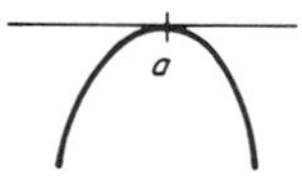 $y^v(a)$ exists
0	0	0	0	Examine $y^v(a)$

[a] Reproduced with permission from G. S. G. Beveridge and R. S. Schechter, *Optimization: Theory and Practice*; Copyright © 1970, McGraw-Hill, New York.

Example 2. Find the maximum value of

$$y = f(x) = 1 + 8x + 2x^2 - \tfrac{10}{3}\partial x^3 - \tfrac{1}{4}x^4 + \tfrac{4}{5}x^5 - \tfrac{1}{6}x^6$$

The first derivative

$$y' = 8 + 4x - 10x^2 - x^3 + 4x^4 - x^5$$

vanishes at $x = -1$ and $x = 2$. Thus, there are two stationary points. To determine which of these is a maximum, the second derivative

$$y'' = 4 - 20x - 3x^2 + 16x^3 - 5x^4$$

is evaluated at $x = -1$ and 2 and found to be zero. The test is therefore inconclusive, and the third derivative is evaluated

$$y''' = -2 - 6x + 48x^2 - 20x^3$$

This odd derivative does not vanish at $x = -1$. Therefore, $x = -1$ is neither the location of a maximum nor a minimum. Indeed, this point is called a *saddle point* or *point of inflection.* Since y''' vanishes at $x = 2$, the fourth derivative is examined:

$$y'''' = -6 + 96x - 60x^2$$

At $x = 2$, $y'''' = -54$, a negative quantity. Thus, a maximum exists at $x = 2$. Thus, an even derivative being negative at a stationary point, if all lower order derivatives vanish, is a sufficient condition for a maximum. ■

Maximum, Minimum, Optimum

In most of the preceding discussion, not much care was taken to distinguish between the twin concepts "maximum" and "minimum." Mathematically, the minimum value attained by a function $f(x)$ is the same as the maximum value attained by the function $-f(x)$. Once this fact is kept in mind, it is possible to talk about maximization or minimization interchangeably, without a loss of generality.

In the ensuing discussion, two other notational abbreviations, namely, Min z and min z, are used. The symbol "Min z" stands for "minimize z," while "min z" stands for "minimum value of z." Similar abbreviations "Max z" and "max z" are used with respect to maximization.

2 OPTIMIZATION WITH EQUALITY CONSTRAINTS

In a number of practical situations, optimizing the value of an objective function has to be carried out over a restricted domain. The problem of enclosing as large an area as possible with a given length of a fence is an example of this category. Here the length of the fence, which is predetermined, acts as a constraint, while the enclosed area serves as an objective function.

Direct Substitution Method

Consider a problem with, say, three variables and one constraint which is posed as

$$\text{Min } z = f(x_1, x_2, x_3) \tag{1}$$

subject to

$$g(x_1, x_2, x_3) = 0 \tag{2}$$

Here Eq. (1) is the objective function and Eq. (2) is the constraint. A straightforward method of solving this problem would be to solve Eq. (2) for, say, x_2 and x_3 in terms of x_1 and to substitute these values of x_2 and x_3 in Eq. (1). This results in an equation for z expressed in terms of only one independent variable, namely, x_1. The calculus method presented earlier can now be applied to find the minimum of f.

Example 3. The erection of a fence is required along the perimeter of a triangular area. For several regulatory reasons, one side of the triangular fence is required to be a units long. Due to material nonavailability, the total length of the fence is also predetermined. It is necessary to find the lengths of the other two sides so that the enclosed area is maximized. Denoting the sides of the triangle by a, b, c and the perimeter by $2s$, the enclosed area A is given by the well-known formula

$$A = \sqrt{s(s-a)(s-b)(s-c)}$$

where $2s = a + b + c$. Therefore, the problem can be stated as

$$\text{Max } A = \sqrt{s(s-a)(s-b)(s-c)}$$

subject to $2s = a + b + c$, $b > 0$, $c > 0$. This problem is very similar to that in Eqs. (1) and (2). Because s and a are constants, the area A is only a function of the variables b and c. Solving for one of these variables, say, for c,

$$c = 2s - a - b$$

and substituting this in the objective function

$$A = \sqrt{s(s-a)(s-b)(s-2s+a+b)}$$

which is only a function of the variable b. Setting

$$dA/db = 0,$$

$$s(s-a)(2s-a-2b) = 0$$

Since $s \neq 0$, $(s - a) \neq 0$, it is necessary that

$$2s - a - 2b = 0$$

or

$$b = s - a/2 \qquad \text{and} \qquad c = s - a/2$$

Consequently, the maximum value of A is $(a/2)\sqrt{s(s - a)}$. ■

The above procedure, called the *direct substitution* method, is not always convenient. If there are a number of constraint equations, the procedure to solve them becomes unwieldy. The direct substitution method sometimes creates a need for the imposition of additional constraints—a self-defeating action because one of the reasons for trying the substitution was to get rid of the constraints.

The Lagrange Multiplier Method

The method of direct substitution is characterized by two steps. In the first step, some of the variables are eliminated (i.e., expressed in terms of the other variables), and in the second step, the resulting objective function is differentiated. However, this method makes an unnatural distinction between those variables that were eliminated and those that were kept in the objective function. From a practical viewpoint, the question of which variables to eliminate has to be resolved by the analyst. The Lagrange multiplier method attempts to eliminate such distinctions by introducing, what appears to be artificial, undetermined constants into the problem.

Without delving in the details, it is sufficient at this point to state the gist of the Lagrange multiplier method. It says that the constrained problem

$$\text{Min } z = f(x_1, x_2, x_3) \tag{3}$$

subject to $g(x_1, x_2, x_3) = 0$ is equivalent to the unconstrained problem

$$\text{Min } \phi = f(x_1, x_2, x_3) + \lambda g(x_1, x_2, x_3) \tag{4}$$

where λ is, as yet, an undetermined quantity, called the *Lagrange multiplier*. Now ϕ is a function of four variables, namely, x_1, x_2, x_3, and λ and a minimum value of ϕ can be found using calculus method. That is, the necessary conditions for a minimum of ϕ are

$$\begin{aligned} &\frac{\partial \phi}{\partial x_1} = \frac{\partial f}{\partial x_1} + \lambda \frac{\partial g}{\partial x_1} = 0, \qquad &\frac{\partial \phi}{\partial x_2} = \frac{\partial f}{\partial x_2} + \lambda \frac{\partial g}{\partial x_2} = 0 \\ &\frac{\partial \phi}{\partial x_3} = \frac{\partial f}{\partial x_3} + \lambda \frac{\partial g}{\partial x_3} = 0, \qquad &\frac{\partial \phi}{\partial \lambda} = g(x_1, x_2, x_3) = 0 \end{aligned} \tag{5}$$

Notice that the last equation is the constraint equation itself.

Example 4. Maximize $A^2 = s(s-a)(s-b)(s-c)$ subject to $a+b+c-2s=0$, where s and a are constants. Here,

$$z = f(b, c) = s(s-a)(s-b)(s-c)$$

and

$$g(b, c) = a + b + c - 2s$$

Therefore,

$$\phi(b, c, \lambda) = s(s-a)(s-b)(s-c) + \lambda(a+b+c-2s)$$

$$\frac{\partial \phi}{\partial b} = -s(s-a)(s-c) + \lambda = 0$$

$$\frac{\partial \phi}{\partial c} = -s(s-a)(s-b) + \lambda = 0$$

$$\frac{\partial \phi}{\partial \lambda} = a + b + c - 2s = 0$$

from which $b = c = (s-a)/2$. ■

The Lagrange multiplier method can also become tedious. If there are a large number of constraints, one has to introduce n Lagrange multipliers, $\lambda_1, \lambda_2, \ldots, \lambda_n$ and one faces the problem of solving the resulting system of equations.

3 PROBLEMS WITH INEQUALITY CONSTRAINTS

There is a basic and profound difference between mathematical abstraction and physical reality. An equality sign in the world of mathematics represents perfect equality. It is a precise mathematical relation with no ambiguity whatsoever. However, an equality is not a physically (or socially) realizable entity. "Liberty and justice for all" is a good slogan but a very difficult proposition. It is a truism that A's liberty has to be curtailed somewhere to keep order in a society. Even in the modern technological world, a constraint such as "room temperature $\leq 68°$F" is easier to realize than one that demands "room temperature $= 68°$F." Thus, inequalities are an inherent part of reality. This is the basic reason for the present interest in optimization problems with inequality constraints. An attempt is made in this section to introduce a number of problems that, when formulated in mathematical language, result in inequalities. This section may be omitted at first reading.

Imagine n cities to be supplied with grain from m granaries located in different parts of a country. The amounts of grain needed at each city are known, as are the supplies at each granary, and the total supply of grain in the granaries is enough to meet the demand. How much should be sent from each granary to each city in order that the total shipping cost be as small as possible? It is assumed that the cost of shipping along each route is proportional to the amount sent but the proportionality constant (cost per ton) differs from one route to another.

Let a_i denote the amount of grain available at the ith granary, b_i the amount required at the jth city, and c_{ij} the unit cost per ton of shipments from i to j. Then the problem is to choose amounts x_{ij}, denoting the amount sent from i to j, which will minimize the total cost, subject to the constraints that the amount of grain shipped from a granary cannot exceed what is in it and that the amount of grain sent to a city must be at least equal to what is needed in the city. Expressed mathematically,

$$\sum_j x_{ij} \leqq a_i \qquad \text{for each granary} \quad i \tag{1}$$

$$\sum_i x_{ij} \geqq b_j \qquad \text{for each city} \quad j \tag{2}$$

where all x_{ij} are nonnegative. The total cost $\sum_i \sum_j c_{ij}x_{ij}$ is to be minimized. Thus, the problem leads to a system of simultaneous linear inequalities.

The problem is further characterized by the fact that one is not only looking for values of x_{ij} that satisfy the inequalities, but also for the solution that minimizes the total cost. Problems of this kind, known as *transportation problems*, are a special case of what is known as the *linear programming* (or LP) problem: the minimization of a linear combination of variables, when the variables must satisfy linear inequalities.

Looking at a special case of the transportation problem leads to some new discoveries. Consider the case in which the a_i and b_j are all 1. The constraints then become

$$\sum_j x_{ij} \leqq 1 \quad \text{and} \qquad \sum_i x_{ij} \geqq 1 \tag{3}$$

An example of such a situation is where i refers to individuals, j refers to possible jobs to which these individuals may be assigned, and c_{ij} is a score that measures the skill of the ith person when doing the jth job. A reasonable way to allocate people to jobs is to favor the allocation where the sum of the scores for the man–job assignments actually made is greatest. This suggests using the transportation-problem format, where we attach the following meaning to the variables x_{ij}: if $x_{ij} = 1$, individual i is assigned to job j; if $x_{ij} = 0$, i is not assigned to j.

Of course, we also want to maximize, rather than minimize, $\sum_i \sum_j c_{ij}x_{ij}$ but this is a trivial difference.

It is natural to ask what interpretation is given to a solution in which the x_{ij} are fractions. The answer is that this will not happen. Although in most linear programming problems the answers will not come out in whole numbers, the special system of the transportation problem has the remarkable property that the answers are whole numbers when the a_i and b_j are whole numbers. Since the x_{ij} are restricted to nonnegative values, it is easy to see that when the a_i and b_j are all 1, this will make each x_{ij} either 0 or 1.

The LP problem occurs in so many different contexts that it is worthwhile, at this stage, to look into the formulation of a simple problem from a verbal description.

Example 5. *A production scheduling problem.* A small job shop has one automatic machine that can be operated a maximum of 7.5 hr per day. It is planned to manufacture both pins and bolts on this machine. A maximum of 1000 bolts per day can be produced due to raw material shortage. Other relevant production and cost data are given in the table below:

Part	Prod. rate (pieces/hr)	Material cost (¢/piece)	Labor & overhead (¢/hr)	Selling price (¢/piece)
Pin	300	2.5	240	5
Bolt	200	3.0	240	7

It is desired to find the quantity of each part to be produced per day in order to maximize profit.

Let

$$x_1 = \text{number of pins produced per day}$$
$$x_2 = \text{number of bolts produced per day}$$

The fact that only 7.5 hr of machine time is available can be translated into the inequality

$$x_1/300 + x_2/200 \leq 7.5 \qquad \text{or} \qquad 2x_1 + 3x_2 \leq 4500$$

The material restriction on the bolts can be written as

$$x_2 \leq 1000$$

As the objective is to maximize profit, the profit z in cents per day can be written as

$$\begin{aligned} z &= (5 - 2.5 - 240/300)x_1 + (7 - 3 - 240/200)x_2 \\ &= 1.7x_1 + 2.8x_2 \end{aligned}$$

All the above information can be collected and compactly written as

$$\text{Max } z = 1.7x_1 + 2.8x_2$$

Subject to the restrictions

$$2x_1 + 3x_2 \leq 4500$$
$$x_2 \leq 1000$$
$$x_1 \geq 0, \qquad x_2 \geq 0$$

The last two inequalities merely emphasize the fact that the production of bolts and pins cannot be negative. ■

Project Scheduling and the Critical Path Method

A problem which is akin to the transportation and quickest path problems, and of great importance in large scale systems is the problem of the determination of a *critical path.* The critical path method (CPM), program evaluation and review technique (PERT), task network scheduling (TANES), and a host of other methodologies with acronyms ranging from ABLE to WRAP are all tools to assist the management in the planning and control of relatively complex projects. Incidentally, there are about 100 variations of PERT or CPM. As is common with many other disciplines, there exists an unclear dichotomy between PERT and CPM. While the goals of both CPM and PERT are essentially the same, the terminology is different. Therefore, it is customary to discuss both PERT and CPM together.

Since planning and proper execution of the operations of large systems is a complex affair, a network or directed graph serves an important visual function. In its simplest terms, a network represents an ordered relationship of all activities necessary to the completion of a project. Therefore, in order to be able to use the network concept, it is necessary to divide the project into several time-consuming *tasks*, *jobs*, or *activities.* It is useful to perform this subdivision so that each activity is independent; that is, it does not rely on other activities of the project for its own progress. In other words, an activity does not depend on other activities for the termination of its task; however, it can depend upon some other activity for its initiation. Thus, an activity is a basic element of the network method.

One diagram frequently used to exhibit the interrelationships in a complicated project is illustrated for a sailboat launching in Fig. 8.3. The circles indicate the jobs (or activities) that must be done; the arrows indicate precedence relations, that is, which jobs must be done before others. From the figure, for example, the boat cannot be masted before it is launched and the necessary rigging attached to the mast. The time required to do each job is indicated by the numbers in the circles.

If we have a sequence of jobs, each one depending only on its immediate predecessor, like the sequence sand hull, paint hull, and launch, then the total time to do the jobs in this sequence is merely the sum of the individual

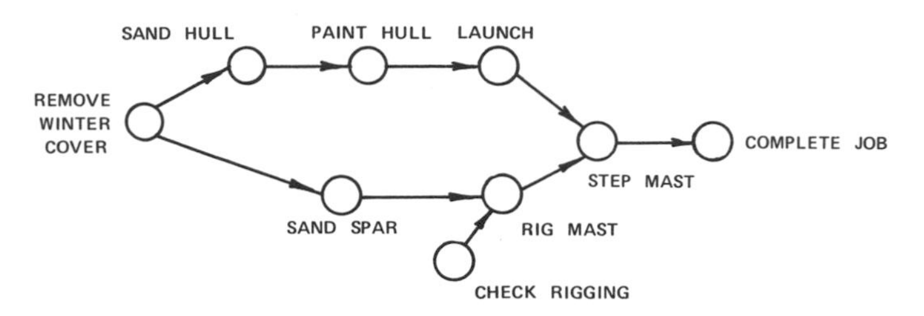

Fig. 8.3 Figure illustrating the precedence relations involved in the task of launching a boat.

job times. However, when there are several branches to the network, jobs may have to wait for each other. In any event, one can verify by enumeration, for example, that the minimum time required to complete the entire project is the largest total time encountered on any path from a starting point to the finish. The path along which this total is obtained is called the *critical path*. The calculation of the critical path in a given network is itself a simple example of a type of problem that goes by the name "dynamic programming."

So far, we have assumed that a fixed time is required to do each job. Actually, it is often the case that by spending more money, a job can be speeded up. Suppose the project is to be finished at a certain definite fixed time T. Which jobs should be speeded up, and how much, so that the project is done on time and the least possible amount of money is spent? Certainly, at first one would want to speed up the jobs on the critical path, but beyond that, the way to proceed is not as clear. Again, we will introduce linear inequalities.

Let x_i be the time at which the ith job is started and t_i the time it takes to execute. Then, if the ith job precedes the jth job, we must have

$$x_i + t_i \leq x_j \tag{4}$$

If we write down all such inequalities, then any values for the x's and t's satisfying the inequalities give us a possible schedule. In addition, if we denote the last job by N, then

$$x_N + t_N \leq T \tag{5}$$

with T as the finishing time referred to above.

Turning now to cost minimization, let us imagine that the cost–time relation for a particular job is given by

$$C(t_i) = c_i - d_i t_i \tag{6}$$

then we must minimize $\sum_i (c_i - d_i t_i)$ subject to the inequalities given above as well as the inequality

$$A_i \leq t_i \leq B_i \tag{7}$$

for all i. Since this is a linear expression, we again have a linear programming problem. Some of the special methods available for the assignment problem can be used to great effect here, making the calculation quite practical for projects involving more than a thousand jobs. The diagramming of complicated projects this way is now very common.

Solving Systems of Equations

Even though the LP formulation owes its origin to situations requiring analysis of competitive situations, the LP method found its way into a variety of places where no competitive situation is immediately apparent. Consider the problem of solving

$$\begin{aligned} 2x_1 + 3x_2 &= 8 \\ 5x_1 + 7x_2 &= 19 \end{aligned} \tag{8}$$

Even though this problem is trivially simple, consider a trial and error type of solution method where one would make a guess of the values of x_1 and x_2. Let the trial value of x_i be y_i for $i = 1, 2$. In general, the trial values do not satisfy the above equation. If y_3 is the residual of the first equation and y_4 that of the second equation, that is, if

$$\begin{aligned} y_3 &\triangleq 8 - 2y_1 - 3y_2 \\ y_4 &\triangleq 19 - 5y_1 - 7y_2 \end{aligned}$$

then, rearranging,

$$\begin{aligned} 2y_1 + 3y_2 + y_3 &= 8 \\ 5y_1 + 7y_2 + y_4 &= 19 \end{aligned} \tag{9}$$

Had the initial guess been a correct guess, the residuals y_3 and y_4 would have been zero; otherwise, it would have been nonzero. If y_3 and y_4 are assumed to be positive (which implies that y_1 and y_2 are also positive), then minimizing $y_3 + y_4$ is tantamount to finding the correct solution of Eq. (8). Thus, solving Eq. (8) can be formulated as

$$\text{Min } z = y_3 + y_4$$

subject to Eq. (9) and

$$y_i \geq 0 \qquad \text{for all} \quad i$$

A Measurement Problem

Consider once again the problem of environmental pollution by trace metals. Consider the problem of measuring the quantity of trace metals in a sample of kidney tissue. One method of making this measurement is by neutron activation analysis. In activation analysis, the tissue material is subjected to neutron bombardment. This bombardment can cause the atoms of some stable elements to become excited, and so they will subsequently undergo a change of state. This change of state for some of the atoms will be via gamma disintegration. These disintegrations are detectable and countable using equipment such as a scintillation counter and multichannel analyzer. This equipment consists of a counter and several (say, 400) memory cells called *channels*. Each channel or memory unit is capable of storing one number supplied to it by the counter.

Each gamma disintegration releases a measurable amount of energy and the counter measures the energy released and classifies or channelizes each disintegration according to the magnitude of the energy released. For instance, if the released energy is less than or equal to 0.01 MeV, it could be assigned to channel 1. If the released energy is greater than 0.01 MeV but not more than 0.02 MeV, it would be assigned to the next channel, that is, channel 2. Thus, at the end of an experiment, cell 1 contains a number equal to the number of disintegrations whose energy had been between 0 and 0.01 MeV, and so forth. Stated differently, the multichannel recorder stores the *gamma ray spectrum*. A typical spectrum is shown in Fig. 8.4. The problem is to determine the quantities of various elements present in the sample under test by analyzing the spectrum.

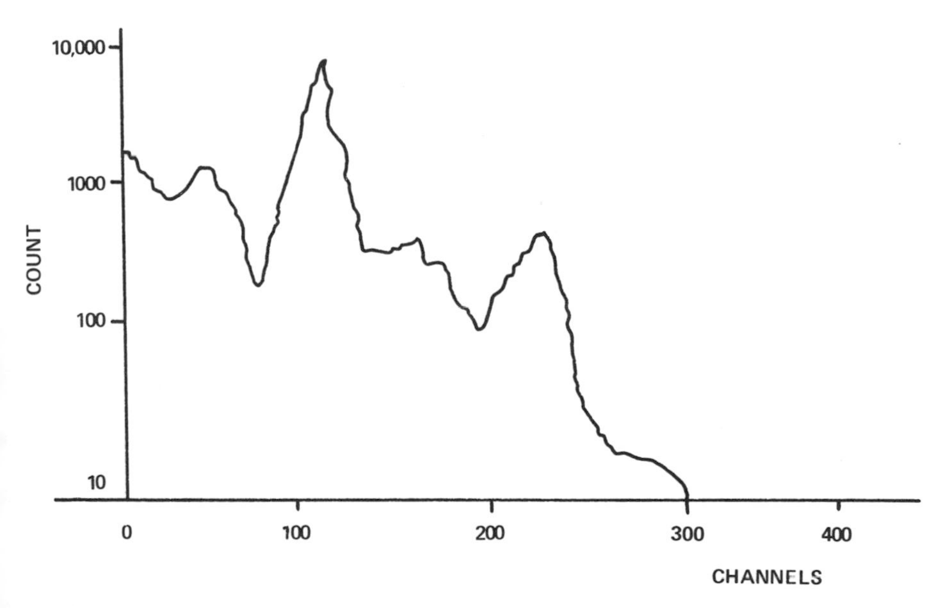

Fig. 8.4 Gamma ray spectrum of a tissue sample known to contain x μg of copper and y μg of sodium.

Analysis of the spectrum involves a comparison of the test spectrum with a set of standard spectra. To formulate the problem in mathematical terms, let

i = an energy range consisting of one or more channels, $i = 1, 2, \ldots, m$

S_i = the total gamma decay count in range i

$c_{i,j}$ = the count in standard spectrum j in range i

$S_{i,j}$ = the count in spectrum under test in range i that is attributable to element in standard spectrum j

Then, one can write

$$\begin{aligned} x_1c_{1,1} + x_2c_{1,2} + \cdots + x_mc_{1,m} &= S_1 \\ x_1c_{2,1} + x_2c_{2,2} + \cdots + x_mc_{2,m} &= S_2 \\ &\vdots \\ x_1c_{m,1} + x_2c_{m,2} + \cdots + x_mc_{m,m} &= S_m \end{aligned} \tag{10}$$

Equation (13) can be quickly solved for x_j in many simple cases. This problem would not have been cited here if this is all that is involved.

Inspection of the entire process reveals that implicitly the subscript i has been chosen to be equal to the subscript j. That is, the number of energy ranges used is assumed to be equal to the number of standard spectra available in the library. This assumption is not always warranted. Second, the data are inherently inaccurate because of the statistical variations in a gamma decay process. These two problems can be resolved in a very simple way. For instance, one can use a regression procedure by trying to fit the data in the least-square sense, that is by minimizing.

$$D = \sum_{i=1}^{n} \left[S_i - \sum_{j=1}^{m} x_i c_{i,j} \right]^2 \tag{11}$$

This procedure works even if $m \neq n$. Thus, this method does not restrict the number of standards and the number of ranges. For instance, one can use each channel as one range, giving a total of 400 ranges and still use, say, only 10 standard spectra for comparison.

Both the least square and simultaneous equation methods suffer from another disadvantage. Both the methods assume that the analyst knows the identity of components in the tissue under test. These and only these should compose the library of standards for a particular analysis. If the analyst does not know the exact components present in the tissue, it is conceivable that one may erroneously use more standards than there are components. This could lead to negative quantities for some of the contributors of the sample.

All these problems can be resolved to some degree of satisfaction by reformulating the problem as

$$\text{Max } D = x_1 \sum_{i=1}^{n} c_{i,1} + x_2 \sum_{i=1}^{n} c_{i,2} + \cdots + x_m \sum_{i=1}^{n} c_{i,m} \tag{12}$$

subject to

$$\begin{aligned} x_1c_{1,1} + x_2c_{1,2} + \cdots + x_mc_{1,m} &\leq S_1 \\ x_2c_{2,1} + x_2c_{2,2} + \cdots + x_mx_{2,m} &\leq S_2 \\ &\vdots \\ x_1c_{n,1} + x_2c_{n,2} + \cdots + x_mc_{n,m} &\leq S_n \end{aligned} \tag{13}$$

The inequalities restrict the counts of the unresolved composite spectrum for m ranges. Maximizing D ensures that the sum of the activities of the components within the ranges will be maximized.

4 THE LINEAR PROGRAMMING MODEL

The similarity of the structure of the mathematical models developed for all the problems in the preceding section is evident. All problems involve a minimization or maximization of a function such as

$$z = \sum_{i=1}^{n} c_i x_i = \mathbf{c}^{\mathrm{T}}\mathbf{x} \tag{1}$$

where $\mathbf{c}^{\mathrm{T}} = [c_1, c_2, \ldots, c_n]$ and $\mathbf{x}$ is a n-dimensional vector. All problems involve m constraints such as

$$\mathbf{Ax} \begin{cases} \geq \\ = \\ \leq \end{cases} \mathbf{b} \tag{2}$$

where $\mathbf{b}$ is an m-dimensional constant vector and $\mathbf{A}$ is a $m \times n$ coefficient matrix.

The objective function in Eq. (1) and the constraints in Eq. (2) are both linear functions of x_i; therefore, models of this type are called *linear programming* (or LP) models. If the objective function is *quadratic* while all the constraints are linear, then the problem is called *quadratic programming* (or QP) models. A typical quadratic objective function is

$$z = \mathbf{c}^{\mathrm{T}}\mathbf{x} + \mathbf{x}^{\mathrm{T}}\mathbf{D}\mathbf{x}$$

where $\mathbf{D}$ is an $n \times n$ constant symmetric matrix. If both the objective function and the constraint set are nonlinear functions of x_i, then the problem becomes a *nonlinear programming problem.* If the coefficients c_i in the objective function or a_{ij} of the constraint set are random variables, the problem is called a *stochastic programming* problem.

A Graphical Solution

Relatively simple LP problems can be solved using graph paper and a straight edge. Obviously graphical methods are useless to solve problems of practical interest as they invariably contain a large number of variables and constraints. However, the graphical method provides invaluable insight into the optimization procedure.

Example 6. Consider the task of getting a solution to the LP problem formulated in Example 5. That is,

$$\text{Max } z = 1.7x_1 + 2.8x_2$$

subject to

$$\begin{aligned} 2x_1 + 3x_2 &\leq 4500 \\ x_2 &\leq 1000 \\ x_1, x_2 &\geq 0 \end{aligned}$$

There are two variables, namely, x_1 and x_2. Therefore, a two-dimensional coordinate plane is sufficient to solve the problem. If each variable is assigned to one of the coordinate axes, then each inequality represents a semiinfinite region in the x_1x_2 plane (see Fig. 8.5). Then *all* the inequalities represent a polygon of the x_1x_2 plane, as shown in Fig. 8.6. This region is termed the *feasible region.* Inspection of Fig. 8.6 reveals that the inside of the polygon $0ABC$ is the feasible region. That is, any point inside, or on the boundary, of the feasible region satisfies the constraints and any point outside the feasible region fails to satisfy the constraints. Thus, search for the values of x_1 and x_2 that maximize z should be restricted to the feasible region. As there are an infinite number of points in a plane region, it is practically impossible to search the entire feasible region. Indeed, such an exhaustive search is not necessary. Plotting $z = 1.7x_1 + 2.8x_2$ for various values of z reveals that the value of z increases as the line is moved away and away from the origin. In fact, z can be seen to attain its maximum when the cost line touches the vertex B of the feasible polygon. If the cost line is moved farther away from the origin, in the hope of increasing z further, then no portion of it falls within the feasible region, thus violating the constraints. Thus, the coordinates defining thc point B satisfy the constraints while maximizing z.

From Fig. 8.6, the coordinates of B are (750, 1000). Thus, the maximum value of z is

$$z = 1.7(750) + 2.8(1000) = 4075$$

■

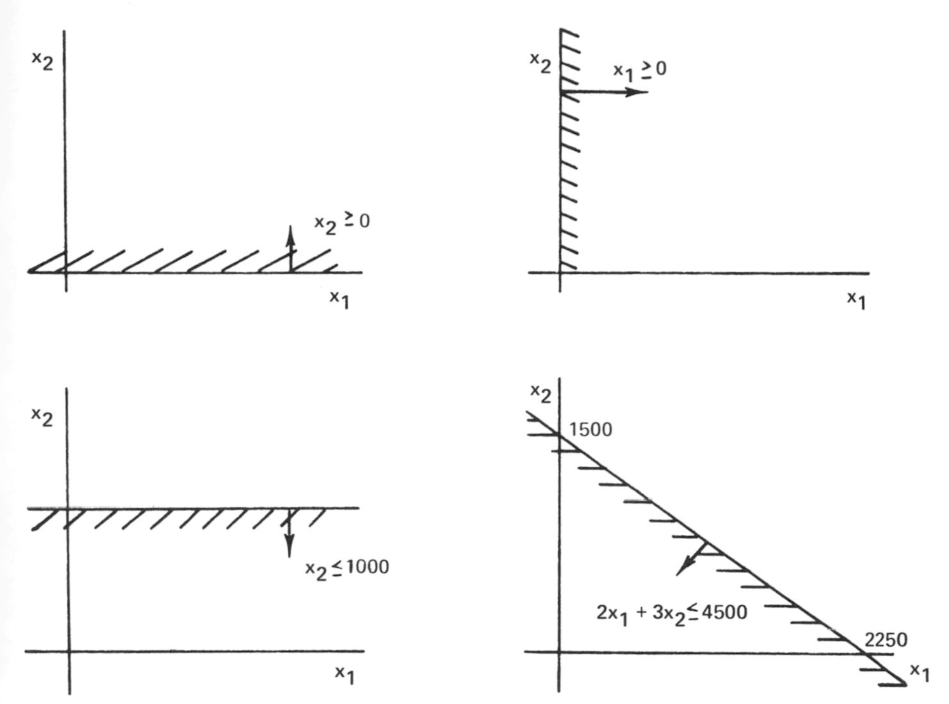

Fig. 8.5 Semiinfinite feasible regions represented by the inequalities in Example 6.

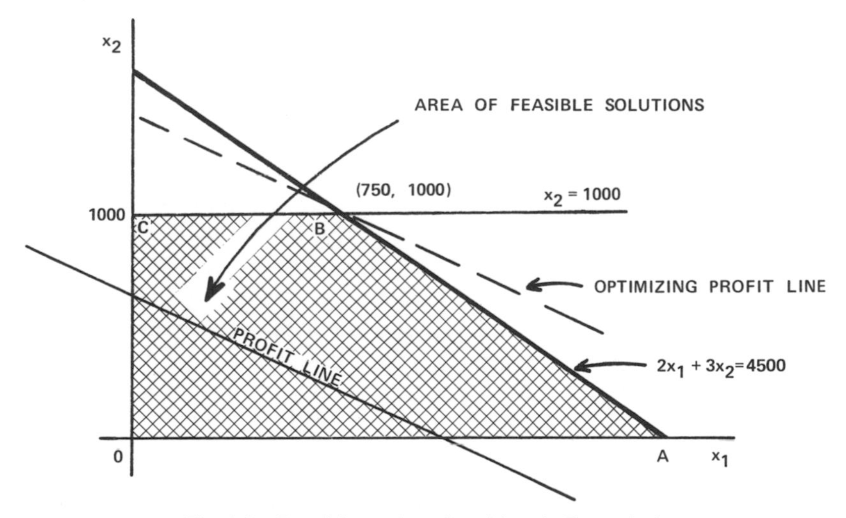

Fig. 8.6 Feasible region of problem in Example 6.

Without further analysis several important characteristics of a LP problem can be summarized.

(1) The variables, such as x_1 and x_2 whose values are required to be determined are called *decision variables.*

(2) The constraints define a region called the *feasible region* (or feasible polygon or feasible polyhedron) and the optimum solution of the LP problem lies on the boundary of the feasible region. Therefore, while searching for an optimum, it is sufficient to confine attention only to the boundary of the feasible region. Indeed, in a majority of cases, it is sufficient to confine the search to the vertices of the feasible region.

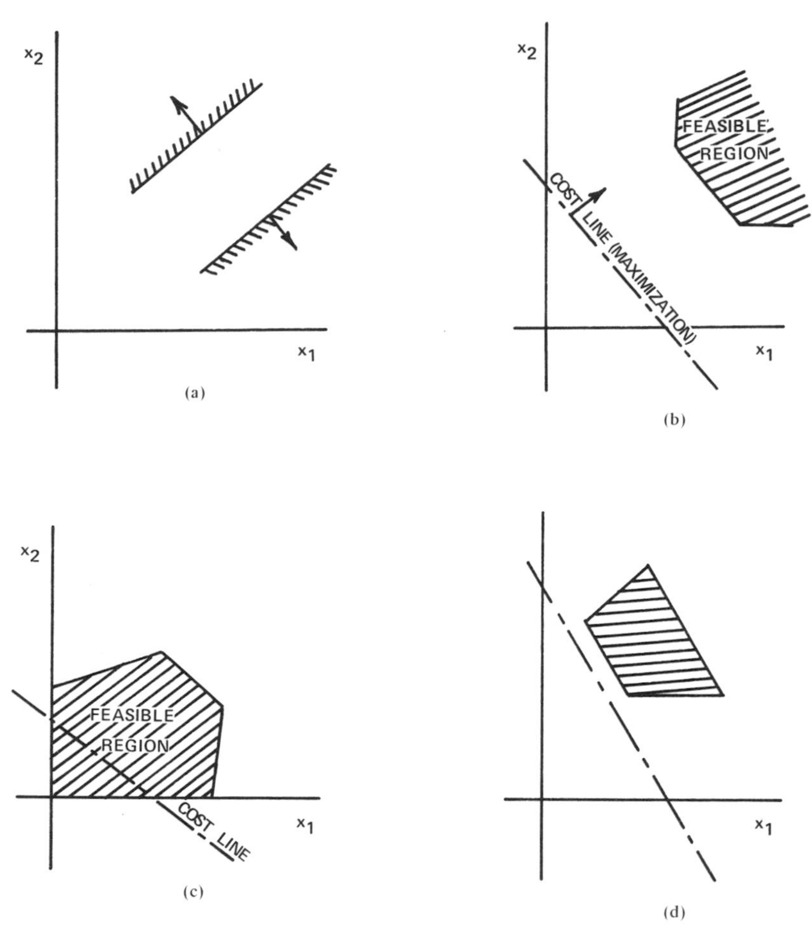

Fig. 8.7 (a) Contradictory constraints. (b) Unbounded area of feasible solutions. (c) Trivial solution. (d) Infinite number of optimum solutions.

Solutions to LP problems do not always exist. If there are contradictory constraints, then there is no feasible area and, hence, no solution exists. (see Fig. 8.7a). Figure 8.7b shows another situation in which a solution does not exist. Here, the area of feasible solutions into which the *cost line* is moving is unbounded. Situations like this may arise where a manufacturer is trying to decide on a price to charge the consumer so that he can maximize his profits. According to the figure, the manufacturer can keep increasing his profits by raising the prices. What the problem neglected to consider was that no one will buy the goods if the price goes beyond a certain level. If this additional consideration or constraint is taken into account, the region of feasible solutions becomes bounded and a solution will exist. A third possible situation is depicted in Fig. 8.7c. This situation may arise when a country wants to reduce the level of pollution to a minimum. The obvious answer is not to allow any pollution. The additional constraint that the country should maintain a specified growth rate would yield a nontrivial solution. Finally, when the optimizing line or cost line happens to be parallel to one of the sides of the feasible region, then there exists an infinite number of solutions. This is shown in Fig. 8.7d.

Barring such specialized situations as shown in Fig. 8.7, it is evident that it is sufficient to confine the search for an optimum to the vertices of the feasible region. Thus, the vertices of the feasible region play a special role in LP and deserve an identity. In LP parlance these vertices are called *extreme points.*

The Standard Form of the Linear Programming Problem

For the purpose of standardization, it is both useful and convenient to convert any given LP problem into a standardized format. In the standard format, the constraints of an LP problem are always written as equalities. For instance, a constraint of the form

$$a_{i,1}x_1 + a_{i,2}x_2 + \cdots + a_{i,n}x_n \leq b_i \tag{3}$$

can always be rewritten as

$$a_{i,1}x_1 + a_{i,2}x_2 + \cdots + a_{i,n}x_n + x_{n+1} = b_i \tag{4}$$

In such cases x_{n+1}, which is assumed to be nonnegative, is called a *slack variable.* If the inequality constraint appears as

$$a_{i,1}x_1 + a_{i,2}x_2 + \cdots + a_{i,n}x_n \geq b_i \tag{5}$$

rather than in the form of Eq. (3), then Eq. (5) is multiplied throughout by -1 to yield

$$-a_{i,1}x_1 - a_{i,2}x_2 - \cdots - a_{i,n}x_n \leq -b_i \tag{6}$$

which can now be converted into the standard form by the addition of a slack variable. Thus, if the original problem contains p inequality constraints, the standard form contains p slack variables.

The optimization of a linear program can either be maximization or minimization. If the objective function is maximization, such as

$$\text{Max } A = c_1x_1 + c_2x_2 + \cdots + c_nx_n \tag{7}$$

it can be converted to a minimization form by multiplying throughout by -1 to yield

$$\text{Min}(-\xi) = -c_1x_1 - c_2x_2 - \cdots - c_nx_n \tag{8}$$

or equivalently as

$$\text{Min } z = -c_1x_1 - c_2x_2 - \cdots - c_nx_n \tag{9}$$

where $z = -\xi$.

Now, consider a linear programming problem stated as follows. Given an objective function (to be maximized or minimized)

$$z = c_1x_1 + c_2x_2 + \cdots + c_nx_n \tag{10}$$

find the values of the decision variables $x_i \geq 0$, $i = 1, 2, \ldots, n$, satisfying the m linear inequality or equality constraints

$$a_{i,1}x_1 + a_{i,2}x_2 + \cdots + a_{i,n}x_n \leq b_i; \qquad i = 1, 2, \ldots, p \tag{11}$$

$$a_{i,1}x_1 + a_{i,2}x_2 + \cdots + a_{i,n}x_n = b_i; \qquad i = p+1, p+2, \ldots, m \tag{12}$$

This problem, stated in the standard format looks as follows. Given the objective function

$$z = c_1x_1 + c_2x_2 + \cdots + c_nx_n + 0.x_{n+1} + \cdots + 0.x_N \tag{13}$$

subject to

$$a_{i,1}x_1 + a_{i,2}x_2 + \cdots + a_{i,n}x_n + x_{n+i} = b_i; \qquad i = 1, 2, \ldots, p \tag{14}$$

$$a_{i,1}x_1 + a_{i,2}x_2 + \cdots + a_{i,n}x_n = b_i; \qquad i = p+1, p+2, \ldots, m \tag{15}$$

Notice that the standard format contains $N = p + n$ variables out of which p are the slack variables. In a more compact notation, this problem can be stated as follows. Given the objective function (to be maximized or minimized)

$$z = \sum_{i=1}^{N} c_ix_i = \mathbf{c}^\mathrm{T}\mathbf{x} \tag{16}$$

where $\mathbf{c}^\mathrm{T} = [c_1, \ldots, c_n, 0, 0, \ldots, 0]$ is an N-dimensional row vector, find the values of the variables x_i, $i = 1, 2, \ldots, N$, satisfying the m equality constraints specified by

$$\mathbf{Ax} = \mathbf{b} \tag{17}$$

where $\mathbf{A}$ is an $m \times N$ coefficient matrix, $\mathbf{x}$ is an $N = (n + p)$-dimensional vector that includes all the original decision variables x_i, $i = 1, 2, \ldots, n$, and the p slack variables, all of which are nonnegative, and $\mathbf{b}$ is an m-dimensional vector. Note that the slack variables are by definition nonnegative whereas the original n decision variables (i.e., x_i for $i = 1, 2, \ldots, n$) are not necessarily so restricted. This situation is rather inconvenient from a mathematical viewpoint. To remove this difficulty all the x_i's are hereafter assumed to be nonnegative (i.e., $x_i \geq 0$ for all i). A majority of the problems arising in practice automatically satisfy this condition. If not, it is relatively straightforward to define a new decision variable (in terms of the old variables) that satisfies the nonnegativity constraint. With this additional assumption, the LP problem can be restated in its *standard format* as follows. Given the objective function (to be maximized or minimized)

$$z = \sum_{i=1}^{N} c_i x_i = \mathbf{c}^{\mathrm{T}}\mathbf{x} \tag{18}$$

$$\mathbf{c}^{\mathrm{T}} = [c_1, c_2, \ldots, c_n, 0, 0, \ldots, 0] \tag{19}$$

find the values of x_i ($i = 1, 2, \ldots, N$) satisfying

$$\mathbf{Ax} = \mathbf{b} \tag{20}$$

$$x_i \geq 0 \qquad \text{for all} \quad i \tag{21}$$

Note that Eq. (20) represents m linear equations in N unknowns. It is assumed that $N > m$, that is, the total number of unknown variables is greater than the number of equations in the constraint set. This assumption is important and makes sense. If $m > N$, there are more equations than unknowns; some of the equations are dependent on others and therefore could be eliminated. If $m = N$, the problem reduces to one of solving a set of simultaneous linear algebraic equations and a unique solution exists (i.e., there is no sense talking about optimization) if $\mathbf{A}$ is nonsingular.

Standard Inequality Forms

There is a standard convention in writing LP problems while using the inequality signs. Consider the minimization problem written as follows:

$$\begin{aligned} x_i \geq 0, \qquad & i = 1, 2, \ldots, n \\ a_{1,1}x_1 + a_{1,2}x_2 + \cdots + a_{1,n}x_n \geq b_1 & \\ a_{m,1}x_1 + a_{m,2}x_2 + \cdots + a_{m,n}x_n \geq b_m & \\ c_1x_1 + c_2x_2 + \cdots + c_nx_n = \text{Min } z & \end{aligned} \tag{22}$$

It is significant to note that the greater than sign is used in all constraints. This orientation of inequality sign is recommended for minimization problems because it is consistent with the fact that the objective function can also be interpreted as having a greater-than relation if z indeed becomes a minimum. Similarly, maximization problems are customarily written with a less-than relation. If this convention is followed, then one has to know the procedure for handling an equality constraint. For instance, a constraint, such as

$$a_{i,1}x_1 + \cdots + a_{i,n}x_n = b_n \tag{23}$$

can always be replaced by the pair of inequality constraints

$$a_{i,1}x_1 + \cdots + a_{i,n}x_n \leq b_n \tag{24}$$

and

$$a_{i,1}x_1 + \cdots + a_{i,n}x_n \geq b_n \tag{25}$$

Depending on the nature of the optimization (i.e., maximization or minimization), one of the above is multiplied by -1.

5 THE SIMPLEX ALGORITHM

One of the more popular algorithms for the numerical solution of LP problems is the so-called *simplex algorithm* originated by Dantzig. At present there are several versions of the simplex method. Only the basic ideas of the algorithm are presented here.

Rewriting the constraint equations of Eq. (4.20) in scalar form, one has

$$\begin{aligned} a_{11}x_1 + a_{12}x_2 + \cdots + a_{1N}x_N &= b_1 \\ a_{21}x_1 + a_{22}x_2 + \cdots + a_{2N}x_N &= b_2 \\ &\vdots \\ a_{m1}x_1 + a_{m2}x_2 + \cdots + a_{mN}x_N &= b_m \end{aligned} \tag{1}$$

It is assumed that $N > m$, all a_{ij}'s are constants, and all $x_j \geq 0$. It was stated earlier that any vector $\mathbf{x}$ whose components $x_j \geq 0$ satisfy all the equations of Eq. (1) constitutes a *feasible solution*. In simplex method, a particular form of a feasible solution plays a central role, namely, a *basic feasible solution*.

Equation (1) contains m equations in N unknowns with $m < N$. If one can pick a nonsingular $m \times m$ submatrix from $\mathbf{A}$, then one can associate one decision variable with each of the m columns of $\mathbf{A}$. This leaves $(N - m)$ decision variables not associated with the columns of $\mathbf{A}$. If all these $(N - m)$ decision variables are set identically equal to zero, then the remaining system of m equations can be exactly solved for the m unknown decision variables.

This solution is called the *basic solution.* The m decision variables that are *not* equated to zero are called *basic variables,* and the $(N - m)$ decision variables that are set equal to zero are called *nonbasic variables.*

A basic solution that is feasible is called a *basic feasible solution.* There is an important theorem that establishes a relation between the basic feasible solution and the extreme point defined in the context of a graphical solution. This theorem states that *every basic feasible solution to the set of equations in Eq. (1) is an extreme point of the feasible region defined by Eq. (1) and vice versa.*

Thus, selecting a basic feasible solution is tantamount to picking a vertex of the feasible region. Not all vertices represent optimum solutions, however. It is necessary to search the vertices in a systematic fashion to find the optimum vertex or the vertex that yields an optimum solution. This is what the simplex method is all about.

From the preceding paragraph, it is clear that the magnitude of effort involved in implementing the simplex algorithm depends on the number of vertices of the feasible region (or equivalently on the total number of basic solutions). For a system of m equations in N variables (with $N > m$), the total number of basic feasible solutions is equal to the number of combinations of m items out of n:

$$C_N{}^m = N!/(N - m)!m! \tag{2}$$

In other words, this is the maximum number of extreme points of the feasible region. Thus, given a basic feasible solution (i.e., given a vertex of the feasible region), theoretically one has to search $C_N{}^m$ vertices to find the optimum. If this is the case, the method is tantamount to exhaustive enumeration. The simplex method takes advantage of another fact.

In a two-dimensional problem (i.e., a problem involving two decision variables), a vertex has two neighboring vertices, three neighbors in a three-dimensional problem, and n neighbors in an n-dimensional problem. Therefore, given a vertex (i.e., a basic feasible solution), one can possibly inspect the n-neighboring vertices and pick one of them that promises the best increase (or decrease) in the objective function. In the worst case, one may end up inspecting all $C_N{}^m$ vertices before being able to locate an optimum. Usually, the optimum is located much faster.

Thus, the simplex algorithm consists of the following steps.

(1) Find any basic feasible solution (bfs) of the constraint equations (i.e., start from any vertex of the feasible region).

(2) From the first bfs advance to the next bfs in the direction that promises the best improvement in the objective function. (This is equivalent to moving from one vertex to another vertex).

(3) Continue moving from one bfs to the next until no further improvement can be found.

The Simplex Tableau

The mechanics of implementing the procedure are first illustrated. Suppose that the given problem is

$$\text{Max } 2x_1 - x_2 \tag{3}$$

subject to

$$\begin{aligned} -3x_1 + 2x_2 &\leq 2 \\ 2x_1 - 4x_2 &\leq 3 \\ x_1 + \ x_2 &\leq 6 \\ x_1, x_2 &\geq 0 \end{aligned} \tag{4}$$

Step 1 Convert the constraint set in Eq. (4) into the standard form using slack variables:

$$\begin{aligned} -3x_1 + 2x_2 + x_3 + 0 \ + 0 \ &= 2 \\ 2x_1 - 4x_2 + 0 \ + x_4 + 0 \ &= 3 \\ x_1 + \ x_2 + 0 \ + 0 \ + x_5 &= 6 \\ x_1, x_2, x_3, x_4, x_5 &\geq 0 \end{aligned} \tag{5}$$

Rewrite the objective function as

$$z = 2x_1 - x_2 + c_3x_3 + c_4x_4 + c_5x_5 \tag{6}$$

where $c_3 = c_4 = c_5 = 0$.

Step 2 As there are three equations in Eq. (5), choose any three of the x_i's as the basic set of variables. Then, the remaining x_i's become nonbasic variables. (Whenever possible, it is convenient to choose the slack variables as the basic set, or to choose one variable from each equation such that the chosen variable has a coefficient $+1$.)

Step 3 Arrange the basic variables, nonbasic variables and the coefficients as shown in Table 8.2. Note that the right side of Eq. (5) appears as the last column and the coefficients c_i appear as the last row of Table 8.2.

Step 4 Locate the most positive entry in the last row and mark it by an arrow. Calculate the ratios of the elements of the column containing the arrow and the last column. In this example, these ratios are $-3/2$, $2/3$, and $1/6$. Take the largest positive ratio and circle that. In this case, the circle appears in the x_4 row. The entry circled is called the pivot. The row and column in which the pivot appears are called the pivotal row and pivotal column.

TABLE 8.2

Initial Tableau for the Simplex Method

Basic set	Nonbasic set x_1	x_2	Basic set x_3	x_4	x_5	b
x_3	-3	2	1	0	0	2
x_4	(2)	-4	0	1	0	3
x_5	1	1	0	0	1	6
c_i	2 ↑	-1	0	0	0	$0 =$ value of z

Step 5 Construct a new table as follows. First replace the label of the pivotal row by the same label appearing on the pivotal column. That is, x_4 is removed from the basic set and x_1 added to it. Second, divide the pivotal row by the value of the pivot. Then, using appropriate row operations with the pivotal row, convert every other entry in the pivotal column equal to zero. The transformed array is shown in Table 8.3.

TABLE 8.3

Second Stage of the Simplex Tableau

	x_1	x_2	x_3	x_4	x_5	b
x_3	0	-4	1	$\frac{3}{2}$	0	13/2
x_1	1	-2	0	$\frac{1}{2}$	0	3/2
x_5	0	(3)	0	$-\frac{1}{2}$	1	9/2
	0	$+3$ ↑	0	-1	1	$3 =$ value of z

Step 6 Repeat Step 4.
Step 7 Repeat Step 5. The result is shown in Table 8.4.

TABLE 8.4

Third (Last) Stage of the Simplex Tableau

	x_1	x_2	x_3	x_4	x_5	b
x_3	0	0	1	$\frac{5}{6}$	4	25/2
x_1	1	0	0	$-\frac{1}{2}$	2	21/2
x_2	0	1	0	$-\frac{1}{6}$	$\frac{1}{3}$	3/2
	0	0	0	$-\frac{1}{2}$	0	$15/2 =$ value of z

Step 8 Repeat Step 4. However, Step 4 cannot be repeated as there is no positive entry in the last row. Whenever a situation like this is encountered, the simplex method is said to have converged.

The optimum values (i.e., the values of x_i's that make z a maximum) are read from the table as $x_1 = 21/2$, $x_3 = 25/2$ and $x_5 = 3/2$. The maximum value of z is found from the table to be 15/2. It is useful to make a parenthetical observation here. Notice that the process of going from Table 8.2 to Table 8.4 is similar to finding the product form of the inverse of a matrix (see Appendix 3).

The procedure described in the preceding example can be stated as a set of rules for any LP problem. Consider the LP problem:

$$\text{Max } z = c_1x_1 + \cdots + c_nx_n$$

Subject to

$$\begin{aligned} a_{11}x_1 + a_{12}x_2 + \cdots + a_{1n}x_n &\leq b_1 \\ a_{21}x_1 + a_{22}x_2 + \cdots + a_{2n}x_n &\leq b_2 \\ &\vdots \\ a_{m1}x_1 + a_{m2}x_2 + \cdots + a_{mn}x_n &\leq b_n \end{aligned}$$

and

$$x_i \geq 0; \qquad b_j \geq 0.$$

Step 1 Add p slack variables and convert the problem into the standard format. Now the problem becomes

$$\text{Max } z = c_1x_1 + \cdots + c_nx_n + c_{n+1}x_{n+1} + \cdots + c_{n+p}x_{n+p}$$

subject to

$$\begin{aligned} a_{11}x_1 + a_{12}x_2 + \cdots + a_{1n}x_n + x_{n+1} &= b_1 \\ a_{21}x_1 + a_{22}x_2 + \cdots + a_{2n}x_n + x_{n+2} &= b_2 \\ &\vdots \\ a_{m1}x_1 + a_{m2}x_2 + \cdots + a_{mn}x_n + x_{n+p} &= b_m \end{aligned}$$

where $x_i \geq 0$, $b_j \geq 0$, and $c_{n+j} = 0$ for $j = 1, \ldots p$.

Step 2 Select m of the x_i's and call them the basis. Choose the slack variables as the basis if all of the m slack variables have a $+1$ as a coefficient (otherwise one has to introduce the so-called *artificial variables*). Arrange the basic variables, nonbasic variables and all other coefficients in the standard form as shown in Table 8.2.

Step 3 Inspect the last column of this table and locate the largest positive entry by an arrow. Calculate the ratios a_{ij}/b_j for all i. Identify the largest positive ratio and circle that a_{ij} which gives the largest positive ratio. This a_{ij} is called the pivot.

Step 4 Prepare a new table with all labels as before except the label on the pivotal row. The pivotal row is now labeled with x_j. Using row operations (i.e., multiplying the pivotal row by an appropriate constant and then adding it to other rows), reduce every entry in the pivotal column, except the pivot, to zero. Reduce the pivot to 1 by dividing the pivotal row by the pivot.

Step 5 Inspect the last row. If all entries of this row are either negative or zero go to Step 6. If not, go to Step 3.

Step 6 Find the optimum values of the basic variables in the last column and the maximum value of z at the intersection of the last row and last column.

Modifications of this algorithm and other specialized procedures are often used to solve special cases such as the assignment problem, transportation problem, shortest route problem, and so on.

6 DUALITY IN LINEAR PROGRAMMING MODELS

The concept of duality is not only important in its own right but it also increases one's understanding of the LP procedures. The dual formulation of an LP model provides a framework for the presentation of important concepts such as shadow prices, opportunity costs, sensitivity analysis, and so forth. For every LP problem, which may be termed the *primal problem*, there exists a corresponding optimization problem called the *dual problem*. The optimum solution of either problem reveals valuable information about the optimum solution of the other problem. Indeed, if the initial simplex tableau of the primal problem contains an $m \times m$ identity matrix, then the solution of either problem by the simplex method yields an explicit solution of the other. Thus, the dual statement of a problem contains all the information of the primal and represents only a different format of problem specification.

There are two different ways in which the dual problem can be formulated: the *symmetric dual* and the *unsymmetric dual*. Formulation of a symmetric dual is illustrated here. If the primal problem is to find $\mathbf{x}$ so as to

$$\text{Max } z = \mathbf{c}^{\mathrm{T}}\mathbf{x} \tag{1a}$$

subject to

$$\begin{aligned} \mathbf{Ax} &\leq \mathbf{b} \\ \mathbf{x} &\geq \mathbf{0} \end{aligned} \tag{1b}$$

then the dual problem would be to find a vector $\mathbf{w}$ that

$$\text{Min } y = \mathbf{b}^{\mathrm{T}}\mathbf{w} \tag{2a}$$

subject to

$$\begin{aligned} \mathbf{A}^{\mathrm{T}}\mathbf{w} &\geq \mathbf{c} \\ \mathbf{w} &\geq \mathbf{0} \end{aligned} \tag{2b}$$

As an illustration consider the problem

$$\text{Max } z = 4x_1 + 5x_2 + 9x_3$$

subject to

$$\begin{aligned} x_1 + \ \ x_2 + 2x_3 &\leq 16 \\ 7x_1 + 5x_2 + 3x_3 &\leq 25 \\ x_1, x_2, x_3 &\geq 0 \end{aligned}$$

If the above problem is considered as the primal, then the dual problem would be

$$\text{Min } y = 16w_1 + 25w_2$$

subject to

$$\begin{aligned} w_1 + 7w_2 &\geq 4 \\ w_1 + 5w_2 &\geq 5 \\ 2w_1 + 3w_2 &\geq 9 \\ w_1, w_2 &\geq 0 \end{aligned}$$

From this it can be seen that the dual formulation is obtained by flipping, so to speak, of the primal problem on its side. That is:

(a) The jth column of the coefficients in the primal problem is the same as the jth row of the dual.
(b) The row of coefficients of the primal objective function is the same as the column of constants on the right side of the dual constraints.
(c) The column of constants on the right side of the primal is the same as the row of coefficients of the dual objective function.
(d) The direction of the inequalities and the sense of optimization are reversed in the primal–dual pair.

In addition to the above observations, the primal and the dual problems are associated with an important theorem of significant practical interest. This theorem, called the *duality theorem*, can be stated as follows. If either the primal or the dual problem has a finite optimum solution, then the other problem also has a finite optimum solution and the extreme values of both the problems are equal, that is, *min* $y =$ *max* z. If either problem has an unbounded optimum solution, then the other problem has no feasible solution. This theorem allows one to switch from the primal to the dual and vice versa.

A Physical Interpretation of Duality

A physical interpretation of the primal problem is relatively straightforward. For instance, the objective function of the primal problem can be interpreted as

$$\text{Max} \sum_{j=1}^{n} \left(\frac{\text{production}}{\text{activity } j}\right)(\text{activity } j) = \text{product}$$

subject to

$$\sum_{j=1}^{n} \left(\frac{\text{resource } i}{\text{activity } j}\right)(\text{activity } j) \leq \text{resource } i \qquad \text{for all} \quad i$$

$$(\text{activity } j) \geq 0 \qquad \text{for all} \quad j$$

Thus, the primal problem poses the following question: with a given productivity c_j and a given upper limit to the availability of each resource (or input b_i), at what level should each activity (or output x_j) be maintained in order to maximize the total production?

Similarly, an interpretation can be given to the dual problem. The major difficulty, however, is that the dual problem carries two variables, namely, w and y, whose meaning is not readily apparent. With this gap in our knowledge, the dual problem can be interpreted as follows:

$$\text{Min} \sum_{i=1}^{m} (\text{resource } i) w_i = ?$$

subject to

$$\sum_{i=1}^{m} \left(\frac{\text{resource } i}{\text{activity } j}\right) w_i \geq \left(\frac{\text{product}}{\text{activity } j}\right) \qquad \text{for all} \quad j$$

$$w_i \geq 0 \qquad \text{for all} \quad i$$

From this it can be seen that the dual constraints will be consistent if w_i are expressed in units of product per unit of resource. Using this notation for w_i, the dual problem becomes

$$\text{Min} \sum_{i=1}^{m} (\text{resource } i)\left(\frac{\text{product}}{\text{resource } i}\right) = \text{product}$$

subject to

$$\sum_{i=1}^{m} \left(\frac{\text{resource } i}{\text{activity } j}\right)\left(\frac{\text{product}}{\text{resource } i}\right) \geq \frac{\text{product}}{\text{activity } j}; \qquad j = 1, 2, \ldots, n$$

and

$$\left(\frac{\text{product}}{\text{resource } i}\right) \geq 0 \qquad i = 1, 2, \ldots, m$$

Thus, the dual problem can be stated as follows: With a given availability of each resource (b_i) and a given lower limit of unit value for each activity (c_j), what level should be assigned to each activity w_j in order to minimize the total resource consumption? The optimal dual variables are variously referred to as *accounting* or *shadow prices.*

The shadow prices define the marginal value of the contribution of each constraint to the objective function. They answer the question: How much can the production (or the output) be increased by relaxing a constraint by one unit? For instance, how many more cars can be produced by increasing the availability of one of the resources by one unit? Conversely, how much will the profit change by tightening the availability of one of the resources? In other words, shadow prices shed some light on the sensitivity of a process to a variation in the variables. The significance of shadow prices lies in their policy implications.

7 DYNAMIC PLANNING WITH LINEAR PROGRAMMING

One of the goals of this chapter is to show how elementary ideas from optimization theory and mathematical programming can be profitably applied to the planning and execution of large-scale complex systems. Planning is a multifaceted operation. It involves a recognition of the problem at hand, a recognition of the mathematical, economic, and statistical attributes of the problem, a recognition of the decision and control variables, and an ability to analyze probable future consequences of competitive decision choices. In a systems approach to planning and decision making, one attempts to decompose a large-scale complex problem into subsystems, each of which is simpler to manipulate. As an illustration of a situation in which such complex decision-making problems arise, let us consider a manufacturing process. The production schedule of a company has to take account of customer demand (tempered by a likelihood of a price cut by competitors), requirements for raw materials and intermediate inventories, the capacities of equipment, the possibility of equipment failure, and other technological restrictions. Other reasons for complexity in real decision-making situations are that the organization may be pursuing inconsistent goals and the economic environment in which the company operates may be uncertain. This type of uncertainty and conflict in goals can also be detected in the management of water resources. In the problem described in Chapter 1, for example, the rules for releasing the water from the reservoir with the goal of maximizing electricity generated may be inconsistent with the goal of minimizing ecological damage. Indeed, the managers responsible for achieving these goals are probably located at two different places in the managerial hierarchy. This requires a coordination of goals whenever a large system is decomposed into several subsystems. Thus, to be successful the systems approach must improve the managerial decision-making process—the improvement being measured by the "cost" of obtaining it.

Mathematical programming (MP) techniques, such as those briefly described in the earlier sections, as well as other MP techniques such as nonlinear, geometric, and dynamic programming techniques, can often be used effectively in the analysis of complex competitive situations. These techniques have been found useful in making better decisions, in attaining better coordination, and in implementing better controls. A basic feature of planning, however, is that it is essentially a dynamic process. The formulations described thus far are essentially static in nature. Nevertheless, a large number of dynamic problems can be solved using LP techniques by using a stagewise approach. In this approach, the system is visualized as consisting of several subsystems connected in cascade. Each subsystem in this cascade can stand for the state of the system at a discrete instant of time. This process is generally referred to as the "time-staged" formulation. Alternatively, the subsystems can stand for segments of a system. As we are, by and large, interested in the flow of some "material" through the system, and because it takes a finite amount of time for this material to move through the system, the segmentation of the spatial dimension can also be visualized as a time-staged formulation. Basic features of these two approaches are briefly demonstrated by means of two somewhat simplified examples. Many of the truly important applications of linear programming, and extensions thereof, have a structure very similar to these examples. No attempt will be made to solve these problems completely; only highlights of the formulation are shown.

A Production Scheduling Problem

Suppose that a company is required to supply S_i television sets over a period of T months. That is, the company has to supply S_1 television sets in the first month, S_2 in the second month, and so forth. Let us divide the production process into several *activities*:

e_i = number of workers engaged in production in month i (this is called the *employment level* activity)

s_i = stock of TV sets on hand at the end of the ith month in excess of current requirements (this is called the *stock level* activity)

x_i = increase in the number of workers at the beginning of month i (this is called the hiring activity)

y_i = decrease in the number of workers at the beginning of month i (this is called the firing activity)

d_i = number of workers idle during month i

The management is required to operate within two basic constraints. First, at the end of month i, there must be at least S_i television sets in stock. This is called the *sales requirement*. Second, the production level in month i is limited by the work force available in that month. This is called the *work force requirement*.

There are two ways of meeting the sales requirement, namely, through the output of workers currently engaged in production and from inventories entering the month i. That is, for $i = 1$, for instance,

(inventories at the end of period 1)
= (production during period 1) + (inventories entering period 1)
− (sales requirement for period 1) (1)

But

(production during period 1)
= (production capacity per worker)(number of workers) (2)

Suppose that each worker can produce four TV sets per month. Then Eq. (1) can be written as

$$s_1 = 4e_1 + 0 - S_1 \tag{3}$$

In the above equation, the inventories entering period 1 are set at zero by assumption. The process can be repeated for $i = 2$ to yield

$$s_2 = 4e_2 + \bar{s}_1 - S_2 \tag{4}$$

Notice the assumptions that went into writing Eq. (4). The production capacity of each worker is assumed to remain unchanged. This is generally true if the manufacturing process is an established one and the workers are all experienced. If this is not true, Eq. (4) can be modified accordingly. Notice also that the second term on the right side of Eq. (4), namely, the inventories entering period 2, is written as $\bar{s}_1$ rather than s_1. That is, the inventories leaving period 1 are not the same as those entering period 2. The assumption here is that there is a spoilage, attrition or unaccountable loss of items in stock. Expressions for $s_3, s_4, \ldots, s_T$ can be similarly written.

TABLE 8.5

A Time-Staged Production Scheduling Problems[a,b]

e_1	s_1	x_1	y_1	d_1	e_2	s_2	x_2	y_2	d_2	e_3	s_3	x_3	y_3	d_3	e_4	s_4	x_4	y_4	d_4	$\cdots$	e_T	s_T	x_T	y_T	d_T		Row
4	−10																			$\cdots$						$= S_1$	1
1		−1	1	1																						$= W_1$	2
	9				4	−10														$\cdots$						$= S_2$	3
−1				−1	1		−1	1	1																	$= O$	4
						9				4	−10									$\cdots$						$= S_3$	5
					−1				−1	1		−1	1	1												$= O$	6
											9				4	−10				$\cdots$						$= S_4$	7
										−1				−1	1		−1	1	1							$=$	8
		$\vdots$					$\vdots$					$\vdots$											$\vdots$			$\vdots$	$\vdots$
																				$\cdots$	4	−10				$= S_T$	$2T-1$
																					1		−1	1	1	$= O$	$2T$
c_1	i_1	h_1	f_1	n_1	c_2	i_2	h_2	f_2	n_2	c_3	i_3	h_3	f_3	n_3	c_4	i_4	h_4	f_4	n_4	$\cdots$	c_T	i_T	h_T	f_T	n_T	Minimize	

[a] e_i employment level; s_i stock level; x_i hiring; y_i firing; d_i idle-workers level; S_i sales requirement; W_i exogenous change in work force.
[b] s_i is assumed to be 10; $\overline{s}_i$ is assumed to be 9.

Now let us look at the work force requirement. Analogous to the previous development, we can write for $i = 1$

(active workers in period 1) + (idle workers in period 1)

= (workers entering period 1) + (workers hired) – (workers fired) (5)

That is,

$$e_1 + d_1 = W_1 + x_1 - y_1 \tag{6}$$

Typically, $W_1 = 0$ if the firm has just started business. For $i = 2$, the work force requirement is

$$e_2 + d_2 = e_1 + d_1 + x_2 - y_2 \tag{7}$$

Equations like (3), (4), (6), and (7) can be written for $i = 1, 2, 3, \ldots, T$. These equations are rearranged as follows:

$$\begin{array}{llllll}
4e_1 - s_1 & & & & & = S_1 \\
e_1 & -x_1 + y_1 + d_1 & & & & = W_1 \\
\quad \bar{s}_1 & & +4e_2 - s_2 & & & = S_2 \\
-e_1 & & -d_1 + e_2 & -x_2 + y_2 + d_2 & & = 0 \\
 & & & & & \vdots \\
 & & & & 4e_T - s_T & = S_T \\
 & & & & e_T \quad -x_T + y_T + d_T & = 0
\end{array} \tag{8}$$

These constraints are arranged compactly as shown in Table 8.5. This table is essentially an expression of the conservation law: "Total uses of resources or products must equal their total supply." The last row of this table shows the unit costs involved.

Now the objective function can be written as

$$\text{Min}\left[\sum_{i=1}^{T} (c_i e_i + i_i s_i + h_i x_i + f_i y_i + n_i d_i)\right]$$

The problem as stated above has all the features of a LP problem and can be solved using one of the standard LP algorithms. However, methods based on dynamic programming are more suitable to solve this problem.

Control of River Pollution

As another illustration of the time-staged formulation where spatial staging takes the place of time staging, consider a problem of pollution control. When organic waste is discharged into a stream, it is oxidized by the oxygen dissolved in the stream. This process reduces the concentration

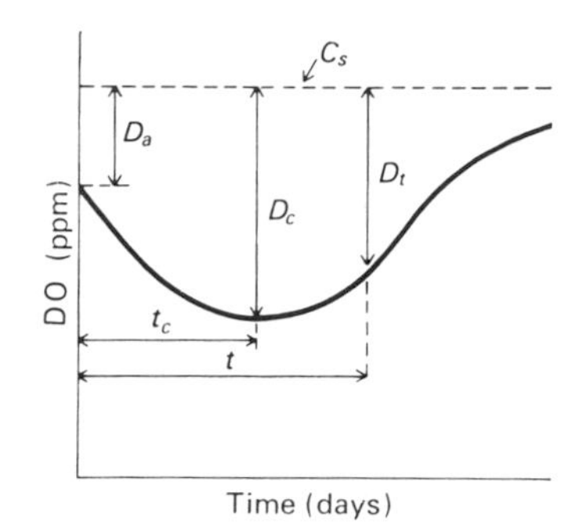

Fig. 8.8 A typical dissolved oxygen sag curve. (Reproduced with permission from J. C. Liebman and W. R. Lynn, *Water Resources Research* **2**, No. 3, 581–591 (1966)

of dissolved oxygen (DO) in the stream. The oxygen is eventually replenished by absorption from the atmosphere. The interplay of these absorption and consumption phenomena often causes the concentration of DO to go through a minimum as shown in Fig. 8.8. Because of the undesirable effects of low oxygen concentration, it is common practice for regulatory agencies to require that the minimum oxygen concentration be kept above some standard value. The standards set by regulatory agencies typically have a feature as shown in Fig. 8.9. Note that the minimum allowable oxygen concentration varies with position from the mouth of the river. As is true for many streams, it is necessary that the waste be treated to remove much of the oxidizable organic matter before it is discharged into the stream. The cost of this treatment depends upon the extent to which the oxygen demand

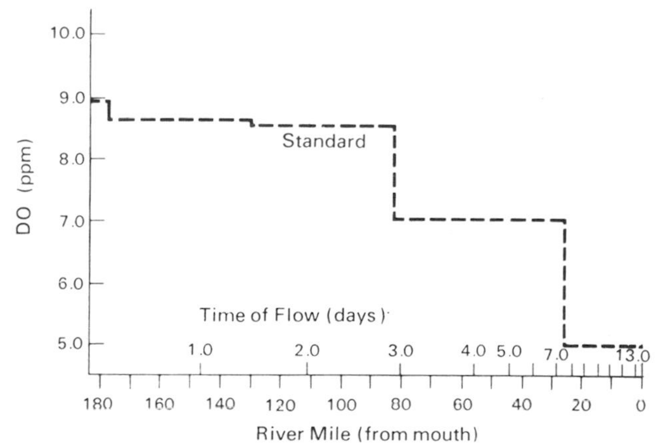

Fig. 8.9 Dissolved oxygen standards. (Reproduced with permission from J. C. Liebman and W. R. Lynn, *Water Resources Research* **2**, No. 3, 581–591 (1966).)

(called the biological oxygen demand or BOD) of the waste is reduced. A typical cost relationship is shown in Fig. 8.10.

The problem faced by a waste treatment plant is the following. If the flow, oxygen content, information about the discharge from the treatment plant, information about the BOD, etc. are given, then how can one minimize the amount of treatment (and so the treatment cost) while keeping the DO level in the lower reaches of the river at or above the minimum standards allowed by the regulatory agencies? The purpose of the present discussion is to show how this problem can be formulated as a mathematical programming problem. A control theoretic formulation (which is essentially a continuous version of the present formulation) and a solution method are presented in Chapter 9.

Consider a river divided into N segments or reaches. For convenience a single waste treatment plant is assumed to exist within each reach. The treatment of water at an upstream plant affects the pollution levels and the level of treatment required at a downstream plant. Hence, the policy of minimizing the cost at each individual plant subject to the standards being met, independently of what is done at other plants, may not yield the lowest overall waste treatment cost. If no conflicts of interest exist among the jurisdictions of the individual plants, it is possible to define an objective function as

$$\text{Min } z = \sum_{i=1}^{N} z_i \tag{9}$$

where z_i is the cost of treatment at the ith treatment plant. This minimization is to be accomplished by an optimal allocation of $p_1, p_2, \ldots, p_N$, where p_i is the percentage of reduction of the BOD at the ith treatment plant. The constraints in this problem are typically a set of equations that describe the relation between waste treatment and minimum DO concentration. These equations are derived from an understanding of the physical, chemical, and

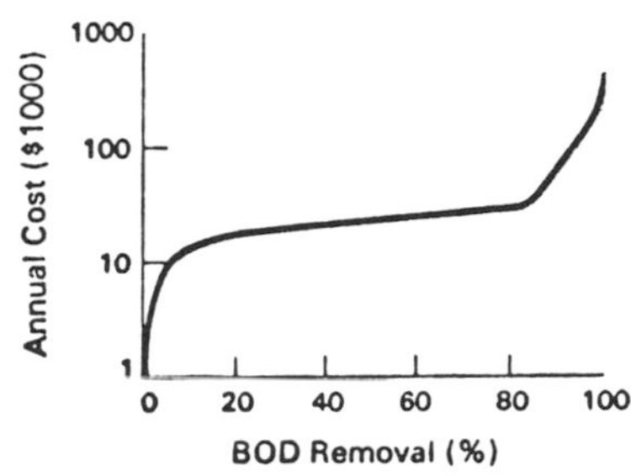

Fig. 8.10 Typical cost curve (Reproduced with permission from J. C. Liebman and W. R. Lynn, *Water Resources Research* **2**, No. 3, 581–591 (1966).)

biological processes that go on in the river. Ignoring these details for a moment, it is possible to write conservation laws among several variables. Toward this end let

f_i = quantity of water flowing in the ith reach of the river

$f_{t,i}$ = tributary flow into the ith reach; all tributary flow is assumed to come at the end of the reach

$f_{p,i}$ = flow from the treatment plant into the ith reach

Then one can write

$$f_{i+1} = f_i + f_{t,i} + f_{p,i+1} \tag{10}$$

Similarly, if one lets

D_i = oxygen deficiency at the beginning of the ith reach

D_S = saturation level of DO

$c_{i,f}$ = DO concentration at the end of the ith reach

c_i = minimum DO concentration in the ith reach

$c_{t,i}$ = DO concentration in the tributary flowing into the ith reach

$c_{p,i}$ = DO concentration in the treatment plant effluent flowing into the ith reach

then, one can write once again

$$D_{i+1} = [D_S - \{f_i c_{if} + f_{t,i+1} c_{t,i+1} + f_{p,i+1} c_{p,i+1}\}]/f_{i+1} \tag{11}$$

Similarly, one can write

$$b_{i+1} = [f_i b_{i,f} + f_{t,i+1} b_{t,i+1} + f_{p,i+1} b_{p,i+1} p_{i+1}]/f_{i+1} \tag{12}$$

where b_i stands for the BOD at the beginning of the ith reach. The subscripts on b carry a meaning similar to the subscripts of c. In addition to the constraints shown above, it is necessary to impose the constraint

$$c_i \geq S_i \tag{13}$$

where S_i is the standard set for the minimum oxygen concentration in the ith reach.

Thus the problem here is to minimize Eq. (9) subject to the constraints given in Eqs. (10)–(13). Notice that all the variables in the constraint equations are not explicitly given. For example, in implementing Eq. (13) one can read off values of S_i from Fig. 8.9. However, to get the values of c_i, one has to use an equation such as

$$c_i = C - (k_a b_i / k_d) \exp(-k_a u) \tag{14}$$

where

$$u = \min[T_i, \theta_i] \tag{15}$$

in which T_i, the time required for oxygen to reach minimum concentration in the ith reach is given by

$$T_i = \left[\frac{1}{k_d - k_a}\right] \ln\left\{\frac{k_d}{k_a}\left[1 - \frac{(k_d - k_a)D_i}{k_a b_i}\right]\right\} \tag{16}$$

and

$$\theta_i = \text{time spent in the } i\text{th reach}$$

In Eq. (16), k_d is the deoxygenation rate constant, k_d is the reaeration rate constant, and b_i is the BOD at the beginning of the ith reach. Equations such as Eq. (14)–(16) are generally obtained by studying the dynamics of river flow and the chemistry of the mixing process. Actual derivation of these equations falls beyond the scope of this work.

EXERCISES

1. Find by direct search the root of

$$y = e^x - x^2 = 0$$

2. Determine the maximum area of a triangle that can be inscribed in a semicircle of diameter $2r$.

3. Graph the following linear constraints. Is the enclosed region convex? If so determine the extreme points of this convex set:

$$\begin{aligned} -2x_1 + 5x_2 &\leq 10 \\ 2x_1 + x_2 &\leq 6 \\ x_1 + x_2 &\geq 2 \\ -x_1 + 3x_2 &\leq 3 \end{aligned}$$

4. Solve the following linear programming problem by the graphical method.

$$\text{Max } z = 3x_1 + 2x_2$$

subject to

$$\begin{aligned} 2x_1 - x_2 &\geq -2 \\ x_1 + 2x_2 &\leq 8 \\ x_1 \geq 0, x_2 &\geq 0 \end{aligned}$$

5. Solve the following linear programming problem by the simplex method:

$$\text{Min } z = x_1 + x_2 + x_3$$

subject to

$$\begin{aligned} x_1 \qquad\qquad - x_4 \qquad - 2x_6 &= 5 \\ x_2 \quad + 2x_4 \qquad + x_6 &= 3 \\ x_3 + 2x_4 - 5x_5 + 6x_6 &= 5 \end{aligned}$$

with $x_i \geq 0$ for all i.

6. Solve the following problem by the simplex method:

$$\text{Min } z = 2x_1 - 3x_2$$

subject to

$$\begin{aligned} 2x_1 - x_2 - x_3 &\geq 3 \\ x_1 - x_2 + x_3 &\geq 2 \\ x_1, \quad x_2, \quad x_3 &\geq 0 \end{aligned}$$

7. Solve the above problem by writing the dual problem first and then solving the dual problem using the graphical method.

8. Consider a discrete function $g(x)$. That is, $g(x)$ is defined for integer values of x. Then $g(x)$ is said to be convex if

$$g(x+1) - g(x) \geq g(x) - g(x-1) \qquad \text{for all} \quad x$$

The function $g(x)$ is concave if

$$g(x+1) - g(x) \leq g(x) - g(x-1) \qquad \text{for all} \quad x$$

(a) Using the above definitions, sketch the shape of convex and concave functions.

(b) If $g(x)$ is the cost of x units of something, then by using a simple argument, show that convex costs occur when each *additional* unit costs at least as much as the previous unit. Similarly, concave costs occur when each additional unit costs no more than the previous unit.

9. Verify the following functions for convexity or concavity. Indicate the range of a, b, x for which each is valid.

(a) $g(x) = ax + b$ (b) $g(x) = ax^2 + b$

(c) $g(x) = \begin{cases} ax + b & \text{for } x \geq 0 \\ -cx + d & \text{for } x \leq 0 \end{cases}$

(d) $g(x) = \begin{cases} 0 & \text{for } \quad x = 0 \\ 13 + 2x & \text{for } \quad x = 1, 2, 3, 4, 5 \\ 28.5 & \text{for } \quad x = 6 \end{cases}$

10. Consider a production planning problem stretching over several time periods $i = 1, 2, 3, \ldots, T$. For period i let x_i be the total man-hours scheduled and r_i be the regular-time wage per man-hour. Also u_i is a given total amount of regular-time man-hours available in period i, $1.5r_i$ is the "time and a half" wage rate per overtime hour, v_i is a given total amount of overtime hours available, and $2r_i$ is the "double-time" wage rate. Write an expression for the total cost $C_i(x_i)$. Is this cost function convex? [Hint: $C_i(x_i)$ resembles one of the $g(x)$ functions of the previous exercise.]

11. Consider a simple project involving the construction of a building. This project can be divided into several jobs or activities as follows:

Job	Estimated time to do the job (weeks)	Description of the job
a	10	Prepare blue prints
b	4	Drive piles
c	3	Erect wooden frame
d	6	Plumbing and wiring
e	3	Painting and carpeting

An essential ingredient in planning with networks is the establishment of a rule of precedence. For the building project, the required predecessor–successor relations to be satisfied are as follows:

$$\begin{array}{llll} a & \text{precedes} & b & \text{and } c \\ b & \text{precedes} & d & \text{and } c \\ c & \text{precedes} & e & \text{and } c \\ d & \text{precedes} & e & \text{and } c \end{array}$$

Using this information, construct a figure such as Fig. 8.3. Determine the critical path by formulating the problem as a linear programming problem. Using a convenient heuristic argument, develop another algorithm to determine the critical path.

12. Figure 8.3 is called an activity on node (or AON) diagram because each node in the figure represents an activity or job. A dual to this representation is the so-called *arrow diagram*, where each activity is represented by an arc. Construct an arrow diagram to the building project in Exercise 11. Discuss the advantages and disadvantages of the arrow diagram and the AON diagram.

13. There are many efficient LP codes currently available in computer libraries. Therefore, if one is good at using LP methods, it is worthwhile to force a problem into a LP format and solve it using one of the standard routines. Demonstrate this idea by considering the equation for multiple linear regression, namely,

$$\hat{y} = \sum_{i=1}^{m} a_i x_i$$

(a) Formulate the problem of finding $a_1, a_2, \ldots, a_M$ as an LP problem. [Hint: Write the objective function as: Min $\Sigma_j |e_j|$, where $e_j = y_j - \hat{y}_j =$ (actual value) − (predicted value).]

(b) Using an LP procedure, estimate the values of a_1, a_2, and a_3 for the following data:

x_1	x_2	x_3	y
1	1	8	6
1	4	2	8
1	9	−8	1
1	11	−10	0
1	3	6	5
1	8	−6	3
1	5	0	2
1	10	−12	−4
1	2	4	10
1	7	−2	−3
1	6	−4	5

14. Modify Table 8.5 by making the following assumptions:

(a) There is no spoilage of items inventoried from one time period to the next.

(b) s_i is measured in units of dozens. That is, $s_i = 1$ unit means $s_i = 12$ television sets.

(c) Introduce the concept of "shrinkage of work force" as follows. For every ten workers left idle in period i, only 9 are available for possible work in period $i + 1$.

15. Consider the production scheduling problem and Table 8.5 once again. Let $T = 4$, $W_1 = 0$, $S_1 = 100$, $S_2 = 200$, $S_3 = 100$, and $S_4 = 160$. Assume that there is no spoilage of items inventoried. Assume that there is no shrinkage effect on workers left idle. Assume that s_i is measured in units of 25's and d_i is measured in units of 10's.

(a) Reconstruct Table 8.5 to reflect these data.

(b) Devise a plan so that every period keeps inventory at a minimum.

(c) Repeat (b) and (c) for $S_3 = 160$ and $S_4 = 240$.

(d) Devise a plan so that the company hires as few workers as possible.

16. Consider the river pollution problem discussed in Section 7. Discuss qualitatively the effect of pumped storage operation on the cost of water treatment. (By pumped storage operation we mean that part of the downstream water is pumped back into an upstream reservoir for later release in order to generate electricity.)

SUGGESTIONS FOR FURTHER READING

1. Excellent books on optimization theory and practice are available. For a treatment of a variety of optimization ideas and illustrative examples, refer to
   G. S. G. Beveridge and R. S. Schechter, *Optimization: Theory and Practice.* McGraw-Hill, New York, 1970.
2. For a thorough treatment of linear programming with applications, consult
   S. I. Gass, *Linear Programming*, 3rd ed. McGraw-Hill, New York, 1969.
3. For a number of realistic applications of mathematical programming, see
   E. M. Beale (ed.), *Applications of Mathematical Programming Techniques.* American Elsevier, New York, 1970.
   A. A. Charnes and W. W. Cooper, *Management Models and Industrial Applications of Linear Programming*, Vols. I and II. Wiley, New York, 1961.
4. A number of topics related to the application of mathematical programming ideas to managerial decision making can be found in
   H. M. Wagner, *Principles of Operations Research with Applications to Managerial Decision Making.* Prentice-Hall, Englewood Cliffs, New Jersey, 1969.
5. Advanced topics related to hierarchial decomposition and aggregation of large scale systems can be found in
   D. A. Wismer (ed.), *Optimization Methods for Large Scale Systems with Applications.* McGraw-Hill, New York, 1971.
6. The river pollution problem discussed in Section 7 is based on
   J. C. Liebman and W. R. Lynn, The Optimal Allocation of Stream Dissolved Oxygen, *Water Resources Res.* **2**, No. 3, 581–591 (1966).

9

Linear Dynamical Structure

INTRODUCTION

Any moderately complex system invariably contains subsystems that are influenced by signals that change with time. What are signals? They are essentially functions of time. Many interesting signals of practical interest change in a predictable manner; so that information is conveyed from the sender to the receiver by transmitting a signal. When such a signal passes through a system, the way in which the signal is altered by the system is of critical importance. Stated differently, the way in which an external signal alters the behavior of a system is of importance in many problems. This behavior of a system, called *response*, is influenced both by the input signal and the characteristics of the system.

The dynamic nature of a system makes the problem more difficult to analyze. Consider an economic system such as an individual business enterprise, say, a restaurant. Suppose that the owner decides on a change in policy on prices or type of food served or time of operation or number of people to employ or some such change. This decision can only be made intelligently if the owner can anticipate the response of the system. That is, what will the restaurant system do if one tinkers with it? Will it result in a reduction of clientele? Will it lead to a profit? Answers to questions like these depend on a number of factors such as the characteristics of the system that encompasses the characteristics of the customers, the organization of the restaurant, geographic location of the restaurant, and several other factors. These problems are not unique to economic systems. Similar problems arise in the study of problems such as the spread of an epidemic, the design of a bridge, building, automobile, or an electrical circuit.

1 PICTORIAL REPRESENTATION OF SYSTEM STRUCTURE

Among the more important contributions that engineers have made to an understanding of complex systems is that of devising highly useful graphical symbolism with which to describe what are essentially mathematical relations. Electric circuit models, block diagrams, signal flow graphs, and analog simulation diagrams belong to this category. Each in its own characteristic way focuses attention on some aspect of the interrelationship among the variables or the elements of a system. Thus, it is possible to study the configuration of a system by temporarily relegating into the background the details of the individual components. This viewpoint makes it feasible to investigate, for example, different configurations, without being especially concerned with the particular physical interpretations given to the individual elements of the configuration. The outcome of such a viewpoint is of fundamental importance, for it means widely differing systems are reducible to a common structural portrayal.

A method of developing diagrams from a physical description of a system is the first step in developing more complex diagrams. Consider a tank of uniform cross-sectional area A being filled from the top at an inflow rate $f_1(t)$ and emptied from the bottom at an outflow rate $f_2(t)$. It is desired to determine a model that predicts the elevation z(t) of water in the tank at any time t. In handling a problem of this kind, it is useful to start with an iconic model of the system (see Fig. 9.1a). Next, it is necessary to identify the relevant variables. In this case, $f_1(t)$ and $f_2(t)$ are relevant independent variables and $z(t)$ is a useful dependent variable. The level $z(t)$ rises or falls according to whether f_1 is greater than or less than f_2. The rate at which $z(t)$ rises, say, therefore depends upon $(f_1 - f_2)$ and A. As derivatives with respect to time represent rates of change, the above flow situation can be translated

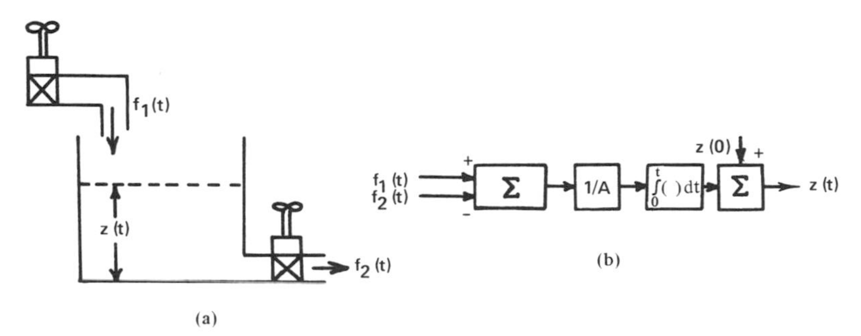

Fig. 9.1 From an iconic model (a) to the block diagram (b).

into

$$\frac{dz}{dt} = (1/A)(f_1 - f_2); \qquad z(t = 0) = z(0) \tag{1}$$

which is a trivially simple differential equation whose solution can be written as

$$z(t) = \int_0^t (1/A)(f_1 - f_2)\, dt + z(0) \tag{2}$$

Differential equations, such as Eq. (1), almost always arise in the study of dynamical systems as they describe the relation between variables and their rates of change.

The information contained in Eq. (2) above can be displayed as shown in Fig. 9.1b. Such figures are, for obvious reasons, called block diagrams. The significance of a block diagram lies not so much in the shapes of the individual blocks but in the mathematical operations represented by the blocks. In Fig. 9.1b, blocks bearing the $\sum$ signs are called *adders* or *summers* and are also often represented in the form of little circles. The adder adds all variables entering into it and produces an output which is an algebraic sum of all its inputs. Therefore, it is necessary to label all incoming arrows into an adder with a positive or a negative sign. The absence of any sign implies a positive sign. The output of this summer constitutes an input to the rectangular box labeled $1/A$, where A is a constant. If the constant value contained in the box is greater than unity, the box is called an *amplifier* with a gain; if the constant is less than unity, it is an *attenuator*. The box labeled with the integral sign is the *integrator*. From this, it is evident that the various boxes in a block diagram stand for various mathematical operations. Indeed, the contents of a box can represent any well-defined mathematical operation. In certain academic disciplines, the usage of certain shapes of blocks has become well entrenched, but it is important to remember that the shape of a block serves only a mnemonic purpose.

Fundamental Couplings

One point that was made in the preceding discussion is that systems can be visualized as collections of blocks or boxes—each representing some function. The argument can be reversed. Suppose that one has a collection of blocks—each capable of performing a certain simple and well-defined operation. Then, it is conceivable that one can build large and complex diagrams using the elementary boxes as building blocks. In order to do this it is necessary to understand the fundamental methods of coupling two or more blocks together. The types of couplings generally fall into four basic categories: (1) the hierarchical coupling, (2) the series or cascade coupling, (3) the parallel coupling, and (4) the feedback coupling. Study of these couplings shed some light on the *topological* and *dynamical structure* of a system.

Hierarchical Coupling

Truly complex systems, almost by definition, evade complete and detailed description. The dilemma in description is basically to seek a compromise between simplicity, a basic prerequisite for human understanding, and the need to take into account the numerous behavioral aspects of a complex system. Hierarchical description is a way to resolve this conflict.

The term *hierarchy* is currently being used to cover a variety of related yet distinct notions. The concept of hierarchy transcends various fields in the sense that it pertains to structural or organizational interrelationships rather than specifics of a phenomenon under study. Central to the hierarchical approach is the belief that systems can be described in terms of their "levels." For instance, a disagreement between two individuals in a company can often be resolved by going to a higher level of the corporation. A dispute regarding the riparian rights of a river than flows through two states is often settled at the "higher" level—a federal authority or arbitration board. Clearly, the relation "is at a higher level than" plays a fundamental role in revealing the hierarchical structure of a system.

The hierarchical structure of a system may be "involuntary" as in, say, man's internal ecology or "voluntary" as the structure of the external society in which he lives. A hierarchical structure appears to be inevitable in any model involving human beings either as individuals or as groups. Indeed, such varied disciplines as physiology, psychology, and sociology may be regarded as various echelons or levels of a hierarchy—if viewed from a proper perspective. For instance, one can study living organisms at the levels of a cell, a tissue, an organ, and the organism. Such a study would fall under the realm of physiology. If one wishes to specialize on the human organism and the forces operating on it, one could climb up the hierarchical pyramid from the human organism to the family, the village, the city, the country, and the whole world level.

Similarly, an industrial automation system can be viewed from three different "levels": (a) physical processing of material and energy, (b) control and processing of information, and (c) economic operation in terms of efficiency and profit. Consider the problem of integrated planning and controlling of a steel making process. From a total systems viewpoint, three main functions to be performed by a planning committee are: production planning, scheduling and coordination of operation, and process control. The top echelon in the hierarchy accepts customer orders, and analyzes and groups the orders for profitability of production and assigns weekly or monthly work loads or production schedules. The next echelon breaks down the production schedules into specific instructions for individual operations. Thus, their main function is *coordination*. The lowest level units are mainly concerned with the control of the process itself.

Hierarchy can also be detected in many naturally occurring systems. Ecological systems (ecosystems, for short), for example, are hierarchical in several respects. The taxonomy (classification) of a species is a hierarchy, the "who-eats-whom" structure is essentially hierarchical. Consider an ecological system in which plants get their energy from the sun, herbivores get theirs from plants, and carnivores depend on herbivores.

In hierarchical structures each block represents a responsibility, function, or activity. The blocks are arranged in levels or echelons. Blocks in lower echelons have a smaller scope within the system and they enter the picture only if finer resolution levels in the space–time specification are contemplated.

The vertical lines connecting various blocks do not necessarily represent physical connections or couplings; they signify relationships. A given block usually incorporates some, but not necessarily all aspects of the subsystems that are represented by blocks below and joined by lines. The goal tree in Fig. 2.1 (page 27) is an example of hierarchical structure. The trophic web shown in Fig. 10.7 (page 382) is another example of hierarchical structure.

Series Coupling

Decisions are seldom isolated activities. Today's decisions influence tomorrow's actions. Many problems in planning involve a number of decisions, each affecting the other. Therefore, it is useful to understand the structure by which a sequence of decision-making stages arise. The structure of interest here is the *serial multistage structure* or the *cascade structure*. In such structures, the output from one stage becomes the input to the next stage as shown in Fig. 9.2. This kind of cascade structure may occur in a variety of different ways. For instance, one may want to take two simple systems, say $\mathscr{A}_1$ and $\mathscr{A}_2$, and build a more complex system by connecting them in cascade. It is also possible that system $\mathscr{A}_1$ is malfunctioning and one may wish to correct the problem by connecting $\mathscr{A}_1$ and a correcting device $\mathscr{A}_2$ in cascade. In

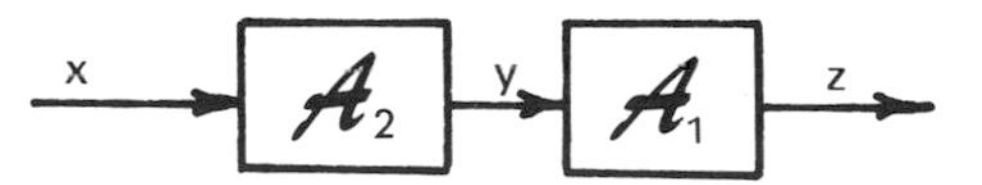

Fig. 9.2 A cascade connection.

Fig. 9.2, $\mathscr{A}_1$ and $\mathscr{A}_2$ represent any abstract oriented objects. The input–output relation for such a cascade connection of oriented objects can be developed as follows:

$$z = \mathscr{A}_1 y, \qquad y = \mathscr{A}_2 x \tag{3a}$$

where it is assumed that the output y is obtained by applying the input x on $\mathscr{A}_2$. Similarly, the operator $\mathscr{A}_1$ operates on y to yield z. Combining the above two relations

$$z = \mathscr{A}_1(\mathscr{A}_2 x) \tag{3b}$$

In the above equations $\mathscr{A}_1$ and $\mathscr{A}_2$ are any valid mathematical operations. They could, for instance, represent coupling matrices, differential operators, or integral operators. If they are indeed matrices, Eq. (3) can be written as

$$z = \mathbf{A}_1 y, \qquad y = \mathbf{A}_2 x \tag{4a}$$

$$z = \mathbf{A}_1(\mathbf{A}_2 x) \neq \mathbf{A}_2(\mathbf{A}_1 x) \tag{4b}$$

Thus, it is clear that in a cascade connection $\mathscr{A}_1$, does not always commute with $\mathscr{A}_2$.

Parallel Coupling

Another common type of interconnection is the parallel or branching structure. Branching structure arises while building decision trees. An example of parallel structure arises if one wishes to correct the performance of a system using the so-called feed-forward compensation as shown in Fig. 9.3. In this figure, the following relations hold:

$$w = \mathscr{A}_1 x + \mathscr{A}_2 x = (\mathscr{A}_1 + \mathscr{A}_2)x \tag{5}$$

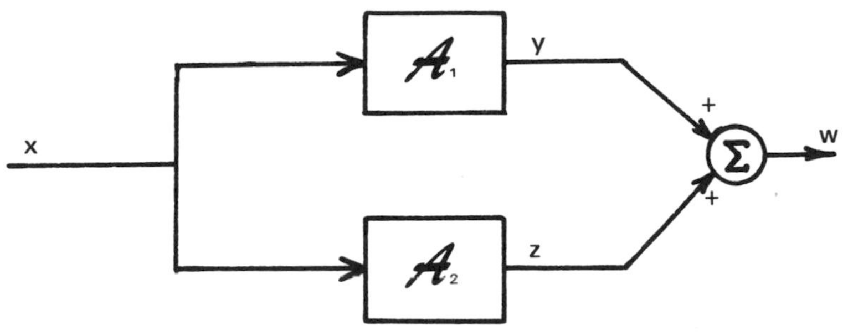

Fig. 9.3 A parallel connection.

Feedback Coupling

Frequently, pictograms such as those in Fig. 9.1 are much too crude, and it is often a convenient guide to both intuition and exposition to conceive more sophisticated representations, when, for example, one wishes to control or regulate the inflow $f_1(t)$ into the tank so that the water level is always maintained at a predetermined level M in the tank. A block diagram reflecting this modification is shown in Fig. 9.4.

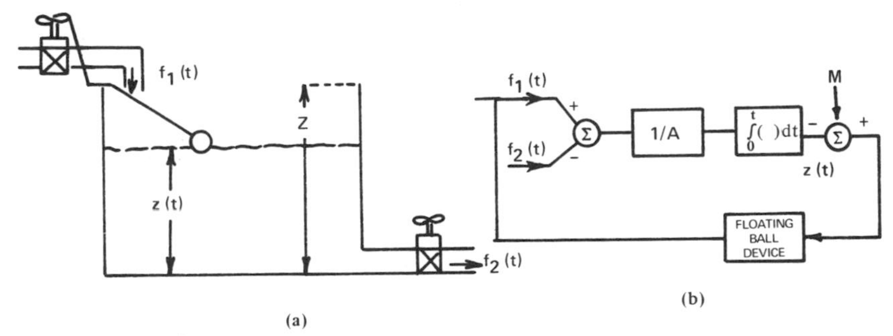

Fig. 9.4 (a) An Iconic model of a water tank with a float. (b) A block diagram model for water level control.

Here, the actual elevation of water level $z(t)$ is monitored by a floating ball and the information about the water level is fed back to the inflow valve. The valve mechanism is arranged such that a flow is maintained as long as the *difference* between M and $z(t)$, that is, $(M - z(t))$ is positive. For this reason, this kind of arrangement is called *feedback*. The closed path is called a *feedback loop*.

The algebra of block diagrams developed for the cascade structure can easily be extended to feedback structures. For example, the simple block diagram with a feedback structure, shown in Fig. 9.5 can be equivalently

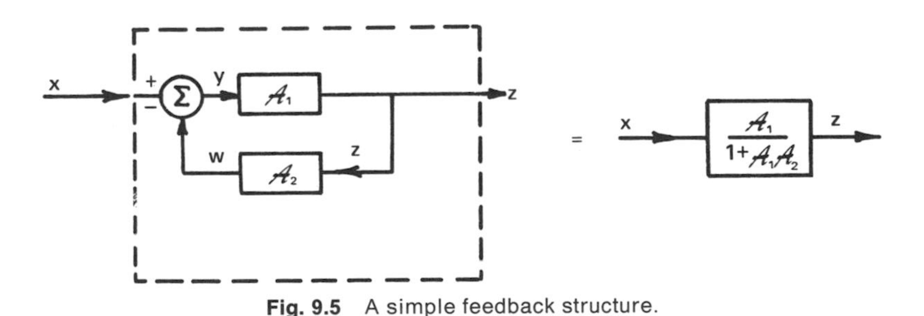

Fig. 9.5 A simple feedback structure.

represented by a single block diagram using elementary analysis. From Fig. 9.5,

$$z = \mathscr{A}_1 y, \qquad y = x - w, \qquad w = \mathscr{A}_2 z$$

Therefore,

$$y = x - \mathscr{A}_2 z, \qquad z = \mathscr{A}_1(x - \mathscr{A}_2 z)$$

which yields

$$z = \frac{\mathscr{A}_1}{1 + \mathscr{A}_1 \mathscr{A}_2} x \tag{6}$$

Now an interesting question can be posed. Suppose that the information in Fig. 9.5 is described by

$$z = \mathscr{A}_3 x \tag{7}$$

rather than by Eq. (6). Then is there feedback? As long as Eq. (7) represents the only source of information one has, one is justified in thinking that the system has no feedback. If, however, the meaning of $\mathscr{A}_3$ is elaborated as in Eq. (6), then one is justified in believing in the existence of feedback. Thus, whether or not there is feedback is really a point of view.

The importance of feedback structure and the role it plays can be revealed by using the concept of feedback to build increasingly complex blocks from relatively simple blocks. For example, an element of infinite gain (i.e., extremely large gain) can be constructed using two unity gain elements and positive feedback, as shown in Fig. 9.6. Now using this infinite gain element as another block, it is possible to build several types of functional blocks. For instance, an infinite gain element with a unit gain element in its feedback path yields a sign inverter, as shown in Fig. 9.7. The input–output relation

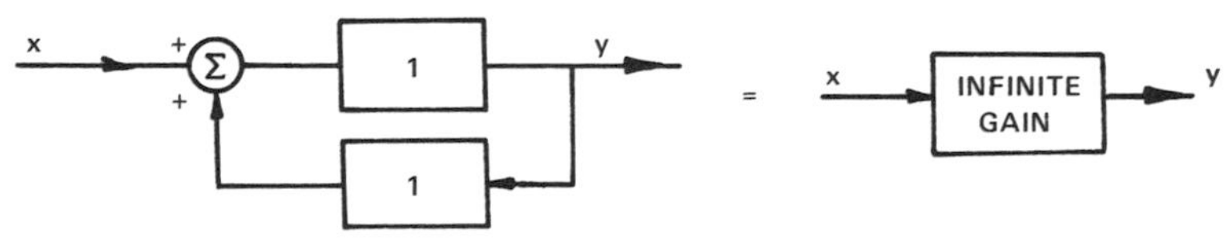

Fig. 9.6 Creation of an infinite gain element using positive feedback.

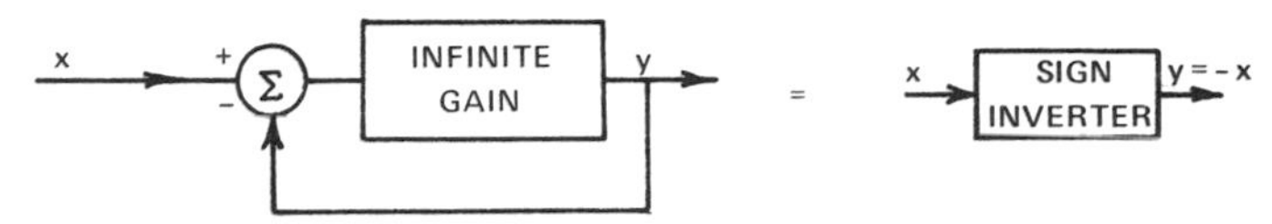

Fig. 9.7 Creation of a simple computational element using an infinite gain element as a building block.

for this configuration is

$$\frac{y}{x} = \frac{\mathscr{A}_1}{1 - \mathscr{A}_1\mathscr{A}_2} = \frac{\mathscr{A}_1}{1 - \mathscr{A}_1} \approx \frac{\mathscr{A}_1}{-\mathscr{A}_1} = -1 \tag{8}$$

or

$$y = -x \tag{9}$$

The infinite gain element plays a fundamental role in analog computation. Table 9.1 shows some of the mathematical operations one can develop using this building block approach.

TABLE 9.1

Construction of Progressively Complex Building Blocks

	$\mathscr{A}_1$	$\mathscr{A}_2$	Type of feedback	Equivalent element
1.	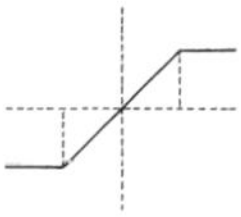	1	+	$y = \operatorname{sgn} x$
2.	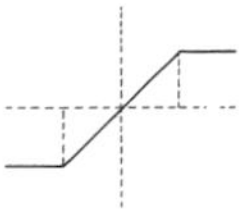	2	+	$y =$ 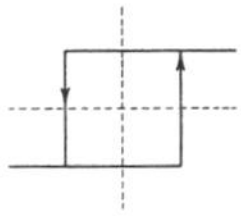
3.	sgn	1	−	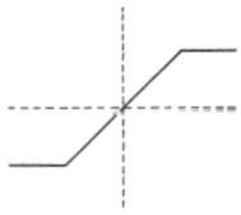
4.	sgn	1	+	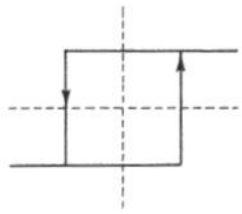
5.	Integrator	1	−	Unit lag
6.	Unit lag	1	+	Integrator
7.	⌊ (the floor function)	1	−	Unit time-lag

Feedback structure is of importance primarily in goal-seeking applications: systems in which the quality of performance is measured by how closely one approaches the desired goal. The impressive success of feedback in controlling both the effects of disturbances and changes in system component characteristics carries with it certain inherent disadvantages. First, feedback requires a more complicated system configuration. Second, feedback tends to make a system unstable (or more sensitive). The stability problem is discussed further in Section 3.

Fascinating examples involving feedback or mutual causal relations can be found in almost all walks of life. Consider a problem from urban ecology. Assume that the attributes of interest are population, sanitation, and some other related areas. Figure 9.8 shows one method of pictorially representing the interactions among these attributes. The arrows indicate causal directions of influence. A plus sign $(+)$ on the directed line segment indicates that a change in the attribute at the root of an arrow causes a corresponding change in the same direction in the attribute at the tip of the arrow. For example, an increase in material standard of living implies an increase in sanitation facilities. As may be noted, some of the arrows form closed paths or loops. If a loop contains an even number of positive or negative signs, it is evident that any change tends to reinforce itself. For example, in the path $P \to M \to C \to P$, an increase in the number of people could lead to an increase in the standard of living, which, in turn, could lead to further migration into the city, which, in turn, increases the number of people in the city. On the other hand, the loop $P \to G \to D \to P$ results in an overall negative influence on P. Here an increase in population causes an increase in pollution (G), which increases the probability of diseases, which, in turn, causes a decrease in population. Thus, any deviation in population gets corrected or counteracted. This kind of counteraction tends to produce *stability*, although the system sometimes oscillates in its attempt to reach stability. The nature of oscillations is influenced by the time lags or time delays involved in the counteraction and the magnitude of counteraction.

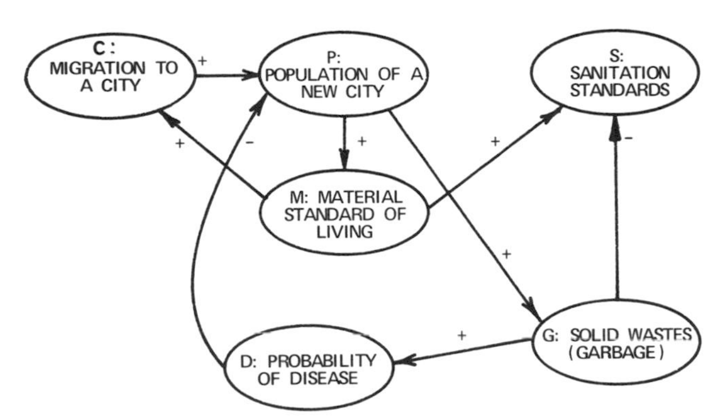

Fig. 9.8 Feedback paths in an urban system.

Oscillatory behavior due to feedback can also be found in economic activities. In 1935, the famous economist, John Maynard Keynes, gave the first satisfactory and adequate explanation of the essential mechanisms on which general economic activity depends. The ideas conceived by Keynes neatly fit the modern concepts of feedback control theory. Keynes' ideas can be arranged as a block diagram, as shown in Fig. 9.9. The starting point of this diagram is the notion that economic activity depends on the rate at which goods are bought. These goods are of two kinds: capital goods and consumer goods. The next step is to recognize that capital goods are another form of investment. The third step is to identify money as the driving force to buy all kinds of goods. This money can be obtained (as profits and wages) by manufacturing the said goods—thus making the cycle complete. However, *all* profits and wages do not automatically go back as investment; some money is saved. If this is all there is to a model, the system would run down to a stop. What keeps the system from halting is new investment. Therefore,

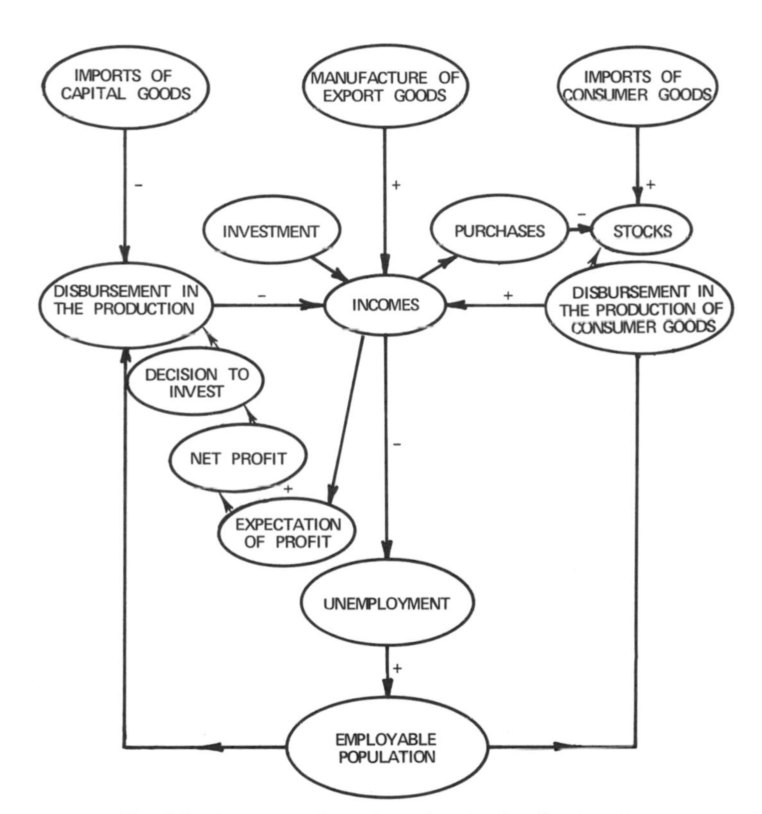

Fig. 9.9 An economic system showing feedback paths.

the level of economic activity and employment depends on the rate of investment. However, the rate of investment depends on the economic activity. Thus, by observing Fig. 9.9 one can observe several feedback loops in an economic activity.

An excellent example of a complex feedback system is the homeostasis mechanism of the human body. Homeostasis means the ability of the body to maintain one or more important signals very close to the desired level. For instance, human body temperature is maintained very close to 98.6°F even if the body is exposed to wide-ranging temperatures. The human body also does a remarkable job in maintaining the level of blood sugar. Many attempts to understand the mechanism of homeostasis were made by using feedback models. One of the major difficulties in modeling the functions of a human body is that each component (or black box) stimultaneously performs a variety of functions. This has the biological advantage of providing several alternative parallel paths (i.e., providing redundancy) by means of which a control action can be accomplished. For instance, muscles used for work are also used for shivering. Blood performs a galaxy of functions.

The complexity attendant in these feedback mechanisms is best illustrated by a specific example. Suppose that a person suffers a sudden loss of blood (called hemorrhage). As a result of the abrupt drop in blood volume, blood pressure falls. As it is vital to maintain a full quota of blood supply to the brain (damage sets in if the brain is deprived of blood for more than 2 or 3 sec), parts of the circulatory system constrict in most of the body to supply the available volume of blood at the required pressure to the brain. The heart and brain vessels are not constricted so as to maintain the supply there. In severe cases, even blood supply to the kidneys falls down to almost zero level. Concurrently, the heartbeat speeds up to as high as 200 beats per minute. This is the immediate body reaction to a hemorrhage. In the hours following the incident, other body reactions rally in: large quantities of body fluids are absorbed from the intestines to restore the blood volume. The patient becomes extremely thirsty, reflecting a need for fluids and salts.

The pituitary, a pea sized gland located at the base of the brain, secretes vasopressin, a hormone that limits the loss of fluid from the body and allows the kidneys to retain salt and water and releases another hormone, ACTH, which stimulates the adrenal glands, located above the kidneys. The adrenal glands secrete cortisol, which simplifies the flow of blood into the veins from the capillaries and also increases the protein concentration in the blood. Thus, the blood composition is slowly restored to its normal level.

The complexity of the problem can be seen by just focusing on one particular aspect: the secretion of cortisol by the adrenal glands. A block diagram for the release of cortisol is shown in Fig. 9.10. The dotted line in the figure indicates that we do not know as yet how changes in blood pressure trigger

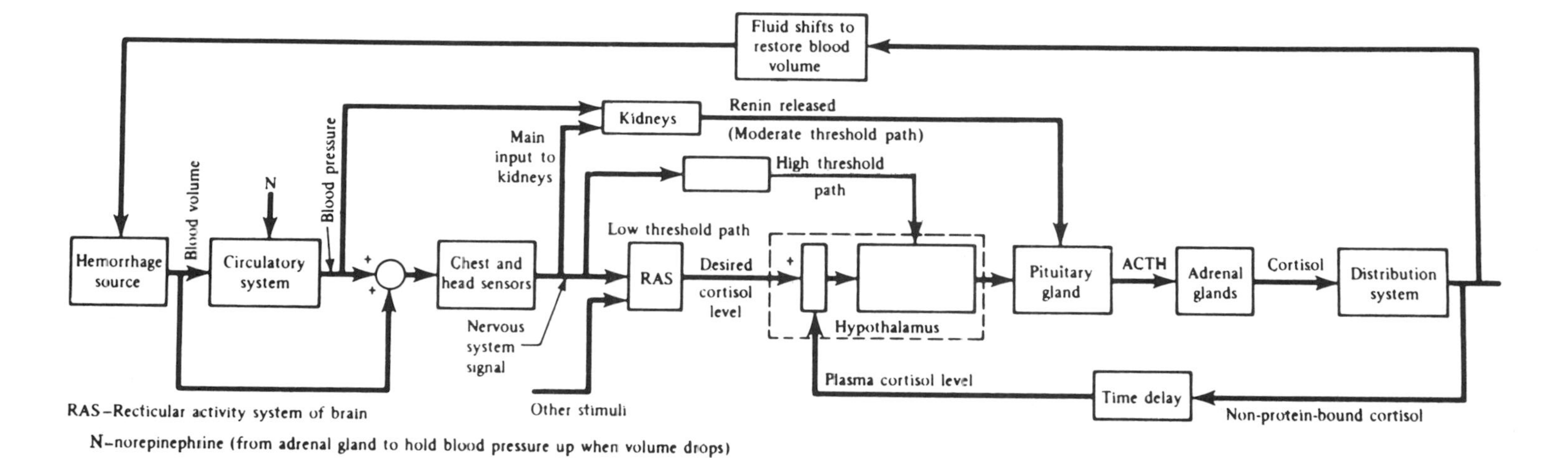

Fig. 9.10 Complete model for cortisol secretion after hemorrhage. (Reproduced with permission from E. E. David, Jr., and J. G. Truxall (eds.), *The Man Made World*, Part III, Polytechnic Institute of Brooklyn, 1968.)

the action of the pituitary gland. Current knowledge indicates the following possibility. Sensors in the head and chest sense the fall of blood pressure and prompt the brain to send a signal to the reticular activity system (RAS) of the brain. The RAS is the portion of the brain that controls sleep, coma, and consciousness. The RAS apparently determines the desired level of cortisol in the blood and sends a message to that effect to the hypothalamus. The hypothalamus compares this desired level with the actual level present in the blood, and the corresponding error signal is sent to the pituitary gland. The pituitary responds by releasing ACTH, and this ACTH triggers the release of cortisol from the adrenal glands. (There is an interesting question that one can ask here. Why did the hypothalamus send its signal to the pituitary rather than a direct signal to the adrenal glands? In other words, what is the need for the intermediate step of ACTH release if all ACTH does is to trigger the adrenal glands? There are several interesting explanations for this mechanism. The student is encouraged to read the literature to find a satisfactory answer.)

The above mentioned model was thought to be correct until the middle 1960s. Recent experiments suggest the inadequacy of the model. Experiments were conducted by deactivating the hypothalamus control and by saturating the feedback element that measures the cortisol level. (This can be done by drugs.) It was found that the body still copes with hemorrhage, indicating alternative paths. Further experiments revealed that an alternative path exists through the kidneys. Indeed, it was found that the kidneys respond both to blood pressure and to the reaction of the pressure sensors located in the chest and head. The response of the kidney is manifested by the release of a chemical substance, renin, which stimulates the pituitary. If this path through the kidneys was also blocked, it was found that there was another path directly from the pressure sensors to the hypothalamus. The block diagram displaying all these paths is shown in Fig. 9.10.

2 TRANSFER FUNCTION MODELS

To a novice, a basic difficulty in handling dynamic systems arises because of a need to handle differential equations, which are more difficult than algebraic equations. Some of this difficulty can be bypassed by the use of Laplace transforms and transfer functions. The role of Laplace transforms in solving differential equations is analogous to the role of logarithms in performing difficult multiplications. To multiply two numbers, one finds the logarithms of the numbers, adds the logarithms, and finds the antilogarithm of the sum that is the required product. Thus, logarithms helped to convert a difficult mathematical operation, such as multiplication, into a relatively simple operation, namely, addition. The Laplace transforms play an analogous role while solving *linear* differential equations. Thus, use of Laplace transforms is restricted to linear systems.

The Laplace transforms $F(s)$ of a function $f(t)$ is defined as

$$L\{f(t)\} = F(s) = \int_0^{\infty} f(t)e^{-st}\,dt \tag{1}$$

where L is the symbol for taking the Laplace transform. The symbol L is read as "Laplace transform of." Given a function $f(t)$, its Laplace transform $F(s)$ can therefore be found by merely evaluating the integral on the right side of Eq. (1). For a majority of functions of practical interest, this integral exists and can be evaluated. For some typical functions, this has been evaluated and the results summarized in Table 9.2. With the help of this table, it is now possible to find the Laplace transform of a linear differential equation, such as Eq. (1.1), which is reproduced here:

$$A\frac{dz}{dt} = (f_1 - f_2); \qquad z(t = 0) = z(0)$$

Taking the Laplace transform on both sides

$$L\left\{A\frac{dz}{dt}\right\} = L\{f_1(t) - f_2(t)\} \qquad \text{or} \qquad AL\left\{\frac{dz}{dt}\right\} = F_1(s) - F_2(s) \tag{2}$$

By virtue of relation (12) of Table 9.2, the above equation becomes

$$A\{sZ(s) - z(0)\} = F_1(s) - F_2(s) \tag{3}$$

TABLE 9.2

Laplace Transform Pairs

	$f(t)$, time domain	$F(s)$, frequency domain
1.	Unit step $u(t)$	$1/s$
2.	Unit impulse $u_1(t)$	1
3.	Ramp function t	$1/s^2$
4.	$e^{\pm \alpha t}$	$1/s \mp \alpha$
5.	$\sin \omega t$	$\omega/(s^2 + \omega^2)$
6.	$\cos \omega t$	$s/s^2 + \omega^2$
7.	$t^n/n!$	$1/s^{n+1}$
8.	$t^n e^{\alpha t}/n!$	$1/(s - \alpha)^{n+1}$
9.	$kf(t)$	$kF(s)$
10.	$f(kt)$	$(1/k)F(s/k)$
11.	$f_1(t) \pm f_2 t$	$F_1(s) \pm F_2(s)$
12.	$f^1(t)$ or $\dot{f}(t)$	$sF(s) - f(0)$
13.	$f^{(n)}(t)$	$s^n F(s) - s^{n-1} f(0) - \cdots - f^{n-1}(0)$
14.	$f(t - \tau)$	$e^{-s\tau} F(s)$
15.	$\int_0^t f(t)\,dt$	$F(s)/s$
16.	$\int_0^t f(t) g(t - \tau)\,dt$	$F(s)G(s)$

which is an algebraic equation in the variable s and can be readily solved for $Z(s)$ to yield

$$Z(s) = \frac{F_1(s) - F_2(s)}{As} + \frac{z(0)}{As} \tag{4}$$

where $Z(s)$ is the Laplace transform of $z(t)$—the elevation of water level in Fig. 9.1. To get $z(t)$, it is now necessary to perform the operation of an inverse Laplace transform. That is,

$$z(t) = L^{-1}\{Z(s)\} = L^{-1}\left\{\frac{F_1(s) - F_2(s) + z(0)}{As}\right\} \tag{5}$$

where the symbol L^{-1} is read as "inverse Laplace transform of." Now one can use Table 9.2 backwards to obtain $z(t)$ from $Z(s)$.

Example 1. Solve $y'' + 3y' + 2y = f(t)$, where $f(t)$ is a unit step function and $y(0) = 0 = y'(0)$. Taking Laplace transforms on both sides,

$$[s^2 Y(s) - sy(0) - y'(0)] + 3[sY(s) - y(0)] + 2Y(s) = 1/s$$

Substituting the values of the initial conditions and solving

$$Y(s) = 1/s(s+1)(s+2)$$

The inverse transform of the above is not available in Table 9.2. However, the right side of the above equation can be expanded in partial fractions as

$$\frac{1}{s(s+1)(s+2)} = \frac{A}{s} + \frac{B}{s+1} + \frac{C}{s+2}$$

where A, B, and C are constants to be determined. The above equation can be rewritten as

$$\frac{1}{s(s+1)(s+2)} = \frac{A(s+1)(s+2) + Bs(s+2) + Cs(s+1)}{s(s+1)(s+2)}$$

Because the denominators are equal, the numerators can be equated:

$$A(s+1)(s+2) + Bs(s+2) + Cs(s+1) = 1$$

The values of A, B, and C can now be evaluated by assigning judiciously chosen values to s. For instance, when $s = 0$, the second and third terms on the left side vanish and

$$A(1)(2) = 1 \qquad \text{or} \qquad A = \tfrac{1}{2}$$

Similarly, when $s = -1$

$$B(-1)(1) = 1 \qquad \text{or} \qquad B = -1$$

When $s = -2$

$$C(-2)(-1) = 1 \qquad \text{or} \qquad C = \tfrac{1}{2}$$

Therefore,

$$Y(s) = \frac{1}{s(s+1)(s+2)} = \frac{\frac{1}{2}}{s} - \frac{1}{(s+1)} + \frac{\frac{1}{2}}{(s+2)}$$

Taking inverse Laplace transforms on both sides

$$y(t) = \tfrac{1}{2} - e^{-t} + \tfrac{1}{2}e^{-2t}$$ ■

A special case of Eq. (4) arises when the initial condition $z(0)$ vanishes. Then the ratio $Z(s)/\{F_1(s) - F_2(s)\} = 1/As$ is referred to as the *transfer function.* Several features of a transfer function are best understood by considering a more general differential equation than the simple one considered so far. Consider an nth order, linear differential equation with constant coefficients, such as

$$\begin{aligned} &y^{(n)} + b_{n-1}y^{(n-1)} + \cdots + b_1 y^{(1)} + b_0 \\ &\quad = a_m x^{(m)} + a_{m-1}x^{(m-1)} + \cdots + a_1 x^{(1)} + a_0 \end{aligned} \tag{6}$$

where $y^{(n)} = d^n y/dt^n$ and $x^{(n)} = d^n x/dt^n$. Using Table 9.2, the Laplace transform of each term can be written as

$$L\{y^{(n)}(t)\} = L\left\{\frac{d^n y}{dt^n}\right\} = s^n Y(s) - y^{(n-1)}(0) - \cdots - y(0) \tag{7}$$

$$L\{x^{(m)}(t)\} = L\left\{\frac{d^m x}{dt^m}\right\} = s^m X(s) - x^{(m-1)}(0) - \cdots - x(0)$$

where $y(0), y^{(1)}(0), \ldots, y^{(n-1)}(0)$ and $x(0), x^{(1)}(0), \ldots, x^{(m-1)}(0)$ are the initial conditions specified with Eq. (6). Substituting Eq. (7) in Eq. (6) and rearranging,

$$\frac{Y(s)}{X(s)} = \frac{a_m s^m + a_{m-1}s^{m-1} + \cdots + a_1 s + a_0}{s^n + b_{n-1}s^{n-1} + \cdots + b_1 s + b_0} + \frac{IC}{s^n + \cdots + b_1 s + b_0}$$

where IC stands for all the terms associated with the prescribed initial conditions. If the system is initially relaxed, that is, initially at rest, all the initial conditions are zero, then

$$\frac{Y(s)}{X(s)} = \frac{a_m s^m + a_{m-1}s^{m-1} + \cdots + a_1 s + a_0}{s^n + b_{n-1}s^{n-1} + \cdots + b_1 s + b_0} \tag{8}$$

which is the transfer function of the system described by Eq. (6).

Notice that one would have obtained Eq. (8) directly from Eq. (6) by replacing every kth derivative in it by s^k for $k = 1, 2, \ldots$. That is, even though

s is defined by the complex number $s = \sigma + i\omega$, s in the transformed domain stands for differentiation with respect to t in the t domain. Because of this, it is customary to interpret s as a differentiation operator and $1/s$ or s^{-1} as an integration operator. Furthermore, this observation enables one to go from transfer function format to a differential equation and vice versa with relative ease.

It should also be noted that in transfer functions representing physical systems, the highest power of s in the numerator cannot exceed the highest power of s in the denominator. A reason for this restriction can be explained as follows. Physical systems are sluggish; they can respond to forcing functions that vary relatively slowly. Stated differently, they act as filters and allow signals with low frequencies to pass through them to appear at the output. High-frequency components of a signal generally do not appear at the output. As the Laplace transform variable $s = \sigma + i\omega$ represents a complex frequency, it can be readily seen from Eq. (8) that as $s \to \infty$, the output $\to \infty$ unless the denominator polynomial is of a higher degree.

It is evident that a transfer function carries with it all the information of a system except information about the initial state of the system. In situations in which the initial state is not very important, a transfer function description is just as good as a differential equation description. Sometimes a transfer function is even more convenient because it involves algebraic equations which are much easier to handle than differential equations. It should be kept in mind that a transfer function carries information only about the behavior of a system; any information about the structural nature of a system is permanently lost once a system is characterized by a transfer function or, for that matter, by a differential equation as well. Thus, if one wants to construct a block diagram by starting from a transfer function, one has many alternatives. Each method leads to a different structural configuration. It is up to the analyst to choose one out of many configurations. Some of the methods of getting block diagrams from transfer functions are described in the example below. The starting point for all these methods in an equation such as Eq. (8) where $n \geq m$.

Example 2. Construct a block diagram to simulate the transfer function

$$Y(s)/X(s) = (s + 1)/(s^2 + 8s + 4) \tag{A}$$

Method 1 This procedure starts by rewriting the above as

$$W(s) = [1/(s^2 + 8s + 4)]X(s) \qquad \text{and} \qquad Y(s) = (s + 1)W(s) \tag{B}$$

The differential equation corresponding to the first of the above equations can be written, by inspection, as

$$\frac{d^2w}{dt^2} = x(t) - 8\frac{dw}{dt} - 4w \tag{C}$$

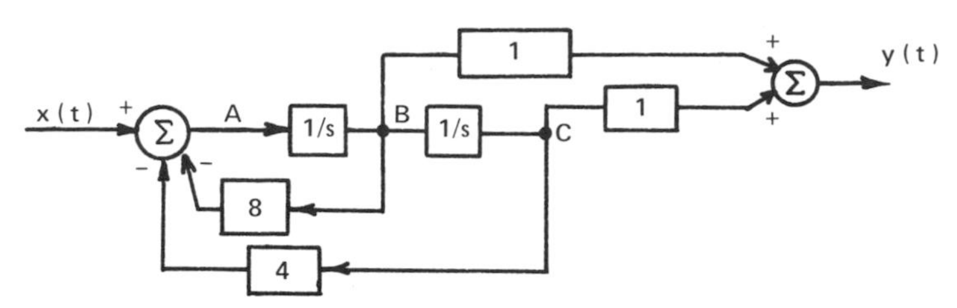

Fig. 9.11 A block diagram to simulate a transfer function by Method 1.

To initiate the construction of a block diagram d^2w/dt^2 is assumed to be known and available at point A in Fig. 9.11. Integrating this twice yields $w(t)$. This operation is shown in the block diagram by passing d^2w/dt^2 through two boxes labeled $1/s$, each representing an integration operation. Now dw/dt is available at the point B and $y(t)$ at the point C. Outputs from the points B and C are tapped and fed respectively to two boxes with gains 8 and 4. The outputs of these boxes, when substracted from $x(t)$, yield d^2w/dt^2, according to Eq. (C). This seemingly artificial way of handling d^2w/dt^2 is called *bootstrapping*. Now $y(t)$ can be obtained from Eq. (B) as

$$y(t) = \frac{dw}{dt} + w(t)$$

This is readily obtained by adding the outputs from points B and C of Fig. 9.11

Method 2 In this method, Eq. (A) is first rearranged as

$$(s^2 + 8s + 4)Y(s) = (s + 1)X(s)$$

Dividing throughout by s^2, the highest power of s, and rearranging,

$$Y(s) = (1/s + 1/s^2)X(s) - (8/s + 4/s^2)Y(s)$$

A block diagram representation of the above equation is shown in Fig. 9.12. Notice how $y(t)$ is obtained by a negative feedback connection. The diagram

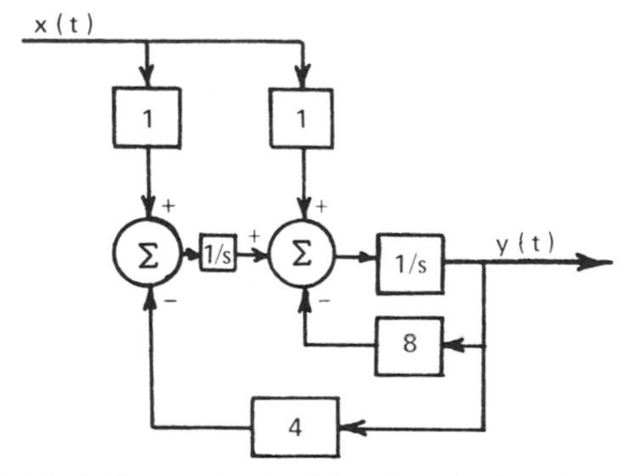

Fig. 9.12 A block diagram to simulate a transfer function by Method 2.

is best understood by traversing each directed path until the output is reached and taking the algebraic sum of the expressions thus obtained. ■

The above two methods are very much similar to the way an *analog computer* is programmed.

Aggregation of Block Diagram Models

It is evident that block diagrams could become large and complex if the system to be studied is large and complex. Drawing a block diagram is only a first step in simulating a system on a computer. The transition from a block diagram to, say, an analog computer simulation diagram is relatively simple. However, if the block diagram has too many blocks, its simulation may become inordinately complex. To be more specific, a typical modern analog–hybrid computer facility can handle about 50 integration operations. Even though digital computers can handle even larger systems with ease, there is always a time–cost trade-off. Due to these reasons, there is always an interest to represent a system of large order by a model of small order. One method of order reduction (a type of aggregation), based on a continued fraction, is presented here.

The gist of the aggregation method is based on two simple observations. For instance, by inspecting Example 2 and the transfer function in Eq. (8), one can conclude that:

(a) The denominator coefficients of the transfer function, that is, b_0, $b_1, \ldots, b_{n-1}$, appear as gains in the feedback paths and the numerator coefficients, that is, $a_0, a_1, \ldots, a_n$ appear as gains in the feed forward paths of the block diagram.

(b) There are as many integrators in the feed forward path, that is, $1/s$ types of terms, as the degree of the denominator polynomial of the transfer function.

These two observations are more or less valid regardless of the form of the block diagram obtained from a transfer function. Therefore, there is no loss of generality if the integration operation is separated out from the gain operation. That is, there is no loss of generality if Fig. 9.13a, say, is redrawn as Fig. 9.13b. In either configuration, the transfer function can be written as

$$\frac{Y(s)}{X(s)} = \frac{(10/s)}{1 + (10/s)(0.5)} \tag{9}$$

$$= \frac{1}{(0.5) + \dfrac{1}{(10/s)}} \tag{10}$$

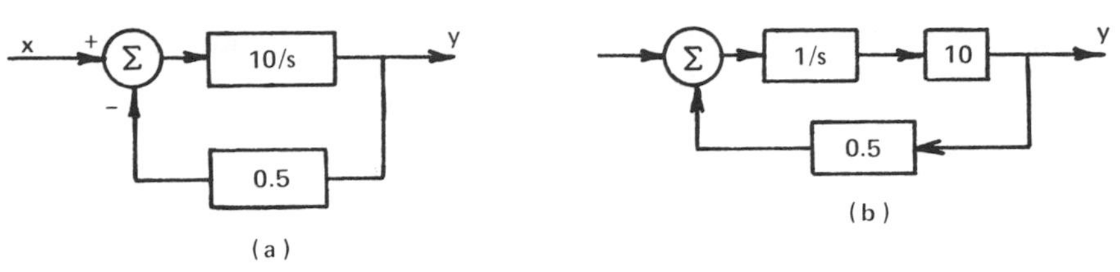

Fig. 9.13 Two different ways of representing a transfer function. (a) Realization of Eq. (2.9); (b) Realization of Eq. (2.10).

Equation (10) is indeed a continued fraction expansion of Eq. (9). Notice that the first quotient of this expansion, namely 0.5, is the gain of the feedback path and the second quotient is the product of $1/s$ and the gain 10 of the feed forward path.

This type of argument can be applied to any block diagram of the type shown in Fig. 9.14 with n feed-forward paths and n feedback paths. In this configuration, it is possible to identify the quotients of a continued fraction. The actual procedure to be followed is given below.

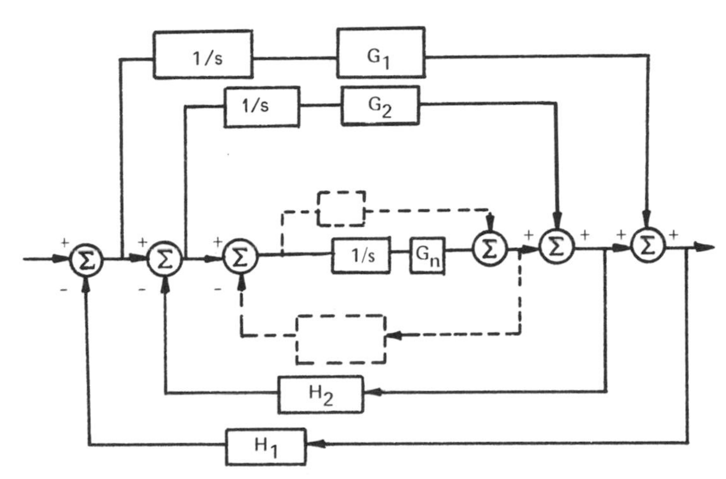

Fig. 9.14 A general block diagram with n forward paths and n feedback paths.

Step 1 Rewrite the given transfer function such that the numerator and denominator polynomials appear in ascending powers of s as in

$$\frac{Y(s)}{X(s)} = \frac{a_0 + a_1 s + \cdots + a_{n-1} s^{n-1} + a_n s^n}{b_0 + b_1 s + \cdots + b_{m-1} s^{m-1} + b_m s^m} \tag{11}$$

where $m \geq n$.

Step 2 Divide the denominator polynomial by the numerator polynomial and using the procedure of synthetic division represent Eq. (11) as follows:

$$\frac{Y(s)}{X(s)} = \cfrac{1}{H_1 + \cfrac{1}{(G_1/s) + \cfrac{1}{H_2 + \cfrac{1}{G_2/s + \cfrac{1}{H_3 + \cdots}}}}} \tag{12}$$

Step 3 Identify the successive quotients of the continued fraction expansion as the transfer functions of the feedback and feed-forward paths, in that order of Fig. 9.14.

The synthetic division terminates after $2m$ quotients are obtained where m is the order of the denominator polynomial. The aggregation method asserts that an approximation to the given transfer function can be obtained by truncating the higher order quotients of the synthetic division and retaining only the first $2k$ quotients where $k \leq m$. How good such an approximation would be is a question left to the student to be worked out as a term project. The following two examples illustrate the ideas presented above.

Example 3. Develop a block diagram to the transfer function

$$\frac{Y(s)}{X(s)} = \frac{s+1}{s^2 + 8s + 4}$$

Step 1 The above equation is rearranged as

$$\frac{Y(s)}{X(s)} = \frac{1+s}{4 + 8s + s^2}$$

Step 2 The process of synthetic division is carried out as shown below:

$$1 + s\overline{)4 + 8s + s^2}(\ 4$$
$$\underline{4 + 4s}$$
$$4s + s^2\overline{)1 + s}(\ 1/4s$$
$$\underline{1 + s/4}$$
$$3s/4\overline{)4s + s^2}(\ 16/3$$
$$\underline{4s}$$
$$s^2\overline{)\tfrac{3}{4}s}(\ 3/4s$$
$$\underline{\tfrac{3}{4}s}$$
$$\underline{0}$$

Step 3 The quotients of the synthetic division are:

$$4, \quad 1/4s, \quad 16/3, \quad 3/4s$$

These values are equated, in that order, to H_i and (G_i/s) for $i = 1, 2, \ldots, 2m$, where m is the order of the denominator polynomial of the transfer function. Thus,

$$H_1 = 4, \qquad G_1 = 1/4$$
$$H_2 = 16/3, \qquad G_2 = 3/4$$

A block diagram for the transfer function is constructed as shown in Fig. 9.15. ∎

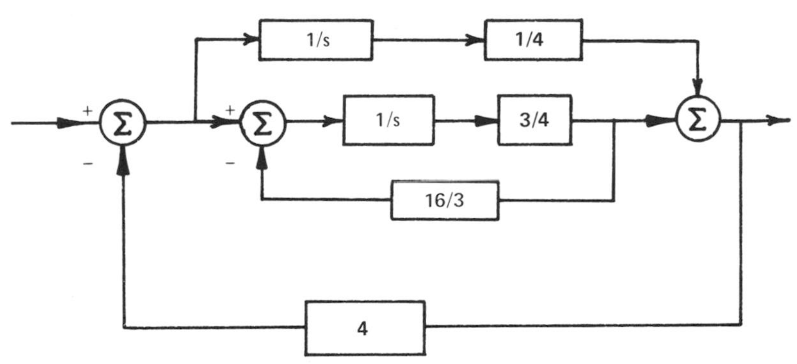

Fig. 9.15 Figure to illustrate the aggregation procedure described in Example 3.

Example 4. A system was originally modeled by the transfer function

$$\frac{Y(s)}{X(s)} = \frac{1441.53s^3 + 78319s^2 + 525286.125s + 607693.25}{s^7 + 112.04s^6 + 3755.92s^5 + 39736.73s^4 + 363650.56s^3 + 759894.19s^2 + 683656.25s + 617497.375}$$

This is a seventh-order system. It therefore needs at least seven integrator blocks. Assume that the computer available has only a capability to simulate two integrator blocks. Therefore, it is required to simplify the given transfer function so that the simplified transfer function can be simulated on the available computer.

After rearranging the numerator and denominator polynomials, the above equation is expanded, using synthetic division, as

$$\frac{Y(s)}{X(s)} = \cfrac{1}{1.016 + \cfrac{1}{\cfrac{4.05}{s} + \cfrac{1}{-0.067 + \cfrac{1}{\cfrac{-3.804}{s} + \cdots}}}}$$

As the simplified model is required to be of second order ($m = 2$), the first four ($2m$) quotients of the synthetic division are retained and remaining terms discarded. That is,

$$\frac{Y(s)}{X(s)} \approx \cfrac{1}{1.016 + \cfrac{1}{\cfrac{4.05}{s}, + \cfrac{1}{-0.067 + \cfrac{1}{\cfrac{-3.804}{s}}}}}$$

$$\approx \frac{0.250s + 1.034}{s^2 + 0.509s + 1.051}$$

This is a second-order transfer function. It can be verified that response of this transfer function approximates that of the original system. ■

Transfer Function Model of a Delay

Delay is an important attribute in system studies and appears in various forms. An automobile driver seeing an obstacle on the road ahead may take as much as 0.3 sec before reacting and moving the foot toward the brake. This is called the *human reaction time.* If a valve on a pipeline is opened, it may take some time before the chemical substance in the pipeline reaches a reaction chamber. This is an example of a *transportation lag.* A real estate entrepreneur, seeing a demand for housing, starts building and it may take as long as three years before the buildings are ready for occupancy.

In all these cases, whatever the name is—dead time, delay, lag, etc.—the system fails to respond immediately. Figure 9.16 shows a pure delay. The input-output relation for this figure is

$$y(t) = x(t - \alpha)$$

Taking Laplace transforms on both sides

$$Y(s) = \int_0^\infty x(t - \alpha)e^{-st}\,dt = e^{-\alpha s}\int_0^\infty x(u)e^{-u}\,du = e^{-\alpha s}X(s)$$

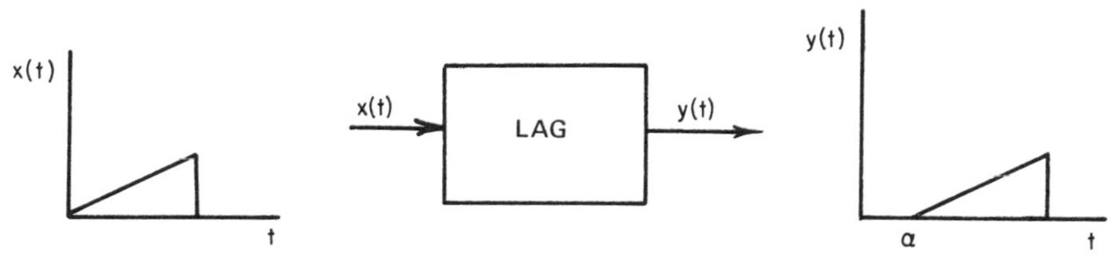

Fig. 9.16 Input–output relation of a lag element.

Therefore, the transfer function of a delay is

$$\frac{Y(s)}{X(s)} = e^{-\alpha s} \tag{13}$$

The havoc that a delay can cause in a feedback loop can be easily illustrated by considering a simple feedback system shown in Fig. 9.17. The overall transfer function of this system is

$$\frac{Y(s)}{X(s)} = \frac{G(s)e^{-\alpha s}}{1 + G(s)e^{-\alpha s}}$$

The exponential term in the numerator simply says that the output is delayed by α units. The exponential term in the denominator causes more problems because the denominator term is the characteristic equation and one is often interested in the roots of the characteristic equation. Indeed, it can be shown that the characteristic equation

$$1 + G(s)e^{-\alpha} = 0 \tag{14}$$

has an infinite number of roots. This illustrates the important fact that even relatively simple systems become difficult to handle in the presence of delays. One method of handling the above situation is to introduce a *sampler* in the feed-forward loop. Discussion of this aspect is beyond the scope of this chapter.

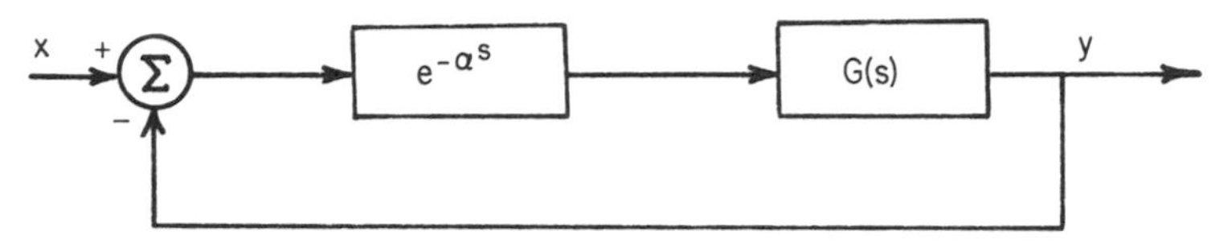

Fig. 9.17 A lag element in a feedback circuit.

Transfer Function Model of a Decision Maker

One of the characteristics of a complex system is the presence of a human being (or society) as a decision-making component. There are two aspects of human decision making that are of interest here. In a broad sense decision making implies selection of one among a set of possible alternatives. In a much narrower sense, decision making implies how an individual tries to perform an assigned task by following a sequence of decision-making stages. The former type of decision making is far more difficult because it involves subjective judgment of values, ordering of preferences, and deciding what is right and what is wrong. In this broad context, the concept of transfer function has never been applied with significant success.

This section deals with the much narrower problem of describing the input–output behavior of a human operator acting as an element in a closed-loop system. Consider the simplified block diagram of such a system as shown in Fig. 9.18. This diagram could represent, for example, the manual control of one axis of an aerospace vehicle. The operator is required to detect a system error visually and to perform a manual movement such that the error is reduced. His response must be consistent both with stability and dynamic performance requirements.

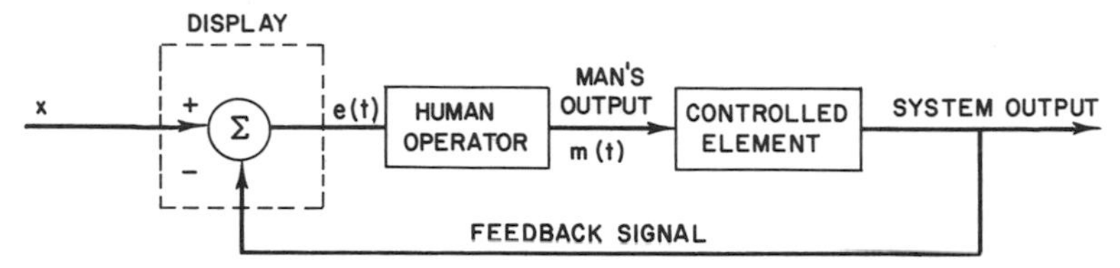

Fig. 9.18 A simple control system with a human operator.

From the standpoint of a system designer, it is extremely desirable that the input–output characteristics of the human operator be expressed mathematically in order to make predictions of system performance possible. However, the construction of an adequate mathematical model of a human operator's behavior, even for a particular task, is extremely difficult. Due to the difficulty of obtaining adequate descriptions of the human operator's behavior in a complex system, nearly all of the work in the past has been done with control systems as simple as that of Fig. 9.18. That is, the operator is assumed to have one or at the most two inputs. Extraneous inputs and feedback loops are generally neglected in the formulation of models. If the display is a subtraction device that presents to the operator only the difference between the input r and the controlled variable c, the task is known as "compensatory tracking." The operator's function is to reduce the error e to zero. If both the reference input r and the output c are displayed, the task is called "pursuit tracking."

Characteristics of the Human Operator in a Control System

The input–output behavior of a human operator in a control system is characterized by the following major features:

(a) *Reaction time*: The behavior of the operator is characterized by the presence of a pure time delay or transport lag, which can be clearly observed in the response to step function inputs.

(b) *Low-pass behavior*: Visual examination (and Fourier analysis) of tracking records reveals that the tracker tends to attenuate high frequencies, the amount of attenuation increasing as the frequency increases.

(c) *Task dependence*: The operator is able to adjust his input–output characteristics in order to perform his control function with a wide range of controlled element dynamics.

(d) *Time dependence*: The dependence of the operator's characteristics on time can be seen in two forms: first, his performance changes with time as he learns, and second, he is capable of sensing changes in environmental parameters and controlled system parameters and adjusting his characteristics accordingly.

(e) *Prediction*: The ability of the human operator to predict the course of a target based on past performance is well known. This ability to extrapolate is important in tracking since it means that tracking behavior is different with "predictable inputs" (such as sine waves or constant frequency square waves) than it is with random or random-appearing inputs. Tracking with a predictable input has been called "precognitive" tracking.

(f) *Nonlinearity*: For certain tasks the operator's behavior appears to be approximately linear, while for other tasks his behavior is nonlinear.

(g) *Determinacy*: A human operator is a nondeterministic system, since his performance is different in successive trials of the same experiment. However, his variability is small in situations where training time is adequate and the task is not considered difficult. Consequently, a deterministic model may be used to describe his performance in a statistical sense.

(h) *Intermittency*: There is a considerable body of evidence which indicates that the human operator behaves as a discrete or sampling system in certain tracking operations.

The various human operator models appearing in the literature are attempts to include the major characteristics listed above into a mathematical description.

A widely used transfer function for a human controller is

$$H(s) = K \frac{T_0 s + 1}{(T_1 s + 1)(T_2 s + 1)} e^{-Ts} = \frac{\text{control}}{\text{error}} = \frac{M(s)}{E(s)}$$

where T is the human reaction time defined earlier. The minimum value of T is usually 0.2 sec; it could be higher if the subject is fatigued, under stress or under the influence of drugs; K is a gain constant which is automatically adjusted by the operator to a suitable value so that the overall system is stable; T_0 is a lead or prediction time constant and is a measure of the extent to which a human operator tries to predict the future error; T_1, T_2 are lag time constants, representing delays in the human processing of signals.

For very simple control tasks, the human operator is relaxed and this condition corresponds to

$$T_0 = 0, \qquad T_1 = T_2 = \tfrac{1}{3}$$

That is, the operator does not bother to predict and behaves in a sluggish way. For complicated tasks, the operator attempts to predict the future and reacts. Thus, T_0 rises to values as high as unity. As the complexity of the task increases, the operator attempts to react more and more quickly, and T_1 and T_2 fall down to as low as $\frac{1}{20}$ sec.

3 THE BEHAVIOR OF SYSTEMS: STABILITY

A system is characterized by (a) its structure and (b) its behavior. Enough has been said about some of the structural features. This section is devoted to an understanding of the behavior of systems. However nicely a system may behave, an analyst would like to know about the chances of its ever displaying a previously unobserved erratic behavior. Certainly jet planes and suspension bridges could be built for a fraction of their current cost if one is willing to let them fall down every few years, but usually a major share of the cost of a system is devoted to making the chances of such a catastrophe sufficiently low.

Because of this inherent fear of the unexpected, one usually observes a system for a period of time before making statements about the repertoire of behaviors of a system. The time one takes to make the said observations depends on a number of things including one's expectations based on past experience. However if one is content with a passive observation, then systems study would not be interesting. This leads to the question of how one can meaningfully intervene and alter the course of behavior of a system.

One of the important indicators of the dynamic behavior of a system is its stability. Once again, the common sense meaning of stability is different from its scientific meaning. To some the word "stable" connotes something that is immobile. The Golden Gate Bridge and the Empire State Building are stable structures, but on a windy day the structures sway appreciably. Thus, stability is not synonymous with immobility. Stability could imply changes within certain limits. Within what limits? With a strong enough wind either structure might be blown down. Thus, the word "stability" in its scientific sense not only implies limits to the changes in the system but also implies limits to the disturbances with which the system is supposed to withstand.

Just as "stability" connotes immobility to some it is suggestive of "goodness" to some others. A fire in a coal mine may burn for days, displaying a high degree of stability, but not much goodness. A given situation can at once be "stable and good," "stable and bad," "unstable and good," and "unstable and bad," depending on the observer's point of view and value structure. A dictatorship in a country A may be stable and good from the viewpoint of country B, whereas it may appear stable and bad to the democratic forces within the country A, it may appear unstable and bad to the deposed leaders, or it may appear unstable and therefore good to the philosophical optimist.

Another widely held misconception is that a system has to be stable to be "useful." It should be remembered that the more a system is stable, the less it is sensitive to external disturbances. That is, a highly stable system could very well be a highly sluggish system. For instance, missiles and torpedos seeking their targets are intentionally designed to be unstable. If a torpedo hits its target before its guidance system enters the unstable region, it has served its useful purpose and the system disintegrates itself before it has a chance to enter the unstable region. Had the missile or torpedo been designed to exhibit stable properties throughout its trajectory, it may end up being sluggish to the point of uselessness.

Historically, the concept of stability has attracted a great deal of attention. The concept originated in mechanics. (A rigid body is said to be in stable equilibrium if it returns to its original position after sufficiently small perturbations; a motion is said to be stable if it is insensitive to sufficiently small perturbations.) This concept is later generalized to include "motions" described by the state variables of a system. It should be mentioned at the outset that important and interesting questions relating to the *existence* of stability and equilibrium arise, especially if the systems are characterized by differential equations. Stability can be analyzed, therefore, by explicit solution of the differential equations describing the system. In essence this involves expressing the solution y of the given differential equation as a sum of exponentials:

$$y = A_1 e^{\lambda t} + A_2 e^{\lambda_2 t} + \cdots + A_n e^{\lambda_n t} \tag{1}$$

If all λ's are negative (or have negative real parts if the λ's are complex numbers), then one can say that the system is stable. This approach is called the *indirect method* of stability analysis as it requires a solution of the equations before questions about stability can be answered. This indirect method not only tells whether a system is stable or not but also, as a by-product, tells the extent of stability or the lack of it. If only a "yes" or "no" answer to the stability question is required, one can apply a direct method such as the Routh or Hurwitz criterion. Many other sophisticated direct methods such as those using root-locus, Nyquist, and Bode plots not only provide a binary answer to the stability question but also give an idea of the degree of stability without requiring us to solve the differential equations describing the system behavior. However, all of the latter four methods are only applicable to linear, time-invariant systems.

The most powerful and widely used method of stability analysis is Liapunov's direct method. This is applicable to nonlinear as well as time-varying systems and does not require an explicit solution of the differential equations. Interestingly enough, this represents a closing of the circle in the quest for stability analysis as the *Liapunov functions* are essentially generalized energy functions which are merely offshoots of the Hamiltonian and the Lagrangian of classical mechanics.

Routh's Stability Test

The Routh stability test is a simple algebraic procedure to determine whether a system is stable or not. In order to apply this test it is necessary to have a transfer function description of the system.

Given a transfer function, such as

$$\frac{Y(s)}{X(s)} = \frac{a_m s^m + a_{m-1} s^{m-1} + \cdots + a_1 s + a_0}{s^n + b_{n-1} s^{n-1} + \cdots + b_1 s + b_0} \tag{2}$$

it is possible to expand it in partial fractions and rewrite it as (if the denominator polynomial has real and distinct roots)

$$\frac{Y(s)}{X(s)} = \frac{K_1}{(s + \alpha_1)} + \frac{K_2}{(s + \alpha_2)} + \cdots + \frac{K_n}{(s + \alpha_n)} \tag{3}$$

Taking the inverse Laplace transform, one can get

$$y(t) = (K_1 e^{-\alpha_1 t} + K_2 e^{-\alpha_2 t} + \cdots + K_n e^{-\alpha_n t}) x(t) \tag{4}$$

Therefore, if the input $x(t)$ is bounded, the output $y(t)$ remains bounded as all the exponential terms are decaying exponentials. If, however, one of the exponentials is of the type $e^{+\alpha_i t}$ then the output goes unbounded even if the input is bounded. When such a situation arises the system is said to be unstable.

Thus, one method of checking stability is to inspect the roots of the *characteristic equation*

$$s^n + b_{n-1} s^{n-1} + \cdots + b_1 s + b_0 = 0 \tag{5}$$

If the characteristic equation has at least one positive root, then that root contributes a positive exponential term to the solution and makes it unstable. In systems terminology roots of the characteristic equation are called *poles* of the transfer function. Sometimes the poles may be complex numbers in the complex s plane. If a transfer function has complex poles, then for stability it is necessary that the poles lie in the left half of the complex plane (i.e., the roots have negative real parts). This test, though simple in concept, is very difficult to apply in practice because it is very difficult to factor polynomials of third or higher degree. Furthermore, there is no need to know the exact location of the poles as long as one can tell whether all the poles are in the left half of the s plane or not. The Routh test accomplishes just this. The exact theory behind the test is based on complex function theory but the mechanics of applying the test are very simple as described below.

To locate the roots of the polynomial

$$b_n s^n + b_{n-1} s^{n-1} + \cdots + b_1 s + b_0 = 0 \tag{6}$$

an array is constructed as follows:

$$\begin{array}{l|llll}
s^n & b_n & b_{n-2} & b_{n-4} & \cdots \\
s^{n-1} & b_{n-1} & b_{n-3} & b_{n-5} & \cdots \\
s^{n-2} & c_1 & c_2 & c_3 & \cdots \\
s^{n-3} & d_1 & d_2 & d_3 & \cdots \\
\vdots & \vdots & & & \\
s^1 & g_1 & & & \\
s^0 & h_1 & & &
\end{array}$$

The first and second rows go to the right until b_1 and b_0 are reached. The constants c_i are calculated from the coefficients in the first two rows as

$$c_1 = \frac{b_{n-1} b_{n-2} - b_n b_{n-3}}{b_{n-1}}$$

$$c_2 = \frac{b_{n-1} b_{n-4} - b_n b_{n-5}}{b_{n-1}}$$

$$\vdots$$

An easy way to remember these formulas is to use the mnemonic device shown below

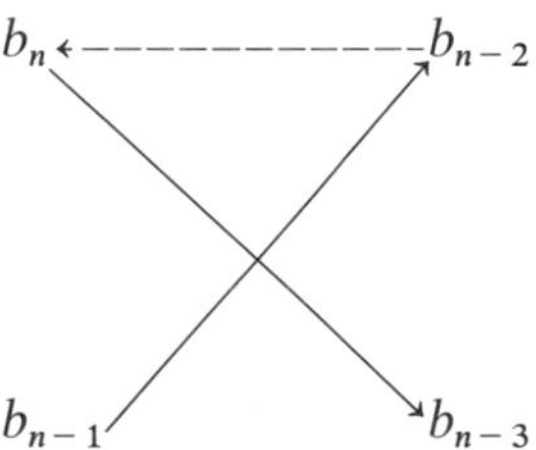

The d_i's are calculated in exactly the same way using the coefficients in s^{n-1} and s^{n-2} rows. The process is repeated until one reaches the last row. The Routh criterion says: *The number of roots with positive real parts is equal to the number of sign changes of the first column in the coefficient array just developed.*

Example 5. The characteristic equation of a system is given by

$$s^5 + 2s^4 + 3s^3 + 6s^2 + 2s + 1 = 0$$

Is this stable? Discuss.

The Routh array is

$$
\begin{array}{c|cccc}
s^5 & 1 & 3 & 2 & 0 \\
s^4 & 2 & 6 & 1 & 0 \\
s^3 & 0 & \frac{3}{2} & 0 & \\
s^2 & \dfrac{0-3??}{0} & & & \\
s^1 & & & & \\
s^0 & & & &
\end{array}
$$

As division by zero is not possible, it is not possible to proceed further. In such cases the zero is replaced by a small positive number, say ϵ, and the procedure is continued:

$$
\begin{array}{c|cccc}
s^5 & 1 & 3 & 2 & 0 \\
s^4 & 2 & 6 & 1 & 0 \\
s^3 & \not{0}\epsilon & \frac{3}{2} & 0 & \\
s^2 & \dfrac{6\epsilon-3}{\epsilon} & 1 & 0 & \\
s^1 & \frac{3}{2}-\dfrac{\epsilon^2}{6\epsilon-3} & 0 & 0 & \\
s^0 & 1 & & &
\end{array}
$$

Now ϵ is allowed to go to zero from the positive side. Then the term in the leftmost column of the s^2 row $\to -\infty$ and of the s^1 row $\to \frac{3}{2}$. Thus, there are two sign changes and the system is unstable.

Alternatively, one can define a new variable, say, $w = 1/s$, and rewrite the given equation as

$$w^5 + 2w^4 + 6w^3 + 3w^2 + 2w + 1 = 0$$

The Routh test can be applied once again to this new polynomial. ■

Example 6. Does the characteristic equation

$$s^4 + 3s^3 + 6s^2 + 12s + 8 = 0$$

represent a stable system?

$$
\begin{array}{c|cccc}
s^4 & 1 & 6 & 8 & 0 \\
s^3 & 3 & 12 & 0 & 0 \\
s^2 & 2 & 8 & 0 & \\
s^1 & 0 & 0 & & \\
s^0 & ? & ? & &
\end{array}
$$

The ϵ method of the previous problem is of no help here. To handle this situation, we look at the row above the row of zeros and define an auxiliary equation as

$$2s^2 + 8 = 0$$

where the coefficients 2 and 8 are taken from the s^2 row of the Routh array. Differentiating this with respect to s, we get the s^1 row as 4, 0. The array can now be completed. If this is done, no sign changes will be found. (Do this!) Therefore, there are no poles in the right half of the complex plane. *However, the system is still considered unstable*! The reason for this situation is that the given polynomial has a pair of complex conjugate roots *on* the imaginary axis which leads to an oscillatory response which by convention is considered unstable. The location of this conjugate pair is obtained by solving the auxiliary equation, whose roots are $s = \pm i2$. ■

The Routh criterion, which is only useful to test the stability of linear systems, only provides a yes or no answer to the stability question. The question of how stable a system is cannot be readily answered by this test. Other tests such as the Nyquist plot and root-locus plot are available for this purpose. Further discussion on this topic can be found in any good book on control theory.

Stability of Nonlinear Systems

What happens if a system is not stable? According to the definition given earlier, the exponential terms in the solution grow unbounded in an unstable system and the system "explodes." However, experience shows that the output rarely grows exponentially or with an exponential envelope. Instead, the output grows for a while and then goes into an erratic waveform. This type of behavior is due to nonlinearities in system components. Thus, before a system "blows up" due to instability, the nonlinearities usually come into play. From this viewpoint nonlinearities are indeed desirable in several cases.

Can we then meaningfully talk about the stability of a nonlinear system? The answer is yes. However, care must be taken to define the precise meaning of stability. Such a definition was first provided by the Russian mathematician A. M. Liapunov. Even a cursory presentation of Liapunov's stability concept is beyond the scope of this work. However, the essence of Liapunov's stability test can be stated as follows.

If a nonlinear (or linear) system is described by the state space model

$$\dot{\mathbf{x}} = f(\mathbf{x}), \qquad \mathbf{x}(t = 0) = \mathbf{0}.$$

then the system is said to be *asymptotically stable* if one can find a function $V(\mathbf{x})$, called the Liapunov function, which has the following properties:

$$\begin{array}{lll} (1) & V(\mathbf{x}) > 0 & \text{for} \quad \mathbf{x} \neq \mathbf{0} \\ & V(\mathbf{x}) = 0 & \text{for} \quad \mathbf{x} = \mathbf{0} \\ (2) & \dot{V}(\mathbf{x}) < 0 & \text{for} \quad \mathbf{x} \neq \mathbf{0} \\ & \dot{V}(\mathbf{x}) = 0 & \text{for} \quad \mathbf{x} = \mathbf{0} \\ (3) & V(\mathbf{x}) \to \infty & \text{as length of vector } \mathbf{x} \to \infty \end{array}$$

That is, the Liapunov function is positive everywhere and its derivative is negative everywhere (except at $\mathbf{x} = \mathbf{0}$). Therefore, if $V(\mathbf{x})$ is visualized as energy in the system, Condition (2) above says that the system is settling back slowly toward an equilibrium.

How do we find $V(\mathbf{x})$ for any given system? No one knows for sure! If the system is linear then one can prove that

$$V(\mathbf{x}) = x_1{}^2 + x_2{}^2 + \cdots + x_n{}^2$$

is a suitable Liapunov function. Maybe there are others. For nonlinear systems, there is no general procedure, but if one can find a $V(\mathbf{x})$ somehow or other, then one can figure out whether a system is asymptotically stable. Inability to find a $V(\mathbf{x})$ does not necessarily mean that a system is unstable. The interested student should refer to advanced books to find answers to all these questions.

4 STATE SPACE MODELS

The utility of Laplace transforms, transfer functions, and other such methods quickly dwindles as a system becomes nonlinear or complex. First, the transfer function concept is difficult to define if a system is nonlinear, noninert, or time varying. Similarly, if a system is described by a partial differential equation, the concept of a transfer function, as it has been defined, is not valid. Second, if the inputs of a system are specified graphically, then the task of determining the Laplace transform of the input signal is tedious, though not impossible. Finally, the computational difficulties of applying these well-known techniques to systems in which there are a large number of input and output variables makes it desirable to look for more general techniques of analysis that are adaptable for automatic computation.

The state variable methods are equally applicable to linear systems, nonlinear systems, discrete systems, time-varying systems, or combinations of these. Indeed, the state variable or state space approach is very general and powerful. Therefore, their application to solve simple, linear, single-input, single-output systems is not warranted. However, this section is mainly concerned with systems characterized by linear, ordinary differential equations. The ideas and procedures discussed here are quite general and are amenable to generalization.

The concept of state is fundamental. It cannot be defined any more than the words "number" or "set" can be defined. An idea of state can be obtained by noting that in order to completely describe a system, one has to include a description of (a) the inputs, (b) the outputs, and (c) the state of the system. In a more formal way, the state of a system is defined as the minimum set of variables, denoted by $x_1, x_2, \ldots, x_n$ specified at time $t = t_0$, which together with the given inputs $u_1, u_2, \ldots, u_m$ determine the state at any future time $t > t_0$.

If the set of input variables is denoted by the input vector **u**, output variables by the output vector **y** and state variables by the state vector **x**, where

$$\mathbf{u} = \begin{bmatrix} u_1 \\ u_2 \\ \vdots \\ u_l \end{bmatrix}, \qquad \mathbf{y} = \begin{bmatrix} y_1 \\ y_2 \\ \vdots \\ y_m \end{bmatrix}, \qquad \mathbf{x} = \begin{bmatrix} x_1 \\ x_2 \\ \vdots \\ x_n \end{bmatrix}$$

then, according to the above definition, the state $\mathbf{x}(t)$ at any time t is related to the initial state $\mathbf{x}(t_0)$ and the nature of the input $\mathbf{u}(t_0, t)$ during the interval (t_0, t). That is,

$$\mathbf{x}(t) = f\{\mathbf{x}(t_0), \mathbf{u}(t_0, t)\}$$

The above equation is very general. It could stand for a difference equation, ordinary linear differential equation or nonlinear differential equation.

Consider the problem of estimating the altitude of a rocket in flight in two-space dimensions. Representing the horizontal axis by r and the vertical axis by z, equations describing the flight of the rocket may be written as

$$\begin{aligned} \ddot{r} &= F \cos \theta \\ \ddot{z} &= F \sin \theta - g \end{aligned} \tag{1}$$

where F is the thrust per unit mass, θ is the direction of thrust, and g is the acceleration due to gravity.

In order to derive a state variable model, it is first necessary to convert the above equations into a system of first-order equations. For this purpose it is convenient to relabel some of the variables as shown below:

$$\begin{aligned} x_1 &\triangleq r, & x_2 &\triangleq \dot{r} \\ x_3 &\triangleq z, & x_4 &\triangleq \dot{z} \\ u_1 &\triangleq F, & u_2 &\triangleq \theta \end{aligned}$$

In terms of the new variables, Eq. (1) becomes

$$\begin{aligned} \dot{x}_1 &= x_2 \\ \dot{x}_2 &= u_1 \cos u_2 \\ \dot{x}_3 &= x_4 \\ \dot{x}_4 &= u_1 \sin u_2 - g \end{aligned} \tag{2}$$

The above set of equations can be compactly rewritten in matrix notation as

$$\dot{\mathbf{x}} = \mathbf{A}\mathbf{x} + \mathbf{B}\mathbf{u} \tag{3}$$

where

$$\mathbf{x} = \begin{bmatrix} x_1 \\ x_2 \\ x_3 \\ x_4 \end{bmatrix}, \qquad \mathbf{A} = \begin{bmatrix} 0 & 1 & 0 & 0 \\ 0 & 0 & 0 & 0 \\ 0 & 0 & 0 & 1 \\ 0 & 0 & 0 & 0 \end{bmatrix}$$

$$\mathbf{B} = \begin{bmatrix} 0 & 0 & 0 & 0 \\ 0 & 1 & 0 & 0 \\ 0 & 0 & 0 & 0 \\ 0 & 0 & 0 & 1 \end{bmatrix}, \qquad \mathbf{u} = \begin{bmatrix} 0 \\ u_1 \cos u_2 \\ 0 \\ u_1 \sin u_2 - g \end{bmatrix}$$

The set in Eq. (3) is called the state variable model or simply the state model.

Note that the state variables x_1 and x_3 define the location of the rocket and the state variables x_2 and x_4 specify the velocity of the rocket. The input or excitation variables are the applied thrust and the direction of the applied thrust. Thus, it is clear that a knowledge of u_1 and u_2 along with a knowledge of the initial state (location and velocity) is sufficient to predict the state of the rocket at any future time.

Note also that solution of Eq. (3) gives the values of all components of the vector **x**. According to the statement of the problem, only the altitude of the rocket is of interest. As $z = x_3 =$ altitude, only the component x_3 of **x** is required. This fact can be written compactly by defining an output y as

$$y = \mathbf{H}\mathbf{x}; \qquad \mathbf{H} \triangleq [0 \quad 0 \quad 1 \quad 0] \tag{4}$$

Thus, the state variable model of this problem is written as

$$\begin{aligned} \dot{\mathbf{x}} &= \mathbf{A}\mathbf{x} + \mathbf{B}\mathbf{u} \\ y &= \mathbf{H}\mathbf{x} \end{aligned} \tag{5}$$

This example has been chosen to demonstrate the mechanics of converting a given problem into a state variable format. A number of questions need to be clarified. For instance: How many state variables are required to describe a system completely? Is there a unique set of variables that qualify as state variables? How does one go about selecting state variables?

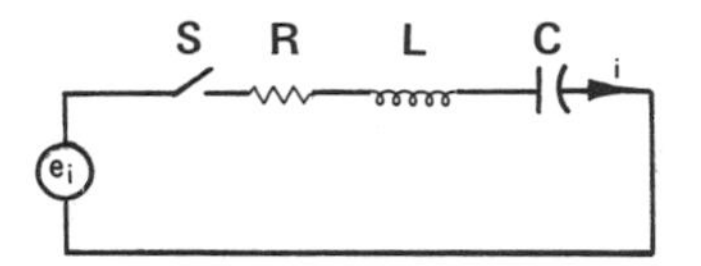

Fig. 9.19 A simple electrical circuit.

Consider a simple electrical circuit as shown in Fig. 9.19. Let the input voltage e_i be in our control and let the current i in the circuit be the quantity of interest. That is, the current i is the output. In Example 4.1 a differential equation describing this system was found to be (note that $i = dq/dt$)

$$R\frac{dq}{dt} + L\frac{d^2q}{dt^2} + \frac{1}{C}q = e_i(t) \tag{6}$$

How do we choose state variables in order to represent the above equation as a state model? One possibility is to let

$$\begin{aligned} x_1 &\triangleq q \\ x_2 &\triangleq \dot{q} \triangleq i \\ u &\triangleq e \end{aligned} \tag{7}$$

In terms of these variables, Eq. (6) becomes

$$\begin{aligned} \dot{x}_1 &= x_2 \\ \dot{x}_2 &= -\,(1/L)x_1 - (R/L)x_2 + u \end{aligned}$$

or in matrix notation as

$$\begin{bmatrix} \dfrac{dq}{dt} \\ \\ \dfrac{di}{dt} \end{bmatrix} = \begin{bmatrix} \dot{x}_1 \\ \\ \dot{x}_2 \end{bmatrix} = \begin{bmatrix} 0 & 1 \\ \\ -\dfrac{1}{LC} & -\dfrac{R}{L} \end{bmatrix} \begin{bmatrix} x_1 \\ \\ x_2 \end{bmatrix} + \begin{bmatrix} 0 & 0 \\ \\ 0 & 1 \end{bmatrix} \begin{bmatrix} 0 \\ \\ u \end{bmatrix} \tag{8}$$

or

$$\dot{\mathbf{x}} = \mathbf{Ax} + \mathbf{Bu} \tag{9}$$

The above state model has one particular drawback. For instance, with the above model, measurement of the initial state $\mathbf{x}_1(0)$ is not convenient because this involves measurement of the charge. However, it is relatively easy to measure the voltage e_i across the capacitor at $t = 0$. Therefore, there is considerable merit in formulating a state model in terms of state variables that are measurable. This can be done as follows.

Instead of starting from Eq. (6), one can start from Kirchhoff's voltage law and write

$$e_R + e_L + e_C = e_i$$

which can also be written as

$$iR + L\frac{di}{dt} + e_C = e_i \tag{10}$$

and

$$i = i_C = C\frac{de_C}{dt} \tag{11}$$

Now Eqs. (10) and (11) are a pair of first-order equations in i and e_C, and so they constitute a state model to the electrical circuit in Fig. 9.19. The above two equations are put in standard format by defining the state variables as

$$x_1(t) = e_C(t) \qquad \text{and} \qquad x_2(t) = i(t)$$

Then, Eqs. (10) and (11) reduce to

$$\dot{\mathbf{x}} = \mathbf{A}_1\mathbf{x} + \mathbf{B}_1\mathbf{u} \tag{12}$$

where

$$\mathbf{x} = \begin{bmatrix} e_C(t) \\ i(t) \end{bmatrix}, \qquad \mathbf{A}_1 = \begin{bmatrix} 0 & 1/C \\ -1/L & -R/L \end{bmatrix}, \qquad \mathbf{B}_1 = \begin{bmatrix} 0 & 0 \\ 0 & 1/L \end{bmatrix}, \qquad \mathbf{u} = \begin{bmatrix} 0 \\ e_i \end{bmatrix}$$

In both Eqs. (8) and (12), the required output, namely, the current i, is obtained by solving the state model. Suppose the required output is e_R, the voltage drop across the resistor, then one has to use the relation

$$e_R = iR \tag{13}$$

to get e_R. Equation (13) can be appended to Eq. (12) as follows. Define

$$\begin{aligned} x_1(t) &= e_C(t) \\ x_2(t) &= i(t) \\ y(t) &= e_R(t) \end{aligned}$$

and

$$\begin{aligned} \dot{\mathbf{x}} &= \mathbf{A}_1\mathbf{x} + \mathbf{B}_1\mathbf{u} \\ y &= \mathbf{H}\mathbf{x} \end{aligned} \tag{14}$$

where $\mathbf{x}$, $\mathbf{A}_1$, and $\mathbf{B}_1$ are as defined before and $\mathbf{H} = [0, R]$ and y is a scalar.

The above procedure vividly illustrates that there is no unique set of state variables to a given problem. Indeed, there exists a number of combinations of variables that can serve as state variables. The important thing to remember is that the set of state variables should be selected in such a fashion that they completely specify the initial conditions (or initial state) of the system. Thus, in order to arrive at the correct state model, one has to ask oneself: What are the variables or components of the system whose initial values carry significant information? If there are any such variables, it is good practice to include them in the set of state variables. Still there is an unanswered question. How does one determine the set of variables whose initial values (i.e., initial conditions) carry significant information? If there are devices or components in a system which are capable of storing energy, then that stored energy (before the switch is closed) carries significant information. In the present example there are two energy storage devices. The inductor L stores electromagnetic energy and the capacitor stores electrostatic energy.

The next question to be answered is the following: How many state variables are required in any given problem? A convenient rule of thumb is: the number of state variables required is equal to the degree of the characteristic equation of the system, or alternatively, one state variable is required for each degree of freedom of the system. Indeed, a convenient way to establish state variables is to construct a block diagram and associate one state variable with each integrator output.

Example 7. Consider the third-order linear system:

$$\frac{d^3y}{dt^3} + 6\frac{d^2y}{dt^2} + 11\frac{dy}{dt} + 6y = u$$

If the state variables are defined as $\mathbf{x} = [y \quad \dot{y} \quad \ddot{y}]^{\mathrm{T}}$, then the state equations become

$$\dot{\mathbf{x}} = \mathbf{A}\mathbf{x} + \mathbf{B}u$$
$$y = \mathbf{H}\mathbf{x}$$

where

$$\mathbf{A} = \begin{bmatrix} 0 & 1 & 0 \\ 0 & 0 & 1 \\ -6 & -11 & -6 \end{bmatrix}, \qquad \mathbf{B} = \begin{bmatrix} 0 \\ 0 \\ 1 \end{bmatrix}, \qquad \mathbf{H} = [1 \quad 0 \quad 0]$$

and u is a scalar.

Another method of arriving at a state model is as follows. Rewrite the given equation using the s operator as:

$$(s^3 + 6s^2 + 11s + 6)y = u$$

Using the partial fraction expansion technique,

$$\left[\frac{1/2}{(s+1)} - \frac{1}{(s+2)} + \frac{1/2}{(s+3)}\right]u = y$$

Associating state variables with the outputs of the integrators, one has

$$\dot{x}_1 = -x_1 + \tfrac{1}{2}u$$
$$\dot{x}_2 = -2x_2 - u$$
$$\dot{x} = -3x_3 + \tfrac{1}{2}u$$
$$y = x_1 + x_2 + x_3$$

or, in matrix notation,

$$\dot{\mathbf{x}} = \mathbf{A}_1\mathbf{x} + \mathbf{B}_1 u$$
$$y = \mathbf{H}_1\mathbf{x}$$

where

$$\mathbf{A}_1 = \begin{bmatrix} -1 & 0 & 0 \\ 0 & -2 & 0 \\ 0 & 0 & -3 \end{bmatrix}, \qquad \mathbf{B}_1 = \begin{bmatrix} \frac{1}{2} \\ -1 \\ \frac{1}{2} \end{bmatrix}, \qquad \mathbf{H}_1 = [1 \quad 1 \quad 1]$$

and u is a scalar. ■

The preceding example illustrates that certain choices of state variables are sometimes desirable as they result in a particularly convenient form for **A**. For instance, if **A** is a diagonal matrix as in $\mathbf{A}_1$ then the system is decoupled and one can solve three first-order equations rather than one third-order equation. Note that the eigenvalues of matrix **A** are -1, -2, -3, which are the same as the eigenvalues of matrix $\mathbf{A}_1$.

The Transition Matrix

The state variables define the state of a system at any time. Suppose the state of a system at time $t = 0$ is known and the state of the system at any other time $t > 0$ is required. To study this problem, consider a simple state model of an undriven system (i.e., $u = 0$, in Eq. (5)):

$$\dot{\mathbf{x}} = \mathbf{A}\mathbf{x}, \qquad \mathbf{x}(t = 0) = \mathbf{x}(0) \tag{15}$$

Taking Laplace transforms on both sides,

$$s\mathbf{x}(s) - \mathbf{x}(0) = \mathbf{A}\mathbf{x}(s)$$

Solving for $\mathbf{x}(s)$,

$$\mathbf{x}(s) = [s\mathbf{I} - \mathbf{A}]^{-1}\mathbf{x}(0) \tag{16}$$

Taking inverse Laplace transform,

$$\mathbf{x}(t) = L^{-1}\{[s\mathbf{I} - \mathbf{A}]^{-1}\}\mathbf{x}(0) \tag{17}$$

Thus, the state of the system at time t, that is, $\mathbf{x}(t)$, is related to the initial state, viz. $\mathbf{x}(0)$, via $L^{-1}\{[s\mathbf{I} - \mathbf{A}]^{-1}\}$. Define the *transition matrix* or *fundamental matrix*, $\mathbf{\Phi}(t)$ as

$$\mathbf{\Phi}(t) \triangleq L^{-1}\{[s\mathbf{I} - \mathbf{A}]^{-1}\} \tag{18}$$

Thus, the state $\mathbf{x}(t)$ is related to the initial state $\mathbf{x}(0)$ via the transition matrix $\mathbf{\Phi}(t)$:

$$\mathbf{x}(t) = \mathbf{\Phi}(t)\mathbf{x}(0) \tag{19}$$

Once the transition matrix can be evaluated, it is an easy matter to calculate $\mathbf{x}(t)$ for any time t.

Example 8. Consider a model described by the linear equation

$$\ddot{y}(k) + 3\dot{y}(t) + 2y(t) = u(t)$$

Define the state variables as $x_1 = y$ and $x_2 = \dot{y}$. In terms of these state variables, the above equation reduces to

$$\dot{\mathbf{x}} = \mathbf{A}\mathbf{x} + \mathbf{B}\mathbf{u}$$

where

$$\mathbf{x}(t) = \begin{bmatrix} x_1(t) \\ x_2(t) \end{bmatrix}, \qquad \mathbf{A} = \begin{bmatrix} 0 & 1 \\ -2 & -3 \end{bmatrix}, \qquad \mathbf{B} = \begin{bmatrix} 0 \\ 1 \end{bmatrix}$$

The transition matrix can be obtained as follows:

$$[s\mathbf{I} - \mathbf{A}] = \begin{bmatrix} s & 0 \\ 0 & s \end{bmatrix} - \begin{bmatrix} 0 & 1 \\ -2 & -3 \end{bmatrix} = \begin{bmatrix} s & -1 \\ 2 & s+3 \end{bmatrix}$$

$$[s\mathbf{I} - \mathbf{A}]^{-1} = \text{adj}[s\mathbf{I} - \mathbf{A}]/\det[s\mathbf{I} - \mathbf{A}] = \begin{bmatrix} s+3 & 1 \\ -2 & s \end{bmatrix} \Bigg/ \det\begin{bmatrix} s & -1 \\ 2 & s+3 \end{bmatrix}$$

$$= \begin{bmatrix} \dfrac{s+3}{(s+1)(s+2)} & \dfrac{1}{(s+1)(s+2)} \\[2ex] \dfrac{-2}{(s+1)(s+2)} & \dfrac{s}{(s+1)(s+2)} \end{bmatrix}$$

Taking inverse Laplace transforms on both sides,

$$\mathbf{\Phi}(t) = \begin{bmatrix} 2e^{-t} - e^{-2t} & e^{-t} - e^{-2t} \\ -2e^{-t} + 2e^{-2t} & -e^{-t} + 2e^{-2t} \end{bmatrix}$$

The procedure presented above is straightforward because the model is only of second order. The task of inverting the matrix $[s\mathbf{I} - \mathbf{A}]$ becomes tedious for higher-order systems. ■

Digital Computer Evaluation of a Transition Matrix

Consider a simple first-order differential equation,

$$\dot{x} = ax; \qquad x(t = 0) = x(0) \tag{20}$$

Solution to this can be written by inspection as

$$x(t) = e^{at}x(0) \tag{21}$$

This result can be generalized to a system of equations. For instance, the solution of

$$\dot{\mathbf{x}} = \mathbf{A}\mathbf{x}; \qquad \mathbf{x}(t = 0) = \mathbf{x}(0) \tag{22}$$

can be written as

$$\mathbf{x}(t) = e^{\mathbf{A}t}\mathbf{x}(0) \tag{23}$$

Comparison of Eq. (23) and Eq. (19) shows that

$$\mathbf{\Phi}(t) = e^{\mathbf{A}t} \tag{24}$$

Therefore, the transition matrix $\mathbf{\Phi}(t)$ can be evaluated on a digital computer using the above identity. The right side of Eq. (24) can be written as a power series:

$$e^{\mathbf{A}t} = \mathbf{I} + \mathbf{A}t + \mathbf{A}^2t^2/2! + \cdots + \mathbf{A}^kt^k/k! + \cdots \tag{25}$$

If this series converges, then the power series can be truncated and $e^{\mathbf{A}t}$ can be approximated by

$$e^{\mathbf{A}t} \approx \mathbf{I} + \mathbf{A}t + \mathbf{A}^2t^2/2! + \cdots + \mathbf{A}^kt^k/k! \tag{26}$$

Whether the sequence in Eq. (26) converges depends upon the eigenvalues of the matrix **A**. The eigenvalues of **A** are determined, for example, by finding the roots of the characteristic equation, viz. $\det(s\mathbf{I} - \mathbf{A}) = 0$. Thus, to ensure convergence, one has to know something about eigenvalues. Finding eigenvalues, or roots of the characteristic equation, is not always easy if the characteristic equation is higher than second order.

5 SYSTEM CONTROL

At the beginning of the first chapter, it was stated that one of the goals in modeling is to learn enough about a system so as to gain control of it and to make it operate in a more desirable way. This is the gist of the control problem. Stated in general terms, the goal of a control problem is to force one or more of the system variables to follow a prescribed pattern of behavior. In a biological system, such as a human being, the goal of the control problem may be to reduce pain and prolong life. In an economic system the factors of interest might be economic growth, inflation, and expenditure of resources. Analogous problems abound in a broad spectrum of processes including electrical, mechanical, industrial, biological, economical, and societal systems.

Engineers have learned to tackle the problem of control using several *ad hoc* and empirical methods. Starting from the time of James Watt's governor in 1769 until 1920 development in this area has been slow. Invention of the radio and the subsequent rise of the electronics industry were a major impetus for the development of servomechanisms and feedback regulation. During World War II there was a tremendous upsurge of interest in the field of control and regulation applied to such problems as radar tracking and gun sighting and firing. Almost all of this work and the work done until the Sputnik era is generally called classical control theory. By and large, classical control theory dealt with linear systems. Some type of linearization was always used while dealing with nonlinear systems. The classical control theory was further characterized by the fact that most of the work was done by first converting the problem into the frequency domain or s domain using Laplace transforms. A number of ingenious methods were devised to study the behavior of systems in the frequency domain using graphical methods such as the Bode plot, Nyquist plot, and root-locus plot.

Demands of modern times made it necessary to formulate problems more realistically and on a firmer mathematical foundation. The first obvious demand was to include nonlinear behavior. Transform techniques such as those based on Laplace transforms, which are valid only for describing linear systems, had to be abandoned in favor of describing systems in the time domain—that is, in terms of differential or difference equations. This is the beginning of the era of modern control theory.

Growth of modern control theory was spurred not only by a compelling need for such theories but also by the fact that modern computers made it possible to solve the problems formulated in terms of modern control theory. The success of this approach in modern industry and aerospace technology is creating optimism, and nowadays several attempts are being made to apply control theoretic ideas to the formidable problems in national economic growth, resource allocation, and investment planning, to mention a few important areas.

Modern control theory assumes that the analyst can specify completely and quantitatively the characteristics of a system and the required system performance. The system characteristics are specified in general by a system of nonlinear differential equations such as

$$\dot{\mathbf{x}} = \mathbf{f}(\mathbf{x}, \mathbf{u}, t) \tag{1}$$

where $\mathbf{x}$ is the state vector, $\mathbf{u}$ is the control vector, and the required performance index is written as

$$J = J(\mathbf{x}, \mathbf{u}, t) \tag{2}$$

Equation (1) merely states that the rate of change of $\mathbf{x}$ depends on $\mathbf{x}$, $\mathbf{u}$, and time—the exact nature of the relation left unspecified. It is sufficient to note at this stage that in a particular problem the right side of Eq. (1) assumes a more specific form. If the problem is linear, it is customary to write Eq. (1) as $\dot{\mathbf{x}} = \mathbf{Ax} + \mathbf{Bu}$, where $\mathbf{A}$ and $\mathbf{B}$ are matrices.

Equation (2), describing the required performance, is also a very general expression. It is important to note that J is a scalar quantity whose value depends upon the vectors $\mathbf{x}$, $\mathbf{u}$, and t. In a typical control problem, one likes to optimize the value assumed by J. In a flight control system, for example, this would mean combining into one scalar quantity the effect of error in maintaining the desired trajectory, the amount of fuel expended, the magnitude of the control action, the time to reach the terminal conditions, and the error in terminal state. Thus, the problem is very much analogous to the multiobjective optimization problem discussed in Chapter 2. In engineering problems, it is relatively easy to combine the various effects. For example, in the case of a lunar trajectory and landing problem, the error in the terminal state would be of crucial importance. In the early days of control theoretic development, people were primarily interested in one aspect of control or another and a number of specialized names were given to various types of problems. These problems are described below.

Terminal Control Problem It is desired to bring the system as close as possible to a given terminal state $\mathbf{x}(t_f)$ within a given period of time. For such a problem, Eq. (2) assumes a form such as

$$J = \int_{t_0}^{t_f} g\{\mathbf{x}_d(t_f) - \mathbf{x}(t_f), \mathbf{u}(t)\}\, dt \tag{3}$$

where $\mathbf{x}_d(t_f)$ is the desired final state and $\mathbf{x}(t_f)$ is the actual final state.

Minimal-Time Control Problem It is desired to reach the terminal state in the shortest possible period of time. For this problem, Eq. (2) would assume the form

$$J = \int_{t_0}^{t_f} dt = (t_f - t_0) \tag{4}$$

The Regulator Problem With the system initially displaced from equilibrium, return the system to the equilibrium state so that the specified performance index is minimized.

The Servomechanism Problem It is desired to cause the state of the system $\mathbf{x}(t)$ to approach as closely as possible a desired state $\mathbf{x}_d(t)$ over a period of time. This is a generalization of the regulator problem.

Minimum Energy Control It is desired to transfer the system from an initial state to a final state with a minimum expenditure of control energy.

Minimum Fuel Control It is desired to transfer the system from an initial state to a terminal state with a minimum expenditure of fuel.

Operations Research Problem It is desired to operate a system so as to maximize the total return or profit.

Thus, the optimal control problem becomes one of finding an optimal control policy $\mathbf{u}(t)$ that minimizes a given performance index. Stated in this manner, the problem is no different in principle from the methods discussed in Chapter 8. However, the present interest is to discuss, without delving too much in the theoretical details, a widely used technique known as Pontryagin's Maximum Principle.

Pontryagin's Maximum Principle

Pontryagin's Maximum Principle is a method based on the calculus of variations and is not too unlike the Lagrange multiplier method discussed in Chapter 8. The Maximum Principle provides only necessary conditions, but not sufficient conditions. That is, if $\mathbf{u}(t)$ is known to be an optimizing value, then it satisfies the Maximum Principle, but not all $\mathbf{u}(t)$ satisfying the Maximum Principle are optimal. The ideas on which the Maximum Principle is founded can be briefly summarized as follows. Consider a system of n differential equations

$$\dot{\mathbf{x}} = \mathbf{f}(\mathbf{x}, \mathbf{u}) \tag{5}$$

where $\mathbf{x} = (x_1 \quad x_2 \quad \cdots \quad x_n)^\mathrm{T}$ and $\mathbf{u} = (u_1 \quad u_2 \quad \cdots \quad u_r)^\mathrm{T}$. The problem is to find an optimal control $\mathbf{u}(t)$ that takes the system from an initial state $\mathbf{x}(t_0) = \mathbf{x}(0)$ to a specified final state $\mathbf{x}(t_f)$ while minimizing

$$J = \int_0^{t_f} f_0(\mathbf{x}, \mathbf{u})\, dt \tag{6}$$

The first step in the process of applying the Maximum Principle is to recognize that Eq. (6) satisfies the differential equation

$$\frac{dJ}{dt} = f_0(\mathbf{x}, \mathbf{u}) \tag{7}$$

The second step is to note that Eqs. (5) and (7) are very similar in structure. Suppose that J is relabeled as x_0. Then Eq. (7) becomes

$$\frac{dx_0}{dt} = \frac{dJ}{dt} = f_0(\mathbf{x}, \mathbf{u}) \tag{8}$$

Now Eq. (8) is appended to one end of Eq. (5), and the augmented system is written as

$$\dot{\mathbf{x}} = \mathbf{F}(\mathbf{x}, \mathbf{u}) \tag{9}$$

where $\mathbf{x} = (x_0 \quad x_1 \quad \cdots \quad x_n)^{\mathrm{T}}$ and $\mathbf{F} = (f_0 \quad f_1 \quad \cdots \quad f_n)^{\mathrm{T}}$. Note that Eq. (9) represents $(n + 1)$ differential equations. In other words, the first two steps are concerned with reformulating the problem so that both the system differential equations and the performance index all appear as one set of equations.

The third step in applying the Maximum Principle is to define a new set of $(n + 1)$ variables

$$\mathbf{p} = (p_0 \quad p_1 \quad p_2 \quad \cdots \quad p_n)^{\mathrm{T}} \tag{10}$$

called the *Lagrange multipliers* or *costate variables* or *adjoint variables*. Now the Maximum Principle states that the control input $\mathbf{u}(t)$ which minimizes the performance criterion J maximizes the value of H, defined by

$$H = \mathbf{p}^{\mathrm{T}}\mathbf{F} = p_0 f_0 + p_1 f_1 + \cdots + p_n f_n \tag{11}$$

Here, the quantity H is called the Hamiltonian. The value of H depends upon $\mathbf{p}$, $\mathbf{x}$, and $\mathbf{u}$ and is therefore written as $H(\mathbf{p}, \mathbf{x}, \mathbf{u})$. Finding the value of $\mathbf{u}$ that maximizes H is a straightforward calculus problem. However, this results in a set of equations in terms of $\mathbf{p}$ and $\mathbf{x}$, where $\mathbf{p}$ is a set of auxiliary variables that was artificially introduced in the derivation process. One way to get rid of $\mathbf{p}$ is to develop a relation between $\mathbf{p}$ and $\mathbf{x}$. It is not too difficult to show that the required equations are

$$\frac{dx_i}{dt} = \frac{\partial H}{\partial p_i}, \qquad x_i(t_0) = x_i(0); \qquad i = 0, 1, \ldots, n \tag{12}$$

$$\frac{dp_i}{dt} = \frac{-\partial H}{\partial x_i}, \qquad p_i(t_f) = p_i(f); \qquad i = 0, 1, \ldots, n \tag{13}$$

There are several important points to be observed in the above pair of equations. Equation (12) represents a set of $(n + 1)$ equations with $(n + 1)$ *initial* conditions. Equation (13) represents another set of $(n + 1)$ equations with $(n + 1)$ *terminal* conditions. That is, the initial conditions of Eq. (13) are not known. Probelms such as these are called *boundary value problems* and are difficult to solve, particularly if one is dealing with nonlinear equations.

As a fourth step, Pontryagin has shown that a necessary condition for optimality is that $p_0 = -1$. Hence, the term with $i = 0$ in Eqs. (12) and (13) is superfluous, and Eqs. (12) and (13) can be rewritten as

$$\frac{dx_i}{dt} = \frac{\partial H}{\partial p_i}, \qquad x_i(t_0) = x_i(0); \qquad i = 1, 2, \ldots, n \tag{14}$$

$$\frac{dp_i}{dt} = \frac{-\partial H}{\partial x_i}, \qquad p_i(t_f) = p_i(f); \qquad i = 1, 2, \ldots, n \tag{15}$$

As a fifth and final step, Pontryagin has shown that a necessary condition for a control $\mathbf{u}(t)$ to be optimal is that the maximum value of the Hamiltonian be constant over the range of time period of interest. That is,

$$H_{\max} = \text{constant}; \qquad t_0 < t < t_f \tag{16}$$

For the particular case where the final time t_f is not a fixed quantity, Eq. (16) becomes

$$H_{\max} = 0 \qquad \text{at} \quad t = t_0 \tag{17}$$

Example 9. At the expense of being trivially simple and somewhat artificial and at the risk of being inaccurate from a physical viewpoint, consider a problem such as the excretion of a drug from a body. This process may be roughly described by an equation such as

$$\dot{x}_1 = -x_1 + u, \qquad x_1(0) = 1$$

where x_1 is the deviation of concentration of a drug from a nominal level in the blood stream and u is the infusion of the drug. This problem is chosen not for its physical significance or accuracy but because it is simple enough to demonstrate the applicability of the maximum principle. We wish to bring the value of x_1 to zero at some final time; that is,

$$x_1(t_f) = 0 \qquad \text{at} \quad t = t_f$$

while minimizing the performance index

$$J = \int_0^{t_f} x_1^2 \, dt$$

and satisfying the constraint $|u(t)| \leq U$. We wish to find out how to infuse the drug—that is, the drug regimen.

The augmented system of differential equations corresponding to Eq. (9) is readily written as

$$\dot{x}_0 = f_0(x_1, u) = x_1^2$$
$$\dot{x}_1 = f_1(x_1, u) = -x_1 + u$$

The Hamiltonian of this system, corresponding to Eq. (11), is

$$\begin{aligned} H &= p_0 f_0 + p_1 f_1 = -f_0 + p_1 f_1 \\ &= -x_1{}^2 + p_1(-x_1 + u) \\ &= -x_1{}^2 - p_1 x_1 + p_1 u = -(\text{a quantity}) + p_1 u \end{aligned}$$

The above quantity attains a maximum value when $p_1 u$ is as large a positive quantity as possible. This can be achieved if p_1 and u are both positive or both negative and $u(t)$ assumes the maximum permissible value. That is, the sign of control signal $u(t)$ is the same as that of p_1 and its magnitude is the maximum allowable value U. That is

$$u(t) = U \operatorname{sgn} p_1$$

where the sgn function is defined in Appendix I. The above equation states that $u(t) = \pm U$, which may be interpreted to mean the drug is either applied at its maximum permissible strength or minimum permissible strength. This type of control is called bang–bang or on–off control. To implement the above control, however, it is necessary to know $p_1(t)$. Using Eq. (14) and (15)

$$\dot{x}_0 = \frac{\partial H}{\partial p_0} = f_0 = x_1{}^2$$

$$\dot{x}_1 = \frac{\partial H}{\partial p_1} = f_1 = -x_1 + u$$

$$\dot{p}_1 = \frac{-\partial H}{\partial x_1} = 2x_1 + p_1$$

The conditions on x_0 and x_1 are

$$x_0(0) = 0, \qquad x_1(0) = 1, \qquad x_1(t_f) = 0$$

and the terminal condition on $p_1(t)$ is

$$p_1(t_f) = 0$$

This last equation is obtained from $\dot{p}_1 = 2x_1 + p_1$ by evaluating it at $t = t_f$ and demanding that $x_1(t_f) = 0$. Thus calculation of $p_1(t)$ is tantamount to solving the two-point boundary value problem

$$\begin{aligned} \dot{x}_1 &= -x_1 + U \operatorname{sgn} p_1; \qquad & x_1(0) &= 1 \\ \dot{p}_1 &= 2x_1 + p_1; & p(t_f) &= 0 \end{aligned}$$

The actual process of solving this nonlinear boundary value problem is beyond the scope of this book. However, it can be done using some of the powerful digital computer algorithms available. ■

In summary, the above example shows how Pontryagin's Maximum Principle can be applied to find the optimum control law. Basically, the power of the technique is derived from the introduction of the auxilary variables and the Hamiltonian. Notice that the solution for the control law, namely, the expression for $\mathbf{u}(t)$, is only a function of time and not of the state $\mathbf{x}(t)$. That is, the manner in which the drug is administered is completely independent of the concentration of the drug in the blood. In other words, how the drug is to be administered is completely predetermined and, therefore, can be preprogrammed. This type of control is called *open-loop control.* The objection to the open-loop control is that if the system is subjected to a disturbance or experiences a change in parameters that was not foreseen, the system will deviate from its optimal trajectory. Then the future optimal trajectory is no longer the precalculated one. Therefore, it is useful to obtain a control law that is a function of the state variables, that is, $\mathbf{u}(\mathbf{x}, t)$. Then if the system is displaced from the optimal path due to some reason, a new optimal path is pursued rather than attempting to return to the original path. This type of control is called the *feedback control.* Often the solution obtained from the Maximum Principle can be transformed into the desired feedback form without too much difficulty.

6 OPTIMUM CONTROL OF RIVER POLLUTION

This section is devoted to a brief and somewhat idealized description of how control theoretic ideas can be used to study the problem of pollution control in surface water bodies such as rivers. It is commonly accepted that the primary indicators of the quality of water in a river are the biological oxygen demand (BOD) and the level of dissolved oxygen (DO). Biological oxygen demand, or BOD, is usually defined as the amount of oxygen required by bacteria while stabilizing decomposable organic matter dumped into water with the help of dissolved oxygen in water. This biodegradation is referred to as *aerobic decomposition.* Dissolved oxygen (DO) is nothing but the amount of oxygen in water that is available for aerobic decomposition. This oxygen enters the water either directly from the air or as a result of photosynthesis in aquatic plants. When organic effluents are discharged into water, the BOD of the added effluents is the total amount of oxygen required for complete aerobic decomposition of the organic matter. To maintain the quality of water, it is therefore necessary to lower the level of BOD and maintain an adequate level of DO.

Consider a section of a river with a continuously treated waste discharged into it at a point (see Fig. 9.20). It is assumed that the concentration of BOD discharged into the river is constant with respect to time. Once the waste enters the river, the BOD decreases as the waste waters mix with fresh river

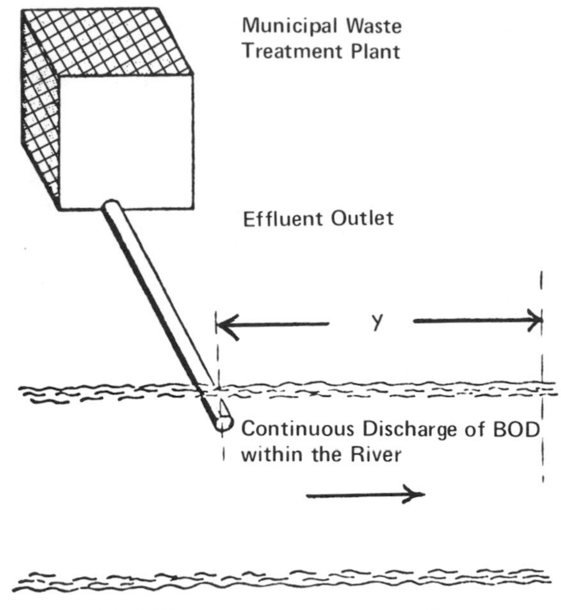

Fig. 9.20 Schematic of a river section.

water. Then if y is the spatial variable measured from the polluting source along the river, then one can indicate the BOD at any point downstream by $B(y)$. Similarly, $D(y)$ indicates the dissolved oxygen at any point y from the source of pollution. With this notation, and several other simplifying assumptions, the equations describing the dynamics of the process can be written as

$$\frac{dB(y)}{dy} = \frac{-K_r}{V} B(y) + \frac{c(y)}{V}; \qquad B(y=0) = B_0$$

$$\frac{dD(y)}{dy} = \frac{K_d}{V} B(y) + \frac{K_a}{V} [D_s - D(y)]; \qquad D(y=0) = D_0 \tag{1}$$

where K_a is the reaeration coefficient, K_r is the BOD removal coefficient, K_d is the deoxygenation coefficient, D_s is the saturation level of the DO, and $u(y)$ is a term that determines how fast or how slow the effluents are dumped into the river at a point y. That is, $u(y)$ is the control term. The quantities B_0 and D_0 are the initial values of BOD and DO respectively at $y = 0$. It is desired to determine $u(y)$ to satisfy certain performance criteria which will be presently described.

Purely for notational convenience, the problem is recast in state variable format by defining $x_1(y)$ and $x_2(y)$ as

$$x_1(y) = B(y) \qquad \text{and} \qquad x_2(y) = D_s - D(y) \tag{2}$$

Now Eq. (1) can be written as

$$\begin{bmatrix} \dot{x}_1(y) \\ \dot{x}_2(y) \end{bmatrix} = \begin{bmatrix} -\beta & 0 \\ \gamma & -\alpha \end{bmatrix} \begin{bmatrix} x_1(y) \\ x_2(y) \end{bmatrix} + \begin{bmatrix} 1 \\ 0 \end{bmatrix} u_1(y), \qquad \begin{bmatrix} x_1(0) \\ x_2(0) \end{bmatrix} = \begin{bmatrix} x_{10} \\ x_{20} \end{bmatrix} = \begin{bmatrix} B_0 \\ D_s - D_0 \end{bmatrix} \tag{3}$$

where $\alpha = K_a/V$, $\beta = K_r/V$, $\gamma = K_d/V$, and $u_1(y) = u(y)/V$. Specification of the initial conditions as $x_1(y=0) = x_{10}$ and $x_2(y=0) = x_{20}$ completes the description of the problem in state space notation.

Selecting Performance Criteria

As discussed in Chapter 2, the selection of performance criteria is not always easy. In this problem, three criteria are chosen merely to illustrate the procedure. One can choose other criteria if the circumstances so demand.

First, every municipality should recognize the riparian rights of the downstream municipalities. That is, when a municipality dumps waste into a river at a point, the deleterious effects of this pollutant on the next section should be minimal. This can be done by requiring that

$$\phi_1 = x_1(y_f) \tag{4}$$

be a minimum where $x_1(y_f)$ denotes the level of BOD at the end of the section of the river under study. Similarly one can ask that

$$\phi_2 = x_2(y_f) \tag{5}$$

be a minimum also. A third objective function is used to reflect the monetary and environmental damage costs. It is not too unrealistic to visualize that these two costs as functions of the concentration of BOD exhibit behavior as shown in Fig. 9.21. Obviously if one increases, the other decreases. Realistically speaking, the cost of environmental damage is difficult to measure

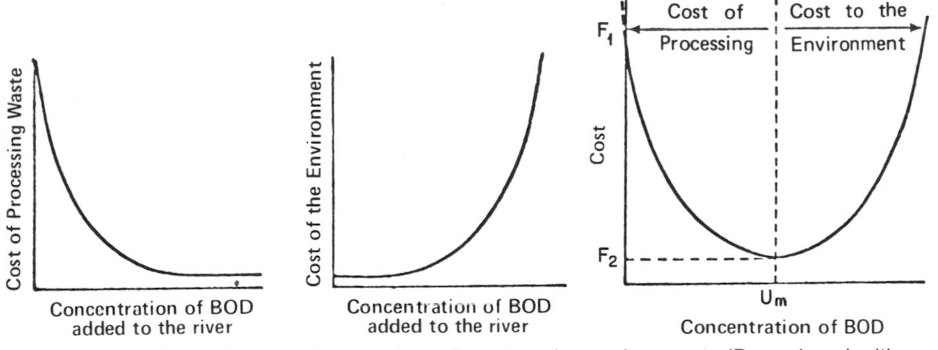

Fig. 9.21 Cost of processing waste and cost to the environment. (Reproduced with permission from Gaurishankar and Lawson, *International Journal of Systems Science* **6**, No. 3, 201–206 (1975).)

and quantify. In Fig. 9.21 it is assumed that cost of processing waste and cost of environmental damage are expressed in the same type of units, such as monetary units. Even then the symmetrical appearance of the curve is deceptive in the sense that a unit decrease in processing cost may or may not result in a unit increase of environmental damage. Nevertheless, these assumptions are made to render the problem mathematically tractable.

The information in Fig. 9.21 can be quantified by approximating the curve by a quadratic in u_1 as $a_0 - a_1u_1 + a_2u_1^2$. Now the third performance index may be defined as

$$\phi_3 = \int_0^{y_f} (a_0 - a_1u_1 + a_2u_1^2)\, dy \tag{6}$$

where a_0, a_1, and a_2 can be expressed in terms of the parameters shown in the figure as

$$a_1 = F_1, \qquad a_2 = \left[\frac{2[F_1 + F_2(u_m - 1)]}{u_m}\right] - F_2, \qquad a_3 = \frac{F_1 + F_2(u_m - 1)}{u_m^2} \tag{7}$$

Now the control problem can be formulated as follows. Find u_1 that minimizes ϕ_1, ϕ_2, and ϕ_3 while satisfying the differential equations shown in Eq. (3). This problem, as stated, is a multiple objective optimization problem. However, the problem will be scalarized by defining a new objective function

$$\begin{aligned} J(\phi_1, \phi_2, \phi_3) &= w_1\phi_1 + w_2\phi_2 + w_3\phi_3 \\ &= w_1x_1(y_f) + w_2x_2(y_f) + w_3 \int_0^{y_f} (a_0 - a_1u_1 + a_2u_1^2)\, dy \end{aligned} \tag{8}$$

where w_1, w_2, and w_3 are weights reflecting the importance of ϕ_1, ϕ_2, and ϕ_3. Who determines these weights and how they are determined is once again a loaded question. Here it is assumed that one can somehow determine these weights.

Formulation of the Hamiltonian

As a first step in calculating the control, one formulates the Hamiltonian H, which was defined in Eq. (5.11) as $\mathbf{p}^T\mathbf{F}$, where $\mathbf{p} = (p_0 \quad p_1 \quad p_2)^T$ and $\mathbf{F} = (f_0 \quad f_1 \quad f_2)^T$. Comparing Eqs. (5.5) and (6.3), it is evident that

$$f_1 = -\beta x_1 + u_1, \qquad f_2 = \gamma x_1 - \alpha x_2 \tag{9}$$

Comparing Equations (5.6) and (6.8), one can conclude that

$$f_0 = w_3(a_0 - a_1u_1 + a_2u_1^2) \tag{10}$$

Therefore, the Hamiltonian for this problem is

$$H = w_3(a_1 - a_1u_1 + a_2u_1^2)p_0 - \beta x_1p_1 + u_1p_1 + \gamma x_1p_2 - \alpha x_2p_2 \quad (11)$$

Choosing $p_0 = -1$, as usual, the value of u_1 that makes H a maximum can be easily calculated. The optimum value of u_1 is obtained by setting $\partial H/\partial u_1 = 0$ to give

$$-a_1w_3p_0 + 2a_2w_3u_1p_0 + p_1 = 0$$

or, with $p_0 = -1$,

$$u_1 = u_1^* = (a_1/2a_2) + p_1(1/2a_2w_3) \quad (12)$$

where the asterisk is used to remind us that u_1 is an optimizing control. In order to implement this control, one has to express u_1 in terms of the state variables. Toward this end, the canonical equations are written as

$$\dot{x}_1 = \frac{\partial H}{\partial p_1} = -\beta x_1 + u_1, \qquad \dot{x}_2 = \frac{\partial H}{\partial p_2} = \gamma x_1 - \alpha x_2 \quad (13)$$

$$\dot{p}_1 = \frac{-\partial H}{\partial x_1} = +\beta p_1 - \gamma p_2, \qquad \dot{p}_2 = \frac{-\partial H}{\partial x_2} = \alpha p_2 \quad (14)$$

Notice that Eq. (13) is the same as Eq. (3) and the coefficient matrix in Eq. (14) is the same as the negative transpose of the one in Eq. (13). Our interest is to solve Eq. (14) for p_1. As Eq. (13) is linear, it can be easily solved using the techniques described in Section 4 to yield, for instance,

$$p_1(y) = -\left(w_1 - \frac{cw_2}{\beta - \alpha}\right)\exp[\beta(y - y_f)] - \left(\frac{cw_2}{\beta - \alpha}\right)\exp[\alpha(y - y_f)] \quad (15)$$

Therefore, the optimum value of $u_1 = u_1^*$ becomes

$$u_1(y) = u_1^*(y) = \frac{a_1}{2a_2} + \frac{p_1}{2a_2w_3}$$

$$= \frac{a_1}{2a_2} - \left(w_1 - \frac{\gamma w_2}{\beta - \alpha}\right)\exp[\beta(y - y_f)] - \frac{\gamma w_2}{\beta - \alpha}\exp[\alpha(y - y_f)] \quad (16)$$

Substituting this in Eq. (3) and solving, one gets

$$x_1(y) = x_{10}\exp(-\beta y) + \frac{a_1}{2a_2}\frac{[1 - \exp(-\beta y)]}{\beta}$$

$$- \frac{K_1}{2\beta}\exp[\beta(y - y_f)] - \exp[-\beta(y - y_f)]$$

$$- \frac{K_2}{\alpha + \beta}\exp[\alpha(y - y_f)] - \exp[-\beta y - \alpha y_f] \quad (17)$$

and

$$x_2(y) = x_{20}\exp(-\alpha y) + \frac{\gamma}{\beta - \alpha} x_{10}\exp(-\alpha y) + \frac{a_1}{2a_2}\frac{[1 - \exp(-\alpha y)]}{\alpha}$$

$$-\frac{K_1}{\alpha + \beta}\exp[\beta(y - y_f)] - \exp[-\alpha y - \beta y_f]$$

$$-\frac{K_2}{2\alpha}\exp[\alpha(y - y_f)] - \exp[-\alpha(y - y_f)] \quad (18)$$

Equation (17) describes the BOD "profile" along the length of the river under optimum conditions of effluent discharge. Similarly, Equation (18) tells the optimum "profile" of $D_s - D(y)$. Equation (16) states how effluents should be discharged along the length of the river. This is more than what we are after. The problem assumed a point source of pollution at $y = 0$, so at $y = 0$

$$u_1{}^*(y = 0) = \frac{a_1}{2a_2} - \left(w_1 - \frac{\gamma w_2}{\beta - \alpha}\right)\exp(-\beta y_f) - \frac{\gamma w_2}{\beta - \alpha}\exp(\alpha y_f) \quad (19)$$

EXERCISES

1. Find the Laplace transform of the following functions using the definition of a Laplace transform as a starting point.

(a) unit step function. (b) e^{-at}. (c) $\cos \omega t$. (d) $\sin \omega t$. [Hint: $e^{i\omega t} = \cos \omega t + i \sin \omega t$.]

2. Find the Laplace transform of $f(t)$, where $f(t)$ is sketched as in Fig. E9.2. [Hint: $L\{f(t - a)u(t - a) = e^{-as}F(s)$, where $u(t - a)$ is a unit step applied at $t = a$.]

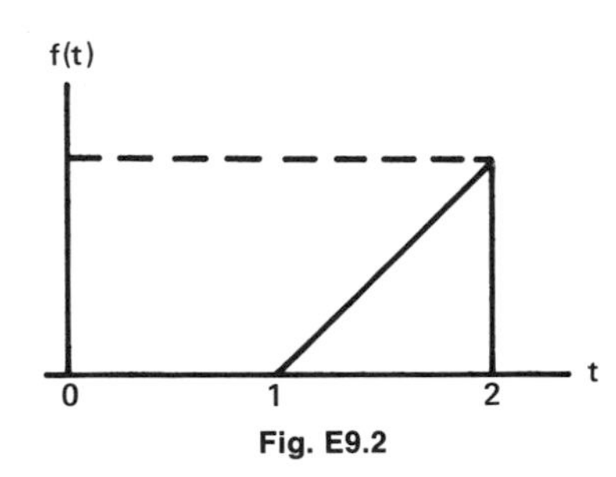

Fig. E9.2

3. Let $f(t)$ be a periodic function of period T. Show that $F(s)$ is given by

$$F(s) = \frac{1}{1 - e^{-Ts}}\{\text{Laplace transform of the first cycle of } f(t)\}$$

4. Solve the following differential equation using Laplace transforms:

$$\ddot{x} + 3\dot{x} + x = m(t); \qquad x(0) = 2, \quad \dot{x}(0) = 4$$

5. Solve the integral equation

$$y(t) = at + \int_0^t y(t) \sin(t - \tau)\, d\tau$$

using Laplace transforms. [Hint: Note that the Laplace transform of the second term on the right side can be obtained using Relation 16 of Table 9.2.]

6. Obtain a single operational expression for the block diagram shown in Fig. E9.6.

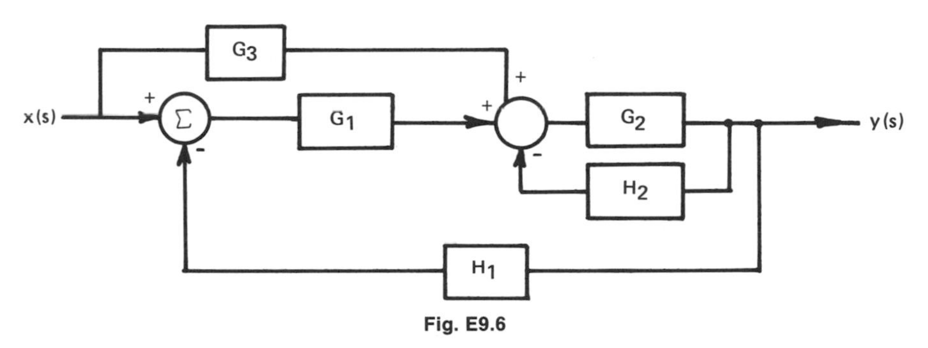

Fig. E9.6

7. Two systems $\mathscr{A}_1$ and $\mathscr{A}_2$ are said to be *inverse* of each other if when connected in tandem, one cancels the effect of the other. Then they are written as $\mathscr{A}_1 = \mathscr{A}_2^{-1}$. Given a system $\mathscr{A}$, show that its inverse $\mathscr{A}^{-1}$ can be realized by interconnecting $\mathscr{A}$, a *summer*, and a scalar with a scale factor -1 as shown in Fig. E9.7.

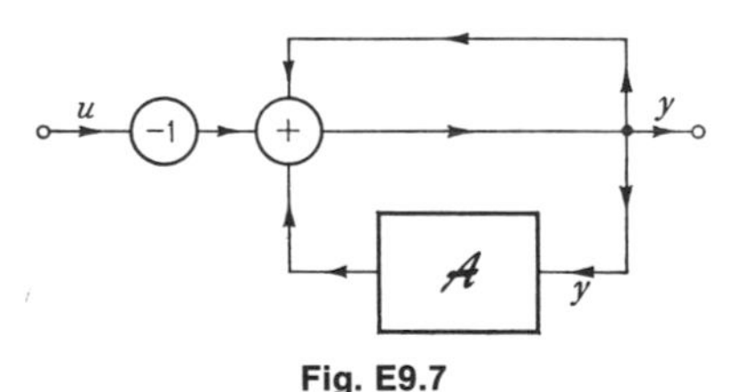

Fig. E9.7

8. Verify the entries in Table 9.1. You may use a graphical procedure if you wish.

9. The characteristic equation of a system is given by

$$s^4 + 3.5 \times 10^3\, s^3 + 3.75 \times 10^6\, s^2 + 1.75 \times 10^9\, s + 0.5 \times 10^{12} = 0$$

Is the system stable? [Hint: Set $s = 10^3\sigma$ and conduct the test on the polynomial in σ.]

10. Conduct the Routh test on the polynomial

$$s^7 + 4s^6 + 5s^5 + 2s^4 + 4s^3 + 16s^2 + 20s + 8 = 0$$

11. Consider a system whose transfer function is given by

$$X(s)/Y(s) = K/(s^3 + 3s^2 + 2s + K)$$

For what range of K is the system stable?

12. Consider a system whose characteristic equation is given by

$$s^4 + 6s^3 + 15s^2 + 18s + 10 = 0$$

Using a Routh array, find the root locations. [Hint: Shift the origin of the s plane in a trial and error fashion.]

13. How many roots of the following polynomial have real parts between 0 and -1?

$$16s^7 + 60s^5 + 23s^4 + 94s^3 + 194s^2 + 256s + 120 = 0$$

14. In labor management negotiations, usually the labor unions want what is called an "automatic cost of living escalator clause," which is tantamount to a positive feedback. A block diagram of such a system is shown in Fig. E9.14. Discuss how additional feedback loops in the form of legislative controls can stabilize the economic system.

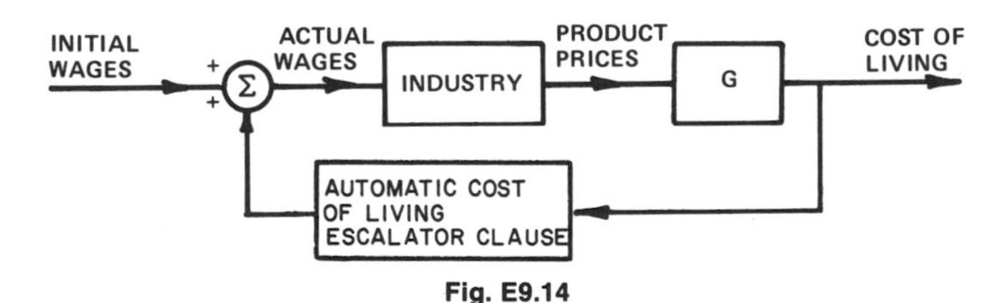

Fig. E9.14

15. Consider an automatic warehousing and inventory control system shown in Fig. E9.15. Find the transfer function $T(s) = C(s)/R(s)$. Suggest a method of determining the sensitivity of $T(s)$ with respect to $G_1(s)$, $G_2(s)$, etc. Which is most sensitive? Discuss.

16. A transfer function model of a system is given by

$$G(s) = 1/(s^2 + 4s + 3)$$

Show that the transition matrix of this model is

$$\mathbf{\Phi}(t) = \begin{bmatrix} 1.5e^{-t} - 0.5e^{-3t} & 0.5e^{-t} - 0.5e^{-3t} \\ -1.5e^{-t} + 1.5e^{-3t} & -0.5e^{-t} + 1.5e^{-3t} \end{bmatrix}$$

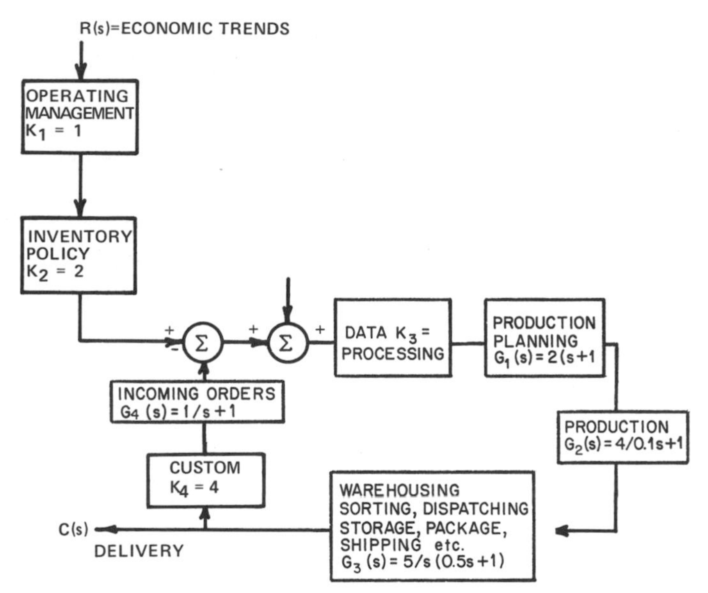

Fig. E9.15 (Reproduced with permission from Shinners, *Modern Control System Theory and Applications* Copyright © 1972 by Addison-Wesley, Reading, Massachusetts.)

17. Consider

$$\ddot{y} + 2\dot{y} + y = \dot{u} + u$$

If $u = \sin t$, find $y(t)$ using the state space method.

18. Let $x_1(t)$ be the number of rabbits and $x_2(t)$ be the number of lynxes in an environment at time t. The lynxes eat rabbits. A linear differential equation model describing $x_1(t)$ and $x_2(t)$ at any time is given by the pair:

$$\dot{x}_1 = kx_1 - ax_2, \qquad x_1(t = 0) = x_1(0)$$
$$\dot{x}_2 = -hx_2 + bx_1, \qquad x_2(t = 0) = x_2(0)$$

Assume that a, b, k, and h are positive numbers.

(a) Let the output y of this system be the number of rabbits x_1. Determine the functions $G(s)$ and $H(s)$ such that $y = G(s)x_1(0) + H(s)x_2(0)$.

(b) Is the system stable if $h = 3$, $k = 1$, $a = b = 2$? How many rabbits will there be as t tends to infinity?

(c) Using the Routh test determine the conditions on a, b, k, and h so that the system is stable.

19. Consider the problem of constructing a model for an epidemic. Let N = total population, s = number of people susceptible to disease, I = number of people infected at any time, and r = number of people who have recovered or are immune. A simple-minded model assumes that the rate of change of r is b times I, and that the rate of new infection is a times the product (sI).

(a) Explain the logic behind the model.

(b) How would you define a stability criterion in this problem? Explain, using examples.

(c) Given $N = 100$, $A = 0.01$, and $B = 0.75$; linearize the system about the operating point $(i, s, r) = (10, 80, 70)$. Is the linearized model stable? Is the linearized model useful?

SUGGESTIONS FOR FURTHER READING

1. For an elementary introduction to the ideas of control and applications of control to problems of every day interest, see
   E. E. David, Jr., and J. G. Truxal (eds.), *The Man Made World*, (in three parts). McGraw-Hill, 1968, New York.
2. A classical reference on Laplace transform theory is
   R. V. Churchill, *Operational Mathematics*, 2nd ed. McGraw-Hill, New York, 1958.
3. For an elementary introduction to control theoretic concepts both from the classical and state space theoretic viewpoints refer to
   J. L. Melsa and D. D. Schultz, *Linear Control Systems*. McGraw-Hill, New York, 1969.
4. For further discussion of state space theory, Liapunov's stability test, Pontryagin's maximum principle, refer to
   P. M. DeRusso, R. J. Roy, and C. M. Close, *State Variables for Engineers*. Wiley, New York, 1965.
   S. J. Citron, *Elements of Optimal Control*. Holt, New York, 1969.
5. If you have detected a similarity between optimal control theory and mathematical programming, then you can find further treatment of these two topics in
   D. Tabak and B. C. Kuo, *Optimal Control and Mathematical Programming*. Prentice-Hall, Englewood Cliffs, New Jersey, 1971.
6. The water pollution problem discussed in Section 6 is largely based on the paper
   V. Gourishankar and R. L. Lawson, Optimal control of water pollution in a river, *Int. J. Systems Sci.* **6**, No. 3, 201–216 (1975).

10

Growth and Decay Processes

INTRODUCTION

Growth and decay processes are of great importance in modeling complex systems. Growth of population and pollution, growth of cancer cells, growth of poverty and unrest are typical problems of contemporary interest. Whenever and wherever there is a change it can always be regarded as a growth or decay process. So wide is the scope of this phenomenon that it is almost impossible to cover all aspects of the problem here.

This chapter begins with a discussion of simple growth processes using discrete time or difference equation models. Then attention will be focused on continuous growth processes described by ordinary differential equations. As nothing can grow forever, models representing growth processes are necessarily nonlinear, the nonlinearity representing the limiting nature of growth. Therefore, growth processes are typically characterized by nonlinear differential equations. Many realistic growth processes are intertwined with other processes, and consequently the equations become coupled. When the number of interacting variables is large, such as in a trophic web, the differential equations may become very difficult to solve. One might not even know the initial conditions to the problem. Thus, it appears that the deterministic approach to modeling is replete with a number of hurdles. However, such deterministic models play a basic role in our understanding of the phenomenon under study; they play the same role as the harmonic oscillator in theoretical physics, the Ising model of quantum field theory, the Kroning–Penny model of lattice vibrations, or the Heisenberg model of ferromagnetism. In spite of their drawbacks all these models provide invaluable insight into the problem. For this reason some simple models of population growth are discussed. Usually several modifications are needed to "fit" these models to realistic situations.

1 DISCRETE GROWTH PROCESSES

Any growth or decay process can be treated as discrete or continuous. The choice essentially depends upon one's viewpoint, the time duration involved, and the resolution level used. Some of the simplest growth processes can be found in everyday experience.

Compound Interest and Amortization

Growth of money in a savings account is one familiar example. Consider a savings account subject to the following regulations:

(1) The annual rate of interest is $r\%$.

(2) The interest is paid and compounded n times a year. That is, a year is divided into n equal parts and interest is paid at the end of each time period.

(3) Deposits can be made at any time, but deposits start earning interest only from the beginning of the next time period. A deposit made on the first day of a time period starts earning interest immediately.

It is desired to construct a model that would predict the balance left in the account at any given time.

Note that $n = 2$ corresponds to semiannual compounding and $n = 365$ corresponds to daily compounding. This problem can be considered as a discrete growth process because the balance in the account changes only at discrete times. Let

$$y(k) = \text{total funds at the end of the } k\text{th period}$$

$$u(k) = \text{total deposits made during the } k\text{th period.}$$

Then, according to the above rules

$$y(k) = y(k-1) + (r/n)y(k-1) + u(k) = (1 + r/n)y(k-1) + u(k) \tag{1}$$

The first term $y(k-1)$ on the right is the principal brought forward from the previous time period, the second term, $(r/n)y(k-1)$ is the interest earned by $y(k-1)$ during the $(k-1)$th time period, and the last term $u(k)$ represents new deposits made during the kth time period.

Example 1. A savings bank pays 5% interest compounded semiannually. Starting with an initial deposit of \$1000, a person makes deposits every six months as follows: \$476, \$355, $-$\$217, $+$\$727. Find the balance after a two-year period.

For this problem $r = 0.05$ and $n = 2$. Therefore, $u(0) = y(0) = 1000$. The input sequence is $u(1) = 476$, $u(2) = 355$, $u(3) = -217$ and $u(4) = 727$. The balance at the end of various periods is

$$y(1) = (1 + 0.05/2)\,y(0) + u(1)$$
$$= (1.025)(1000) + 476 = 1501$$
$$y(2) = (1.025y(1))(1501) + 355 = 1893.53$$
$$y(3) = (1.025y(2))(1893.53) - 217 = 1723.87$$
$$y(4) = (1.025y(3))(1723.87) + 727 = 2493.96$$

■

Consider the problem of a person who borrowed d dollars to purchase a house. This debt is to be paid back by a sequence of periodic payments where a portion of each payment reduces the outstanding principal, while the remaining portion goes toward payment of interest. If the person wishes to make N equal payments of p dollars, what should be the value of p?

To solve this problem, let

$y(k)$ = outstanding debt after the kth period
$u(k)$ = amount of the kth payment
r = interest rate per payment period (usually a constant)

Then

$$y(k) = y(k-1) + ry(k-1) - u(k) = (1 + r)\,y(k-1) - u(k) \qquad (2)$$

Note that

$$u(k) = p; \qquad k = 1, 2, \ldots, N \text{ and } y(0) = d.$$

An easy way to find a solution to the above equation is to evaluate it for various values of k as follows:

$$y(1) = (1 + r)\,y(0) - u(1) = (1 + r)d - p$$
$$y(2) = (1 + r)\,y(1) - u(2) = (1 + r)[(1 + r)d - p] - p$$
$$= (1 + r)^2 d - (1 + r)p - p$$

Similarly

$$y(3) = (1 + r)^3 d - (1 + r)^2 p - (1 + r)p - p$$

or, in general

$$y(k) = (1 + r)^k d - (1 + r)^{k-1} p - (1 + r)^{k-2} p - \cdots - p$$
$$= (1 + r)^k d - p \sum_{i=0}^{k-1} (1 + r)^i; \qquad k = 1, 2, \ldots, N$$
$$= (1 + r)^k d - \frac{p(1 + r)^k - 1}{r}; \qquad k = 1, 2, \ldots, N \qquad (3)$$

The last equality is obtained by virtue of the mathematical identity

$$1 + \alpha + \alpha^n + \cdots + \alpha^{n-1} = (\alpha^n - 1)/(\alpha - 1) \tag{4}$$

In the present problem $k = N$ and $y(N) = 0$. Therefore,

$$y(N) = 0 = (1 + r)^N d - p \frac{(1 + r)^N - 1}{r}$$

or

$$p = \frac{r(1 + r)^N}{(1 + r)^N - 1} d \tag{5}$$

A block diagram of the process shown in Eq. (2) can be drawn as shown in Fig. 10.1. Note that z^{-1} is an operator, analogous to the Laplace transform variable s, operating on $y(k)$ to produce $y(k - 1)$. That is, z^{-1} delays the input by one time period.

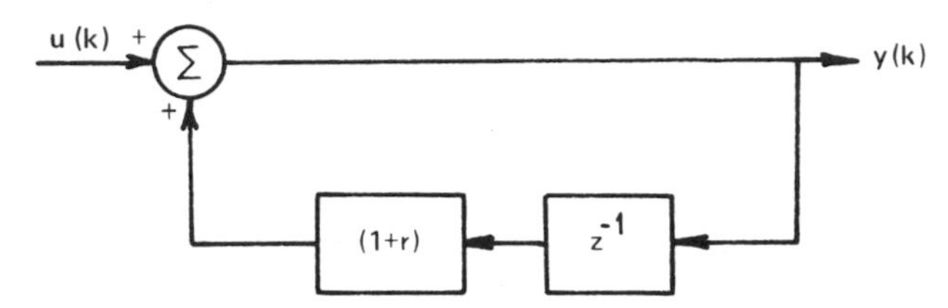

Fig. 10.1 A block diagram representation of a discrete growth process.

Growth of Rabbit Populations

Consider a simple model of rabbit population. Assume

(1) A pair of rabbits (one male and one female) is born to each pair of adult rabbits at the end of every month.
(2) A newborn pair produces its first offspring at two months of age.
(3) Once paired, a pair of rabbits keep producing other rabbits forever according to the above rules.

Let

$$y(k) = \text{number of rabbit pairs at the end of } k\text{th month}$$

At the end of the kth month, the number of newborn pairs reaching reproductive stage is $y(k - 2)$. Therefore,

$$y(k) = y(k - 1) + y(k - 2) \tag{6}$$

Let $y(0) = 1$, $y(-1) = 0$. That is, a newborn pair of rabbits was placed in the test area initially.

Then,

$$y(1) = y(0) + y(-1) = 1, \qquad y(2) = y(1) + y(0) = 2$$
$$y(3) = 3, \qquad y(4) = 5$$

A general solution to this problem is given by

$$y(k) = \frac{1}{\sqrt{5}}\left[\left(\frac{1+\sqrt{5}}{2}\right)^{k+1} - \left(\frac{1-\sqrt{5}}{2}\right)^{k+1}\right]; \qquad k = -1, 0, 1, 2, \ldots \quad (7)$$

The sequence of numbers generated by the above equation is the famous Fibonacci sequence.

2 CONTINUOUS GROWTH

One means of determining how large a population will be at various times in the future is to find the population growth equation that most accurately describes the situation at the present time, then assume that this equation holds good for all time. Let $N(t)$ denote the number of individuals in a population (or population density, in some cases) at time t. This number N is of course an integer, and it will change over time by integral amounts. Nonetheless, if N is large, it can be considered as a continuous variable.

Let us begin with an assumption. If a population is introduced into an environment, where it is not already found, growth occurs (if it occurs!) initially at a rate that is directly proportional to the size of the population at that time. The mathematical translation of this statement is the equation

$$\frac{dN}{dt} = kN \quad (1a)$$

where k is a constant of proportionality often referred to as the *growth constant*. Let us also assume that at time $t = 0$, that is, at the beginning of the experiment, we started with a population of size N_0. This can be written as

$$N(t = 0) = N(0) = N_0 \quad (1b)$$

What do equations of this kind mean? What light do they shed on the time history of the growth process? To answer these questions, we have to look for a function or functions that not only satisfy Eq. (1a) but also the condition stipulated in Eq. (1b).

Recalling that $d/dt(e^{bt}) = be^{bt}$ for any constant b, it is clear that $N = e^{kt}$ is a possible solution of Eq. (1a). Moreover, since Eq. (1a) is homogeneous, $N = Ce^{kt}$ is a solution for any constant C. Because we know what N should be at time $t = 0$, this constant C can be evaluated by using knowledge available through Eq. (1b). Hence, as a candidate for the solution of Eq. (1a) we have

$$N = N_0 e^{kt} \tag{2}$$

After a solution to an equation is obtained, a good analyst should conduct two tests, namely, (a) Is this a correct solution? (b) Is this the only solution? The first question can be answered by differentiating Eq. (2) once and verifying that Eq. (1a) is, indeed, satisfied. An answer to the second question can be obtained by resorting to a rigorous mathematical proof which is beyond the scope of this work. An alternative way is as follows. Since $N(t)$ is the size of the population at time t, and thus never zero, let us rewrite Eq. (1a) as

$$\frac{1}{N}\frac{dN}{dt} = k \tag{3}$$

Integrating both sides between the limits 0 and t, and letting η represent a dummy variable of integration

$$\int_0^t \frac{1}{N}\frac{dN}{d\eta}\,d\eta = \int_0^t k\,d\eta$$

That is,

$$\ln N]_0^t = kt \qquad \text{or} \qquad \ln N(t) - \ln N_0 = kt$$

or

$$\ln\frac{N(t)}{N_0} = kt \qquad \text{or} \qquad N(t) = N_0 e^{kt} \tag{4}$$

which is identical to Eq. (2). Thus, there is only one solution. The base of the logarithms used in the above analysis only affects the scale (of a graph). Therefore, it makes no difference whether the logarithms are to base 10 (log) or to base e (ln).

A surprisingly large number of growth phenomena (or decay phenomena) in physical, chemical, biological, medical, and social sciences follow Eq. (4) or equivalently the differential equation (1). The constant k contains non-dimensional parameters involving all factors that influence the growth in question.

Example 2. The study of the dose–response relationship is fundamental in pharmacology. Consider the problem of determining the dose level of a drug. Let y_0 be the initial concentration of a drug and $y(t)$ be the concentration at any time t. Assume that the drug follows a simple mechanism in disappearing from the body and assume that the rate of change of concentration of the drug in the blood is directly proportional to the amount present in the blood. This yields the differential equation

$$\frac{dy}{dt} = -ky$$

The assumptions made are not too unreasonable. In fact, it is known that penicillin follows roughly the above differential equation.

The above equation is solved to get

$$y = y_0 e^{-kt} \qquad \text{or} \qquad y/y_0 = e^{-kt}$$

From the above equation it is easy to see that, at time $t = 1/k$, the concentration of the drug will be $1/e$ of the initial concentration y_0. That is, approximately 63.2% of the drug has disappeared from the system by time $1/k$. The constant k depends upon the drug under study and can be experimentally determined.

In practice, ingestion of medicine is not a one-shot operation. More realistically, a constant amount y_0 is added to the body at intervals of time T. Then the concentration of the drug at time T is given by

$$y(T) = y_0 e^{-kT} + y_0$$

where the first term shows the amount of drug remaining in the body from the first dose and the second term shows the amount added at this time. At time $2T$, another dose is added and the expression is reduced by e^{-kT}. Thus, immediately after time $t = 2T$, the concentration of the drug is

$$y(2T) = (y_0 e^{-kT} + y_0)e^{-kT} + y_0$$

Repeating this process, the total concentration of the drug immediately after time $t = nT$ is given by

$$\begin{aligned} y(nT) &= y_0 + [y_0 + y_0 e^{-kT} + \cdots + y_0 e^{-(n-1)kT}]e^{-kT} \\ &= \frac{y_0(1 - e^{-(n+1)kT})}{1 - e^{-kT}} \end{aligned}$$

As $t \to \infty$, the concentration approaches $y_0/(1 - e^{-kT})$, which in turn is approximately y_0.

Using such an analysis, a pharmacologist can study the nature of the accumulation of drugs, the procedures to be adopted to maintain a uniform concentration of a drug over a period of time, and any other therapeutic measures. However, waiting for t to go to infinity is impractical. To overcome this, an initial large dose may be given and the required concentration maintained by booster doses. If the initial dose is

$$y_i = y_0(1 - e^{-kT})^{-1}$$

and repeated doses of y_0 are given at intervals of length T, the amount of drug disappearing from the system is computed as before. Now $y(nT)$ is computed by replacing the coefficient of $e^{-(n-1)kT}$ by y_i. Thus,

$$\hat{y}(nT) = y_0 + [y_0 + y_0 e^{-kT} + \cdots + y_i e^{-(n-1)kT}]e^{-kT}$$

which when simplified yields

$$\hat{y}(nT) = \frac{y_0(1 - e^{-nkT})}{1 - e^{-kT}} + y_i e^{-nkT}$$

Replacing y_i by the expression specified above and simplifying, we get

$$\hat{y}(nT) = y_0/(1 - e^{-kT})$$

which is a constant if k and T are constant. Thus, an initial dose of concentration y_i and booster doses of concentration given at intervals of T maintain the concentration of the drug at a constant level $\hat{y}$. ■

Example 3. The concept of body burden has been utilized by many in studying the effect of environmental pollutants on human health. Body burden is defined as the level of such a pollutant in the organs, blood, plasma, or other body fluids of an organism, which produces or is capable of producing damage or significant interference with the body functions. One way to measure the body burden of a substance is to measure the amount stored, taking into account the uptake and excretion from various sources.

Suppose that a person is exposed to a substance S and let

$x(t)$ = amount of S excreted through urine and faeces in time $(0, t)$

$y(t)$ = amount of S consumed via food and beverages in $(0, t)$

$z(t)$ = amount of S per unit volume of blood at time t

$u(t)$ = amount of S inspired via lungs in $(0, t)$

$v(t)$ = amount of S stored in body tissues in $(0, t)$

Assuming that uptake and excretion through other sources are negligible, one can write

$$z(t) = z(0) + \alpha[y(t) + u(t) - x(t) - v(t)]$$

where α is a constant.

Differentiating both sides with respect to t

$$\frac{dz}{dt} = \alpha\left[\frac{dy}{dt} + \frac{du}{dt} - \frac{dx}{dt} - \frac{dy}{dt}\right]$$

It is generally true that the rate of change of the substance S in a tissue is proportional to the total amount of S in the blood. That is,

$$\frac{dv}{dt} = bz$$

where b is some constant. Let

$$w = y + u$$

Combining the preceding three equations,

$$\frac{dz}{dt} + cz = \alpha\left(\frac{dw}{dt} - \frac{dx}{dt}\right)$$

where $c = \alpha b$.

Solution of the above equation gives the value of z as a function of t. As we do not readily know of a method of handling this equation, two possible things can be done. As a first step, one can make assumptions that lead to a mathematical simplification. For instance, the right side of the above equation vanishes if $w - x$ is a constant. As w is the total amount of uptake and x is the excretion, $(w - x)$ represents the excess of uptake over excretion. Thus the use of $(w - x) =$ a constant is a special case. This case leads to the solution

$$z(t) = z(0)e^{-ct}$$

An analog computer can be used to study the more general case and the special case discussed above can be used as a test to validate the analog computer model.

Example 4. The field of mathematical economics is a prolific source of problems leading to differential equations. Consider the problem of building steel mills in a developing country. As steel is required to build a steel mill, the amount of steel a country can produce depends on the amount it has in stock. Let

$$S_a(t) = \text{quantity of steel available at time } t$$
$$S_c(t) = \text{capacity of mills at time } t$$

Assuming that all other ingredients (including manpower and knowhow) are available in abundance, the rate of steel production is directly proportional to the existing mill capacities. This statement yields the equation

$$\frac{dS_a}{dt} = k_a S_c; \qquad S_a(0) = c_1 \tag{A}$$

where c_1 is the stockpile at some initial time. From this it is clear that the rate of steel production can be increased by increasing the steel-producing capacity, that is, by building a lot of mills. But this requires steel. Suppose a certain fraction of the stockpile available is allocated for building more mills. (How the remaining fraction is utilized is inconsequential at the present time.) The rate at which new mills can be built can be assumed to be directly proportional to the steel available in the stockpile. This gives

$$\frac{dS_c}{dt} = k_c S_c; \qquad S_c(0) = c_2 \tag{B}$$

where c_2 is the initial plant capacity. Since part of the steel produced goes back to construct more mills, Eq. (A) must be modified to

$$\frac{dS_a}{dt} = k_a S_c - bS_a; \qquad S_a(0) = c_1 \tag{C}$$

(An interesting problem in planning is to determine the quantity of steel to be allocated for increasing the plant capacity that will maximize $S_a(t)$, the amount of available stockpile at a specified future time T.)

An unrealistic feature of the foregoing analysis is the assumption that the rate of increase in capacity is directly proportional to the rate of allocation of the stockpile to this effort. In practice, construction of steel plants is not that simple. In addition to raw materials such as coking coal, limestone, or iron ore, successful construction and operation requires good transportation facilities (TF), manpower (MP), and an imaginative administration (IA). Therefore, it is more realistic to rewrite Eq. (B) as

$$\frac{dS_c}{dt} = \min[k_c S_a, \text{TF}, \text{MP}, \text{IA}]; \; S_c(0) = c_2 \tag{D}$$

Here the notation min $[a, b, c, d]$ denotes the minimum of the quantities a, b, c, and d.

This is a typical *bottleneck process*, where the rate of production depends upon the scarcest resource or the weakest link. Processes of this kind occur commonly in many other industrial and economic complexes.

3 LIMITS TO GROWTH

After a population has been in a new environment for some time and the density of population has increased to a high level, competition for food and other resources becomes severe. This is usually reflected in dropping rates of birth and survival. (This phenomenon is almost universal except in the human species!) Then the population is said to be limited by environmental resistance. For any given population of animals or plants, and for any given environment, there is a certain maximum number M of members that the environment can support. As the population grows more and more dense, the growth rate becomes smaller and smaller as N approaches M. This kind of population growth can be described reasonably well by a variety of nonlinear differential equations.

From an elementary viewpoint, a change in the size of a population can occur as a consequence of births and deaths. If b is the birth rate and m is the mortality rate, then

$$\frac{dN}{dt} = kN; \qquad N(0) = N_0 \tag{1}$$

can be written as

$$\frac{dN}{dt} = (b - m)N; \qquad N(0) = N_0 \tag{2}$$

As it stands, there is nothing new in Eq. (2); it merely elucidates the meaning of the symbol k in Eq. (1). But now it is not difficult to see that both b and m may themselves depend on the size of the population N. For example, if the food supply is limited, it is reasonable to expect that m will increase as N increases. If this increase is assumed to have a direct proportionality, then $m = m_1 N$, and

$$\frac{dN}{dt} = (b - m_1 N)N = k(N)N; \qquad N(0) = N_0 \tag{3}$$

which is a nonlinear equation because the growth rate k is a function of N. In view of the heuristic argument given above, one expects $k(N)$ to decrease as N increases, perhaps approaching zero as N approaches the said maximum M. If N is such that $(b - m_1 N) > 0$, then the population increases. When $(b - m_1 N) = 0$, that is, when $N = b/m_1$, it is reasonable to expect that the population cannot increase anymore. These predictions can be quickly verified.

Setting $M = b/m_1$, Eq. (3) can be rewritten as

$$\frac{dN}{dt} = m_1(M - N)N \tag{4}$$

Separating the variables,

$$\frac{dN}{(M - N)N} = m_1 \, dt \tag{5}$$

The left side can be decomposed, using partial fractions, to yield

$$\frac{1}{M}\left(\frac{1}{N} + \frac{1}{M - N}\right) = m_1 \, dt \tag{6}$$

Integrating both sides with respect to 1 within the limits 0 to t,

$$\left[\frac{1}{M}(\ln N - \ln(M - N))\right]_0^t = m_1 t \tag{7}$$

or

$$\frac{N}{M-N} = \frac{N_0}{(M-N_0)} e^{Mm_1t} = \frac{N_0}{(M-N_0)} e^{bt} \tag{8}$$

$$N(t) = \frac{[MN_0/(M-N_0)]e^{bt}}{1+[N_0/(M-N_0)]e^{bt}} \tag{9}$$

Now let us plot this $N(t)$ as a function of t. This can be done as follows. For small values of t, e^{bt} is small, and the second term in the denominator of Eq. (9) can be neglected compared to the first term. Thus, for very small values of t, Eq. (9) can be approximated by

$$N(t) \approx \frac{MN_0}{(M-N_0)} e^{bt} = Ae^{bt} \tag{10}$$

Therefore, $N(t)$ shows exponential growth. As t increases, the second term in the denominator becomes large compared to the first term, and $N(t)$ can be approximated by

$$N(t) \approx \frac{MN_0}{(M-N_0)} e^{bt} \bigg/ \frac{N_0}{(M-N_0)} e^{bt} = M \tag{11}$$

The resulting shape of the curve is shown in Fig. 10.2. This S-shaped curve is called the *logistic growth* curve or *sigmoid growth* curve. This phenomenon was first observed by the Belgian sociologist P. V. Verhulst in connection with the growth of human populations. For this reason, Eq. (3) is often referred to as the Verhulst equation.

The above analysis is based on several simplifying assumptions. In practice, the environmental factors influencing the growth of a population appear in

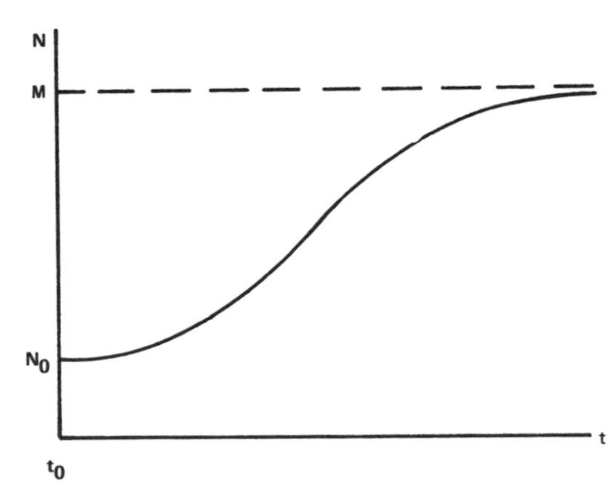

Fig. 10.2 A logistic growth curve.

many forms. The availability of food is of course important. Also of equal importance are the degree of spatial clumping or distribution of food and the manner in which these clumps are arranged (a large number of small clumps or a small number of large clumps). Climatological factors, such as fluctuating temperatures, affect the reproduction, growth, and survival rates of many species. Other environmental factors such as a reduction of dissolved oxygen in water due to pollution could eliminate schools of fish, and defoliation of forests would destroy the natural habitat of big cats and warblers. When all such environmental factors are included, the mathematical discussion becomes very complicated. In addition to this, if some of the environmental factors interact, it makes the task of data collection and interpretation extremely difficult. For example, deer are more likely to die due to starvation if (a) their population density is too high, (b) if the lowest foliage on tall trees is too high for them to reach, and (c) if snow covers the low vegetation. In this case, it is necessary to understand the interaction of the three factors in order to truly model the environment.

There are other pitfalls in the collection and interpretation of environmental data. For instance, suppose that the growth and survival rates of a species are measured at three different sites. Suppose that despite the densities being different at all three sites, the growth and survival rates are identical. Does this mean that growth and survival rates are independent of density? To answer this question affirmatively or negatively, first it is necessary to ascertain whether any of the density effects were masked by other enviromental effects. A solution to this problem can be obtained by carefully designing an experiment in which different sites and different densities are used as two of the factors. That is, the experiment should try to sort out the effects of density and environmental factors by taking samples at several different density locales for each type of site.

This discussion sheds some light on the magnitude of the data collection effort. Detailed consideration of this aspect of the problem leads us too far afield. However, some insight can be gained by estimating the data requirements of the problem under discussion. First, there must be adequate replication of space and time to explore the effects of various site and weather combinations. As the weather is constantly changing at each site, data should be collected for an adequate number of years and at an adequate number of sites. How many sites are to be observed? For how long? What attributes of weather are relevant? What is the extent of resolution required? There is no cookbook recipe to answer these questions. This is one of the major decisions that a systems analyst should make. Assume that there are n independent attributes required to be measured. As a rule of thumb, one would never attempt to plot a curve on a two-dimensional plane with less than six experimentally observed points. Then, ideally 6^n sets of data points are required to faithfully study the relation among the n independent variables.

For large n, 6^n becomes impractically large. Then, using Youden squares, confounding, or other tricks of experimental design, n should be trimmed down to a reasonable size.

The "rate constants" appearing in the preceding models of population growth are likely to be affected by changes in temperature, humidity, age distribution of the species, and several other ecological factors that are external to the system under study. Thus, there is considerable merit in considering the "rate constants" as functions of time or even random variables. Any treatment of the latter case is beyond the scope of this work. Two simple linear cases involving time dependence are treated in the sequel.

In many real situations, the quantity k in Eq. (1) is not likely to remain a constant. Seasonal fluctuations, or other factors could conceivably cause variations in k. This can be taken into account by rewriting Eq. (1) as

$$\frac{dN}{dt} = k(t)N \tag{12}$$

This is a homogeneous differential equation with variable coefficients. A general way to solve Eq. (12) is to use numerical techniques in conjunction with a computer.

Sometimes, the growth process can be modeled by a nonhomogeneous equation of the type

$$\frac{dN}{dt} = kN + Q(t) \tag{13}$$

The quantity $Q(t)$ can be thought of as an exogenous factor that disturbs an otherwise normal growth pattern. $Q(t)$ may represent a sudden natural calamity or may represent the hunting habits of an omnivorous predator like man or a seasonal factor. Equation (13) is a nonhomogeneous equation with constant coefficients. This equation can once again be solved, as before, by using a computer.

4 COMPETITION AMONG SPECIES

Thus far, the discussion has centered on growth processes that are essentially influenced by environmental factors. There exist numerous examples of assemblies which consist of a number of elements that influence each other through competition or cooperation. Some of the more important cases are: populations of biological species; political parties; business enterprises; coupled reacting chemical components in the atmosphere, in bodies of water, and in organisms as a whole or a part; components of the nervous system; and elementary excitations in fluids. In all these problems such terms as population, environment, species, and niche take on a very special meaning.

By appropriate modifications, the following discussion primarily aimed at animal or plant species, can be adapted to suit the needs of any of the above listed problems.

Nowadays, the term "ecological niche" is widely used to refer to any set of environmental conditions that permits a population to survive permanently and with which the population is capable of interacting. Consider two environmental factors X_1 and X_2. The upper and lower values of these factors limit specific organisms in their survival and reproduction. If a species S_1 exists within the boundaries $X_{L1} \leq X_1 \leq X_{H1}$, $X_{L2} \leq X_2 \leq X_{H2}$, then the rectangle bounded by these values is the ecological niche θ_1 for species S_1. This simple concept can be easily generalized to n environmental factors.

Consider two species S_1 and S_2 whose ecological niches are θ_1 and θ_2, and whose populations, respectively, are $N_1(t)$ and $N_2(t)$. The differential equations for the growth of these two species can be written, analogous to Eqs. (3.3) or (3.4) as

$$\begin{aligned} \frac{dN_1(t)}{dt} &= (b_1 - m_1 N_1)N_1 = m_1(M_1 - N_1)N_1; \qquad N_1(0) = N_{10} \\ \frac{dN_2(t)}{dt} &= (b_2 - m_2 N_2)N_2 = m_2(M_2 - N_2)N_2; \qquad N_2(0) = N_{20} \end{aligned} \tag{1}$$

whose solution, during the initial growth stage, can be written as

$$N_1(t) \simeq N_{10}e^{b_1 t}, \qquad N_2(t) \simeq N_{20}e^{b_2 t} \tag{2}$$

Therefore, the ratio of $N_1(t)$ to $N_2(t)$ is

$$N_1/N_2 = (N_{10}/N_{20}) \exp(b_1 - b_2)t \tag{3}$$

In modeling, this kind of mathematical analysis is not an end in itself. The analyst must keep in his mind, constantly, the implication of the equations that present themselves before him. For instance the above analysis indicates that the species S_1 and S_2 can coexist only if $b_1 \equiv b_2$ *precisely*. If $b_1 > b_2$, species S_1 slowly dominates S_2 and drives the latter to extinction. Similarly, if $b_1 < b_2$, then S_1 goes to extinction. In nature, precise mathematical equality of the reproduction rates of two different species is so unlikely, that this possibility can be ignored. This leaves us with a philosophically unpalatable principle that two nearly equally competent competitors cannot coexist. This principle is sometimes termed *competitive exclusion principle*, or Volterra–Gause principle.

It is evident that the competitive exclusion principle leads to conclusions that are contrary to conventional wisdom. Competing political systems, industrial systems, and competing species in the same ecological niche are known to coexist—peacefully or otherwise. If one begins with the assumption that every species competes with every other until only the best survives, then one has also to explain the diversity of flora and fauna that exist today!

Many explanations can be put forward to explain the coexistence of different species.

(i) *Geographical isolation*: Herbivores of Australia (kangaroos) did not compete with European herbivores (rabbits). Therefore, when man attempts to disrupt the geographic isolation of species, he must be careful to prevent the unwitting working of the exclusion principle. A less efficient company may be able to coexist with a more efficient one if they are separated by considerable distance and the transportation costs are heavy. (As a historical anecdote, it is interesting to note in passing that an ingenious American enterpreneur in the steel industry attempted to bypass this kind of protection available to smaller companies on the West Coast by introducing the now famous "Pittsburgh plus" pricing system.)

(ii) *Ecological differentiation*: In biology, ecological differentiation is a necessary rule for survival. Many species adapt themselves to their immediate surroundings by suitable specialization. In business, product differentiation plays an analogous role. Although two species may compete for all available resources, one may be more efficient (due to specialization) than the other in exploiting some of the resources.

(iii) *Interbreeding and mergers*: If two competing species are so closely related genetically that they can interbreed, one group, instead of replacing the other, simply merges with it. Mergers similarly prevent extinction in economics.

This discussion suggests the possibility that either S_1 and S_2 occupy slightly different niches or S_1 and S_2 are distinguishable. It is also possible to question some of the assumptions that went into the derivation of Eq. (1). For instance, Eq. (1) shows no interaction between the two species. If the interaction occurs primarily as competition for the same food, Eq. (1) can be modified to read

$$\begin{aligned} \frac{dN_1(t)}{dt} &= m_1(M_1 - N_1 - \alpha_{12}N_2)N_1; \qquad N_1(0) = N_{10} \\ \frac{dN_2(t)}{dt} &= m_2(M_2 - N_2 - \alpha_{21}N_1)N_2; \qquad N_2(0) = N_{20} \end{aligned} \tag{4}$$

In Eq. (1) M_1 and M_2 were the limiting size (or carrying capacity) of the populations. In the presence of another species, this limiting size has to be smaller. The extent to which the limiting size has to be decreased obviously depends upon the quantity of resource the other species is consuming. Equation (4) results if this dependence is assumed to be directly proportional to the size of the competing species.

Since competition of the two species depends upon the probability of contact between individuals, a possible method of defining α_{ij} is

$$\alpha_{ij} = \sum_e p_{ie}p_{je}/p_{ie}^2 = \beta_i a_{ij} \tag{5}$$

where index e stands for the eth environment, and p_i and p_j are the probabilities for the appearance of individuals of ith and jth species, respectively, in the eth environment. The fraction $\beta_i = 1/\sum_e p_{ie}^2$ is often used as a measure of the *width of the niche*. Evidently, $\alpha_{ij} = 1$ with equal niche width for both the species and $\alpha_{ij} < 1$ if species S_i has a wider niche but S_j is more specialized.

A first step in understanding a growth process described by a nonlinear differential equation of this kind would be to locate the *equilibrium points* of the process. A system is said to be in equilibrium when the rates of change are equal to zero. The present system will be in equilibrium if

$$\frac{dN_1}{dt} = 0 \qquad \text{and} \qquad \frac{dN_2}{dt} = 0 \tag{6}$$

This implies from Eq. (4) that

$$m_1 N_1(M_1 - N_1 - \alpha N_2) = 0 \qquad \text{and} \qquad m_2 N_2(M_2 - N_2 - \beta N_1) = 0$$

or either

$$N_1 = 0 \quad \text{or} \quad M_1 - N_1 - \alpha N_2 = 0$$

and

$$N_2 = 0 \quad \text{or} \quad M_2 - N_2 - \beta N_1 = 0$$

That is, either

$$N_1 = 0 \quad \text{or} \quad N_1 = M_1 - \alpha N_2 \qquad \text{and} \qquad N_2 = 0 \quad \text{or} \quad N_2 = M_2 - \beta N_1$$

The case $N_1 = 0$ and $N_2 = 0$ is trivial and has no practical interest. The second case gives a more interesting condition

$$N_1 = M_1 - \alpha N_2 = q_1 \qquad \text{and} \qquad N_2 = M_2 - \beta N_1 = q_2 \tag{7}$$

Here q_1 and q_2 merely represent the equilibrium values of x and y. These two conditions for equilibrium are plotted as shown in Fig. 10.3. The lines thus obtained are called *saturation lines* because for any point p on this line the x coordinate of p gives the limit to which the species S_1 can grow if at that time S_2 has a population level given by the y coordinate. For an equilibrium in the growth process both dN_1/dt and dN_2/dt should vanish. Therefore, it is useful to plot both the equilibrium lines on the same graph sheet. As the equilibrium lines may have different slopes for differing values of M_1, M_2, α, and β, a merger of the two graphs could lead to four distinct cases as shown in Fig. 10.4.

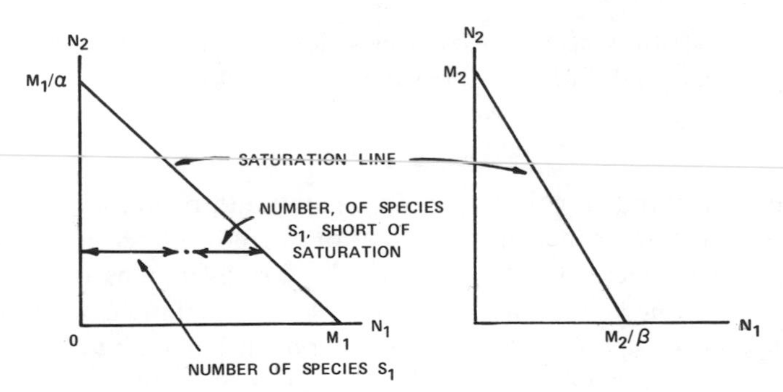

Fig. 10.3 Equilibrium condition in a growth process.

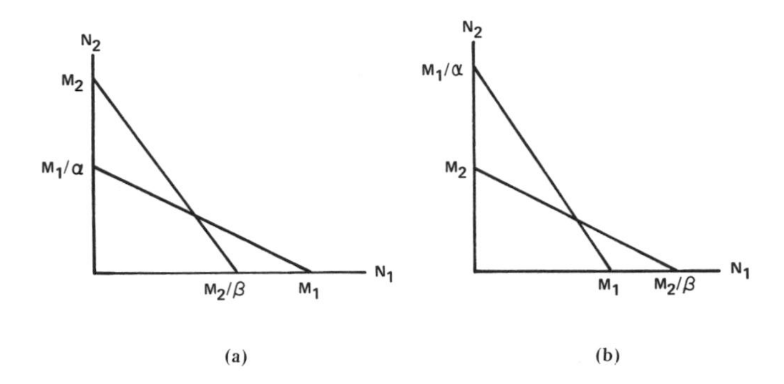

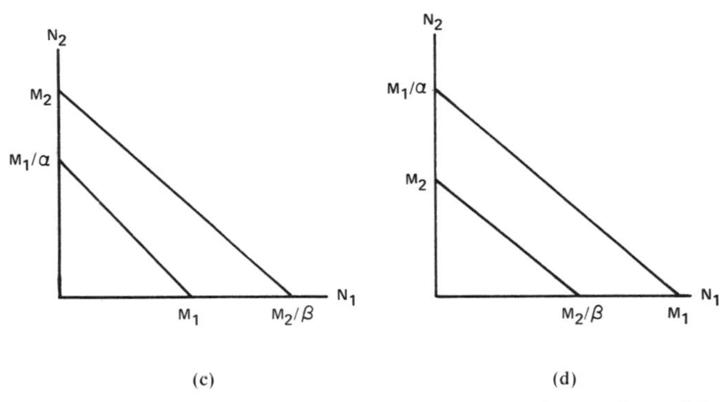

Fig. 10.4 Four different configurations of the saturation lines: Case (a); Case (b); Case (c); Case (d).

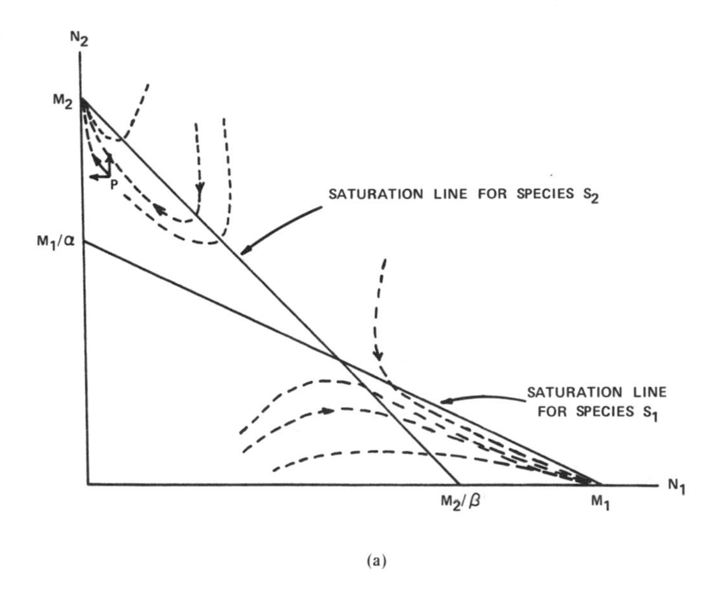

(a)

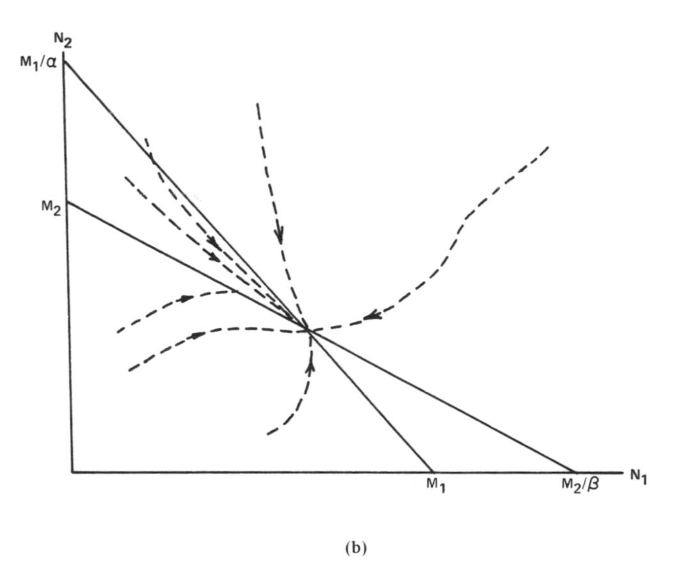

(b)

Fig. 10.5 Trajectories of population levels of two interacting species. (a) For the configuration shown in Fig. 10.4a; (b) for the configuration shown in Fig. 10.4b.

The growth pattern of the populations of species S_1 and S_2 in each case can be obtained by considering typical points in the xy plane for each case. This procedure is demonstrated below.

Case a. $M_2 > M_1/\alpha$ and $M_1 > M_2/\beta$. Stated differently

$$\alpha > M_1/M_2 \qquad \text{and} \qquad \beta > M_2/M_1$$

At point p, in Fig. 10.5a, both the species are not at saturation. In fact, species S_1 is "oversaturated." As this cannot happen in practice, x has to decrease and y increase to reach an equilibrium. The fact that x has to be smaller and y larger is shown by drawing two vectors at p_1, one pointing to the left and one to the top. That is, the vectors are drawn in the direction pointing toward the saturation lines of the respective species. The *resultant* of these two vectors gives the direction of the path along which the population levels of the two species should move to attain equilibrium. The procedure is repeated for various other points. The final growth patterns are shown by the dashed lines. Inspection of these lines reveals that Case (a) leads to a growth condition of *unstable equilibrium.* For this set of conditions, one species always gets extinct and the other reaches saturation. A similar graph for Case (b), is shown in Fig. 10.5.

Coevolution of Host–Parasite Systems

If two species S_1 and S_2 exist in the same environment, several patterns of growth can be observed. These patterns can be represented symbolically and descriptively as

$+ +$	synergism	$0\ 0$	neutrality
$+ 0$	metabiosis	$0\ -$	antagonism
$+ -$	competition	$-\ -$	mutually negativistic

where a $+\ -$ means that one species gets hurt in the presence of the other. For example a $+\ -$ situation could represent a host–parasite system or a predator–prey system. This competition need not be among animal species. Growth of usage of one product may lead to a decline in the usage of another product.

To keep the analysis simple, a specific host–parasite system is examined in very elementary terms. If there are no parasites to kill a host, the host population will expand. (For simplicity, let us assume that all other factors that influence the growth of a population remain unchanged.) Let us further assume, as before, that the rate at which the host population increases is proportional to the size of the population itself. This leads to

$$\frac{dh}{dt} = ah; \qquad a > 0 \quad \text{if} \quad p = 0 \tag{8}$$

where $h = h(t)$ is the size of the host population and $p = p(t)$ is the size of the parasite population at time t.

Similarly, if there are no hosts, the parasites die due to lack of nutrition and the decay of parasite population follows:

$$\frac{dp}{dt} = -bp; \qquad b > 0 \quad \text{if} \quad h = 0 \tag{9}$$

The negative sign on the right side is indicative of decay.

However, if both hosts and parasites are present, the mortality of hosts due to parasite attacks becomes an important factor. The exact relation describing the pattern of attack on the hosts by parasites depends on a wide variety of factors. However, it is reasonable to assume that the attacks are directly proportional to the encounters between the two species and the encounters themselves are proportional to the product hp. With this in mind, Eqs. (8) and (9) can be modified to

$$\frac{dh}{dt} = ah - chp, \qquad \frac{dp}{dt} = -bp + dhp \tag{10}$$

where $a, b, c, d > 0$.

This set of simultaneous, first-order, nonlinear ordinary differential equations is often referred to as Volterra's model for a host–parasite system. The solution of these equations will reveal the dynamic interdependence of two species. As these equations are nonlinear, it is indeed not very easy to solve them analytically.

As before, the equilibrium points of the system can be obtained easily by equating the derivatives on the left side to zero. This gives

$$\frac{dh}{dt} = 0 = ah^* - ch^*p^*, \qquad \frac{dp}{dt} = 0 = -bp^* + dh^*p^* \tag{11}$$

where h^* and p^* are the values of h and p at equilibrium. Now

$$\begin{aligned} h^*(a - cp^*) &= 0 \qquad \text{or} \qquad h^* = 0, \quad p^* = a/c \\ p^*(dh^* - b) &= 0 \qquad \text{or} \qquad p = 0, \quad h^* = b/d \end{aligned} \tag{12}$$

Thus, in the hp plane there are two *equilibrium points*: (0, 0) and $(b/d, a/c)$. The nature of the solution near the equilibrium point E defined by $(b/d, a/c)$ has several interesting properties.

In order to examine the nature of the solution near E, it is useful to consider another coordinate system centered at E itself. This shift in coordinates can be accomplished by defining a new coordinate system in terms of the old one as follows

$$u = h^* - b/d, \qquad v = p^* - a/c \tag{13}$$

Differentiating Eq. (13) with respect to time and using Eqs. (10) and (13), one gets

$$\begin{aligned}\frac{du}{dt} &= \frac{dh^*}{dt} = ah^* - ch^*p^* = -cuv - \frac{bc}{d}v \\ \frac{dv}{dt} &= \frac{dp^*}{dt} = -bp^* + dh^*p^* = duv + \frac{ad}{c}u\end{aligned} \tag{14}$$

The above equations describe the dynamic behavior of the populations when viewed from the new origin or the uv coordinate system (see Fig. 10.6).

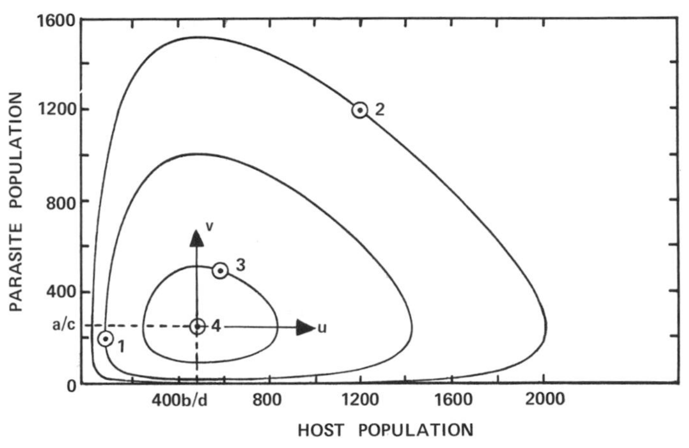

Fig. 10.6 Plot of hosts versus parasites for four sets of initial conditions: (1) $P_0 = 200$, $H = 100$; (2) $P_0 = 1200$, $H_0 = 1200$; (3) $P_0 = 500$, $H_0 = 600$; (4) $P_0 = 250$, $H_0 = 500$.

If the host and parasite populations are exactly equal to b/d and a/c, respectively, there would be an equilibrium. Consider what happens if the populations are disturbed from this equilibrium point by a very small amount; say, the new population is now at the point p—not too far from E. As long as p is sufficiently close to E, the values of u and v are small and the product uv is still smaller and can be neglected from Eq. (14) as a first approximation. This leads to

$$\frac{du}{dt} = \frac{-bc}{d}v, \qquad \frac{dv}{dt} = \frac{ad}{c}u \tag{15}$$

This set of first-order linear equations can be written as

$$\frac{d^2u}{dt^2} = -(ab)u \tag{16}$$

whose solution can be written as

$$u(t) = A \sin(\omega t + \phi_0) \tag{17}$$

where $\omega = \sqrt{ab}$, and A and ϕ_0 are arbitrary constants. These constants can be determined if two additional conditions, say, the initial values of host and parasite populations, are given. An analogous expression is valid for $v(t)$. The sinusoidal nature of the solution near the equilibrium is indicative of the well-known and often observed fact that host–parasite populations fluctuate in cycles.

Multispecies Communities

In a system with more than two competing species, Eq. (1) can be generalized to read

$$\frac{dN_i}{dt} = k_i N_i \left(1 - \frac{N_i}{M_i}\right) + \gamma_i \sum_{j=1}^{N} a_{ij} N_i N_j; \qquad i = 1, 2, \ldots, N \tag{18}$$

The first term $k_i N_i$ describes the behavior of the ith species in the absence of others. As long as $k_i > 0$, the ith species in this model is postulated to grow in an exponential (Malthusian) manner. When $k_i < 0$, there will be exponential decay. The term in the parenthesis $(1 - N_i/M_i)$ is the Verhulst term, and it accounts for the saturation effects. The influence of this term is to damp out fluctuations in the population of species i. Such damping effects may not always be visible because of the enormous time scales involved in the evolution process. To simplify analysis, this term is often neglected. The quadratic term describes the binary interactions of one species with all other species. A positive a_{ij} tells how rapidly encounters between the ith and jth species will contribute to an increase in N_i and vice versa. A vanishing a_{ij} is indicative of no interactions. A unit decrease in host (or prey) population generally does not correspond to a unit increase in parasite (or predator) population [see Eq. (4.5)]. The equivalence number γ_i subsumes in it, among other things, a scale factor to account for this asymmetry between host and parasite numbers. Under the above assumptions, one can say

$$a_{ij} = -a_{ji} \qquad \text{and} \qquad a_{ii} = 0 \tag{19}$$

Equilibrium conditions for the above multispecies community model can be studied by equating $dN_i/dt = 0$ for all i. As an aid in illustrating this concept, consider two simple trophic web configurations as shown in Fig. 10.7. The solid arrows indicate the flow of material or energy from one species to another. The dotted arrows represent flow from an infinite reservoir of food or energy. It is evident from an inspection that the configuration shown

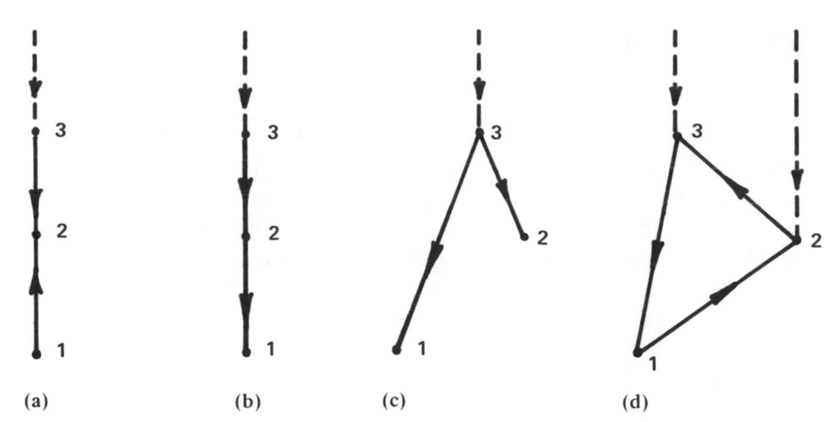

Fig. 10.7 Graphs showing different trophic web structures involving three interacting species. Only configurations (a) and (b) are discussed in the text. (Adapted, with permission, from N. S. Goel, S. C. Maitra, and E. W. Montroll, *On the Volterra and Other Nonlinear Models of Interacting Population*. Academic Press, New York, 1971.)

in Fig. 10.7a quickly degenerates because species 1 would perish as it is eaten by species 2 and there is no source of food or energy available for the first species to thrive on. In the configuration shown in Fig. 10.7b, for example, species 3 has an infinitely large source to thrive on. Therefore, it has an opportunity to reach close to its saturation level. Therefore, N_3/M_3 plays a significant role in Eq. (18). If species 1 and 2 are assumed to be far away from their saturation levels, then N_1/M_1 and N_2/M_2 are small enough to be negligible, and Eq. (18), under steady state conditions, can be approximated by

$$\begin{aligned} 0 \quad + a_{12}q_2 + a_{13}q_3 \qquad\qquad &= k_1/\gamma_1 \\ a_{21}q_1 + \quad 0 \quad + a_{23}q_3 \qquad\qquad &= k_2/\gamma_2 \\ a_{31}q_1 + a_{32}q_2 - (k_3/M_3\gamma_3)q_3 &= k_3/\gamma_3 \end{aligned} \tag{20}$$

where the q_i stand for the equilibrium values of N_i for all i.

Several other simplifications can be made to this problem by observing Fig. 10.7b. For example, as there is no link between species 1 and 3, one can set $a_{13} = a_{31} = 0$. As a positive a_{ij} is indicative of increasing values of N_i, a positive a_{ij} should correspond to an arrow from species j to species i in the trophic web. Therefore, a_{12} and a_{23} are positive and a_{21} and a_{32} are negative. Now, Eq. (20) can be written as

$$\begin{aligned} a_{12}q_2 \qquad\qquad\qquad &= -k_1/\gamma_1 \\ a_{21}q_1 + \quad 0 \quad + a_{23}q_3 \qquad\qquad &= -k_2/\gamma_2 \\ 0 \quad + a_{32}q_2 - (k_3/M_3\gamma_3)q_3 &= -k_3/\gamma_3 \end{aligned} \tag{21}$$

The solution of this equation is

$$q_2 = -k_1/(\gamma_1 a_{12}) = +k_1/\gamma_1 a_{21}$$
$$q_3 = M_3\left[\frac{k_1\gamma_3 a_{32}}{k_3\gamma_1 a_{21}} + 1\right] \quad (22)$$
$$q_1 = -\frac{k_2}{\gamma_2 a_{21}} + \frac{a_{32}}{a_{21}} M_3\left[\frac{k_1\gamma_3 a_{32}}{k_3\gamma_1 a_{21}} + 1\right]$$

As $q_i < 0$ are of no practical interest, only positive solutions will be considered now. Toward this end, notice that all γ_i are positive. The fact that species 1 will perish in the absence of species 2 makes the coefficient k_1 negative. Because a_{21} is positive, q_2 is seen to be positive. Because a_{32} is negative, q_3 can be positive if the condition $(k_3/\gamma_3) > a_{32}q_2$ is satisfied. Similarly, a condition for the positivity of q_1 can be derived.

The Diversity and Stability of Multispecies Communities

Consider a complex ecological system with various species of plants and animals. Some of the species, such as the vegetation, get their energy from primary sources such as the sun. The other species depend on plants and animals for their food. This interdependence leads to competition for resources such as sunlight, land area, soil moisture (for plants), and vegetation (for herbivores). This type of interdependence can be pictorially shown by means of a trophic web. Several aspects of the organization of such trophic webs are generally discussed in terms of diversity and stability.

Diversity refers to the richness and variety of species in a community. Indeed, the concept of diversity can be applied, in a much broader context, to the distribution of the elements of any set among its subsets. For example, the concept of diversity can be used to describe a unispecific population distributed in year classes, particles counted in size classes, or components of a chemical mixture counted in terms of molecules of different kinds. Diversity obviously depends upon the analyst's ability to discriminate: the individuals in an ecosystem, for example, can be classified as to species, as to genotypes, or even as to the kinds of DNA.

Stability is a poorly defined concept in ecology. Most of the stability concepts currently used are those that were borrowed from physical sciences. When one talks about local or global stability of an ecological system, one normally implies that the number of individuals in a species either explodes or the species goes extinct. Thus, stability indicates something about the persistence (or sensitivity) of a population to external disturbances and sheds some light on the probability of extinction of a species.

What is the relevance of these studies to complex systems in general? The relevance occurs at two levels. Increasing human populations and increasing affluence are imposing ever increasing demands on the ecological systems that fulfill the food, fiber, and aesthetic needs of humanity. As long as these demands are modest and the rates of change are small, the natural systems have sufficient resilience (or stability) to absorb considerable abuse and punishment. Thus, a basic understanding of the stability properties of an ecological system is rewarding in its own right. Second, an understanding of how a natural system copes with external disturbances helps humanity to understand the stability properties of the society. For instance, most of the preceding analysis can be quickly adapted to study the dynamics of armament races among countries. The same ideas can also be used to study the propagation of information, such as a technological innovation, in a society. Similar ideas are also being widely used to study the growth patterns of cities and to develop strategies for the advancement of less developed countries.

5 GROWTH PROCESSES AND INTEGRAL EQUATIONS

Differential equations are basic for an understanding of the dynamic behavior of any system. There is a vast body of knowledge about differential equations. However, for computational purposes their structure is awkward. As was seen and as will be seen again, differentiation is an undesirable operation from a computational viewpoint. Because of this, integrators became the basic building blocks of analog computers, and integration was taken as a basis in the numerical methods of solving differential equations. Conceptually, also, the differential equation approach to system description often misleads students by an apparent reversal in the direction of the causal process. For instance, in

$$\frac{dy}{dt} = x \qquad \text{or} \qquad py = x \tag{1}$$

the cause or forcing function is x and the effect or response is y. This cause–effect sequence cannot be represented in a block diagram format using the differential operator p or, equivalently, the Laplace transform variable s, while preserving the predecessor–successor relation of cause and effect. For instance, Fig. 10.8 represents Eq. (1) as a block diagram. According to conventional block diagram notation, the output of the block should represent the effect, whereas Fig. 10.8 shows x, the cause, as the output.

Instead of leaving this situation at this stage, one can look at the problem constructively. What is the meaning of a reversal of cause and effect? It

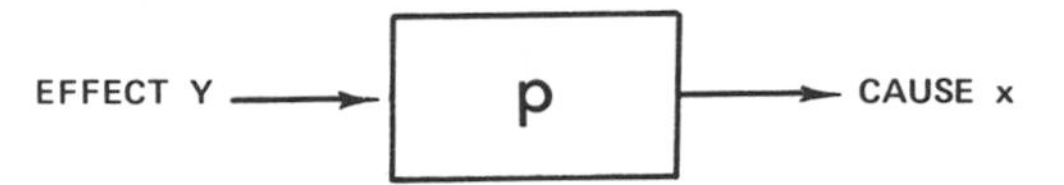

Fig. 10.8 A block diagram showing the relation $py = x$.

means the effect (or response) precedes the cause (or excitation) if the box contains the operator p. That is, a differentiator has clairvoyance! It can predict the future! From this viewpoint a differentiator is an important element. For instance, given the present value of a population and some past history, a differentiator can predict the future values of a population. (This is precisely one of the reasons why differential equations are so important as modeling tools.)

The importance and need for this predictive capability can be illustrated by using another example. In an antiaircraft fire control system, the guns must be aimed at the probable location of the target aircraft when the bullets reach the point of hit, not at the location of the aircraft at the instance of firing the gun. In other words, the system must predict the behavior of the motion of the target aircraft. This task is accomplished as follows. The flight behavior of the target aircraft is studied for a small period of time, say, from t_0 to t_1 (see Fig. 10.9). Then one assumes that the target aircraft cannot deviate from its present course instantaneously (due to inertia). That is, the target aircraft must fly during (t_1, t_2) in the same general direction as it did during (t_0, t_1). That is, the target aircraft is assumed to fly along the direction of the tangent at t_1 (see Fig. 10.9).

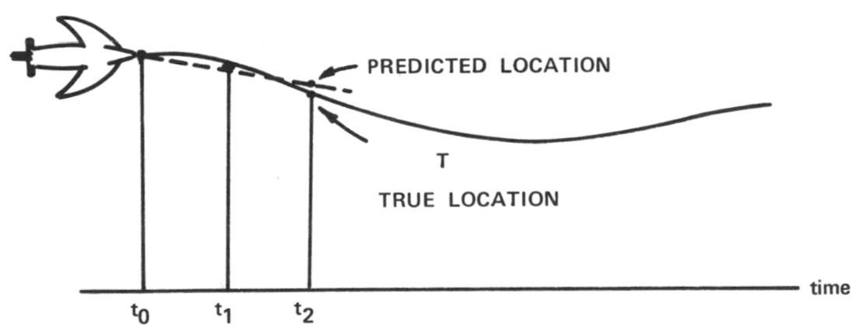

Fig. 10.9 Flight pattern of an aircraft.

The difficulty with the reversal of cause-effect sequence can be avoided by rearranging Eq. (1) as

$$y = \int x \, dt \qquad \text{or} \qquad y = (1/p)x \tag{2}$$

Using the integral operator $1/p$, or equivalently $1/s$, Eq. (2) can be shown in block diagram language as in Fig. 10.10. Thus, use of operator $1/p$ or $1/s$

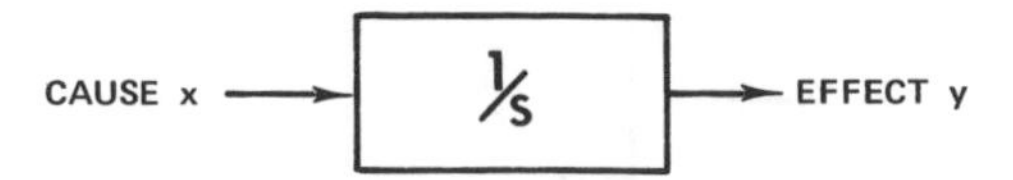

Fig. 10.10 A block diagram showing the relation $(1/s)x = y$.

preserves the causal structure of a problem in its block diagram. Furthermore, computationally integration is desirable as it is a smoothing operation. Conceptually, integration is a deductive process, a process easier to visualize than differentiation, which is essentially an inductive generalization. Thus, intuitively, there is some merit in considering the possibility of representing dynamical systems in terms of integral equations. Mathematically also, integral equations are more concise. Whereas a description of a differential equation is not complete without a further specification of initial conditions or boundary conditions, no such specifications are necessary for an integral equation; all the required information is contained in one equation. A differential equation only describes the *local behavior* of a variable, say, y. Starting from a given point x, a differential equation permits the construction of many possible solutions in a stepwise fashion. The boundary or initial conditions are then invoked to choose a particular solution. Inasmuch as these auxiliary (i.e., initial or boundary) conditions are a determining feature, it would be useful to formulate the system equations for y in such a manner as to include all the conditions explicitly in one equation. Such a formulation must relate $y(x)$ not only to the values of y at the neighboring points but also to all points in the region that influence y. An integral equation is an equation of this type. Since it contains all the information, sometimes it is more convenient to represent the dynamics of a system this way.

An Integral Equation for Population Growth

Consider, once again, the problem of population growth. For simplicity consider only the population of the females of the species. Let

$b(t)$ = female births at time t (strictly speaking, $b(t)\,dt$ = births during the time interval $(t, t + dt)$)

$l(u)$ = probability (or fraction) of the females living to age u

This female population can give birth to male or female offspring. As only the female offspring is of interest, one has to consider only a fraction of the total births. Therefore, let

$m(u)\,du$ = probability (or fraction) of having a girl child in the next du years for a woman who has reached the age u

With this notation, it can be clearly seen that

$b(t-u) =$ female births that occurred u years ago from the present time t

$b(t-u)l(u)m(u)\,du =$ The births occurring at the present moment t as a result of female births that occurred u years ago

Then, the total births at the present time t are

$$b(t) = \int_{\alpha}^{\beta} b(t-u)l(u)m(u)\,du + g(t) \tag{3}$$

where the integration (or summation) is taken over all females of child bearing age, say, from α to β, and $g(t)$ is a term that allows for births to those females who are already alive at time $t = 0$. It is clear that

$$g(t) = 0 \qquad \text{for} \quad t > \beta$$

Equation (3) is called an *integral equation* of the Volterra type.

Thus, it is clear that a growth process can be formulated as a matrix equation (in the discrete case) or a differential or integral equation in the continuous case. Indeed, many other problems in chemotherapy, telephone traffic theory, beam vibrations, hereditary mechanics, and radiative transfer, among others can be formulated as integral equations.

The above integral equation for population growth can be solved by using Laplace transforms. Defining the maternity function $\varphi(u)$ as

$$\varphi(u) = l(u)m(u) \tag{4}$$

Equation (3) becomes

$$b(t) = \int_{\alpha}^{\beta} b(t-u)\varphi(u)du + g(t) \tag{5}$$

Define

$$L\{\varphi(u)\} = \Phi(s) \qquad L\{b(t)\} = B(s) \qquad L\{g(t)\} = G(s) \tag{6}$$

Using the convolution theorem, the Laplace transform of the first term on the right side is $B(s)\Phi(s)$. Therefore, in the s domain,

$$B(s) = B(s)\Phi(s) + G(s) \tag{7}$$

or

$$B(s) = \frac{G(s)}{1 - \Phi(s)} \tag{8}$$

In the above equation, $\Phi(s) = 1$ is the characteristic equation. Indeed,

$$\begin{aligned}\Phi(s) &= L\{\varphi(u)\} = L\{l(u)m(u)\} \\ &= \int_\alpha^\beta e^{-su} l(u)m(u)\, du\end{aligned} \tag{9}$$

Therefore, the characteristic equation is

$$\int_\alpha^\beta e^{-su} l(u)m(u)\, du = 1 \tag{10}$$

If $s_1, s_2, \ldots,$ are the roots of this characteristic equation, then Eq. (8) can be written, using partial fractions, as

$$B(s) = \frac{Q_1}{s - s_1} + \frac{Q_2}{s - s_2} + \cdots \tag{11}$$

Taking inverse transforms

$$b(t) = Q_1 e^{s_1 t} + Q_2 e^{s_2 t} + \cdots \tag{12}$$

Note that the value of Q_i depends upon $G(s)$ or, equivalently, $g(t)$—the births arising out of initial age distributions.

An outstanding question that has not been answered thus far is about finding the roots of Eq. (10). As the function in s on the left side of Eq. (10) is monotone decreasing, it can have only one real root; all other roots must be complex conjugate pairs. Based on this model for females from the United States in 1971, the first three roots were found to be

$$\begin{aligned} r_1 &= 0.0074 \\ r_2, r_3 &= -0.0394 \pm j\, 0.2406 \end{aligned}$$

Therefore, for this case

$$\begin{aligned} b(t) \approx\ & Q_1 \exp(0.0076t) + Q_2 \exp(-0.0394 + j\, 0.2406)t \\ & + Q_3 \exp(-0.0394 - j0.2406)t \end{aligned} \tag{13}$$

Using this equation, one can project the female population of the United States beyond 1971. Inspection of the above expression reveals that $b(t)$ exhibits a wave nature. Such waves are called population waves. More discussion on population waves and their effect on planning can be found in the last section of this chapter.

Integral Equations and Renewal Processes

The Volterra type integral equation discussed above has many other applications. To maintain generality, Eq. (5) is rewritten as

$$b(t) = b_0(t) + \int_0^t \varphi(t-u)b(u)du = b_0(t) + \int_0^t \varphi(u)b(t-u)du \tag{14}$$

or still more generally as

$$b(t) = f(t) + \int_0^t K(t, u)b(u)\,du \tag{15}$$

Equations of this type are called *renewal equations*, and the process modeled by it is called a *renewal process*. The name implies that "life starts afresh" every time the solution hits a low level. In an inventory problem, the growth of stock more or less repeats itself every time the "on-hand" stock hits the zero level. But there are subtle practical differences in application and the student should be aware of them. If the equation describes the variation of a population and if the population level hits zero, the species becomes extinct—there is no chance for renewal! For this reason, integral equations describing population behavior are rarely referred to as renewal equations.

A renewal equation may describe, for example, the rise and fall of water levels in a reservoir impounded by a dam. In such problems, there is considerable practical interest in knowing when the reservoir would first fill up, overflow, or dry out. In other words, there is significant practical interest in the asymptotic behavior of the solution of the renewal equation. Treatment of this problem, though quite interesting, falls beyond the scope of this work.

6 THE DISCRETE EVENT APPROACH

The growth models used thus far to describe real ecological systems are unrealistically simple. In fact, birth, death, and even growth are complicated due to the randomness of natural phenomena. Furthermore, N_i are discrete quantities—not continuous variables. How can differential equations, with deterministic assumptions, describe such phenomena? How can one take into account the complex breeding habits and peculiar life cycles of different species? How can one include in the model spatial and temporal variations of populations, immigrating currents, herding instincts, feeding habits, attack mechanisms, searching times for prey, variations in season, climate, geography, and topography? How does one deal with symbiotic and other kinds of interactions? In modeling problems of this kind one is faced with a stunning, even paralyzing, array of variables and data. While modeling for the management of ecological systems, one has to include control and policy variables as well. Activities such as stocking a lake with a designated species periodically, spraying, and several others are essentially discrete in nature. The "event"-oriented modeling technique originated in this context.

In the discrete event approach, a model is visualized as a set of interacting components. Each component is thought of as a set of processes. Each process is akin to a computer program—it lays down a sequence of activities to be carried out. External conditions can intervene, inhibit, or change the course of an activity.

Other "discrete"-oriented modeling techniques, such as those based on cellular automata, are coming into vogue. Discussion of these and related matters falls beyond the scope of this work.

7 POPULATION PLANNING

This section is essentially a qualitative introduction to the very general idea concerning population planning. The major purpose of this section is to provide the reader with some basic information. The serious student can pick up portions of this topic for more intensive study.

The concept of population planning is intimately related to the notion of population dynamics. Population dynamics, in turn, is concerned with changes in population density resulting from interactions between a population and its environment insofar as these are reflected in changes in reproduction, mortality, and migration. Why do we want models to describe the dynamics of a population? In a very general way, models are required to predict the numbers and densities of a given population at any time. This predictive capability can, in turn, be used for planning.

Using a very coarse resolution level, the problem of population planning can be divided into three subproblems: (1) the population of human beings, (2) the population of animals, and (3) the population of plants. In fact, these three problems are interrelated, and the study of these interrelations may be visualized as a problem falling in the realm of ecology. Ecology is a science that deals with the relations between an organism—possibly groups of organisms or a part of an organism—and its environment (insofar as it is able to survive due to its ability to maintain itself in this environment, thereby making use of its adaptability). Thus, the topic of population planning, to be meaningful, should always be considered in its proper ecological context. In an ecological system space, time, matter, energy, and information are the basic resources. Unfortunately, ecology is remarkably short on long runs of data on these five basic resources. This lack of data is a perennial source of trouble in many population planning studies. This fact should be kept in mind at all times regardless of whether one is planning human populations, animal populations, or plant populations.

Human Populations

Treatment of the problem dealing with any aspect of planning human populations is a delicate matter at best. A little reflection reveals that the planning of human populations is intimately related to the economy and technology of a society. Scientific and technological developments since the time of Malthus have evidently broadened the range of choice open to mankind with respect to population planning, and this range will continue to expand. By the same token, improvements in technology, such as communication and transportation, prevented large-scale annihilation of populations subjected to famine, pestilence, and disease and unwittingly made it possible for more and more people to live at mere subsistence level than was possible before. Thus, there is a vast difference between the notions of (a) the maximum number of people that could exist on earth at a bare subsistence level and (b) the minimum number of people that could exist under maximum conditions of satisfaction.

The Concept of Optimum Population

It was recognized from antiquity that the optimum population necessarily depends upon factors such as natural resources, economic productivity, and political philosophy. Questions such as "Under a given set of conditions, what population density is most advantageous to a given area?" were posed by Confucius of China and by Kautilya of India. The latter articulated this concept in writing in his book *Arthasastra* (the theory of economics). The Greek philosopher Plato acknowledged in his *Laws* the impact of temporal and spatial distributions of populations on his vision of the "city–state." Almost all these early thinkers suggested the use of demographic migrations as measures of population distribution—not control. However, it was Malthus who introduced the concept of a *limit* to population. He stated in effect that the available resources indeed are a limiting factor on population, not vice versa. Contrast this with the Marx–Engel theory, which asserted that the institutional setup of a society is responsible for overpopulation rather than limited resources. Now it appears that both these mechanisms are in action. It is being recognized that in the present social circumstances reformatory measures such as enlightened tax structure, social security programs, educational campaigns, and retirement programs should form an integral part of any population planning project. It is also being recognized that the goal should be not so much aimed at avoiding "over"-population but at a policy that will secure adjustment between population and natural resources—that is, to secure maximum "efficiency." This de-emphasis on "over" comes about primarily because it is so difficult to define the term "over" in "overpopulation."

Optimality is a relative concept in the sense that it depends upon the criterion used. One suggested criterion is the *social welfare function.* Another classical criterion that was popularized by the works of Malthus and John

Stuart Mill was *efficiency of production.* It should be kept in mind that the latter criterion was suggested in an age that preceded the industrial revolution and in an atmosphere replete with economic theorizing and preoccupation.

The conventional wisdom of these times was stated by Edwin Cannon: Under certain circumstances the productiveness of industry in a country or countries may be effected by an increase or decrease of the population of the country or countries in question. It is not true that an increase of population must always diminish the productiveness of industry, or that a decrease of population must always increase it. The truth is that the productiveness of industry is sometimes promoted by an increase of population and sometimes by a decrease of population.

Population theorists then attempted to tie in the "economic theory of population" with the classical subsistence theory of wages. The fallacy of this approach as pointed out by Ricardo, is that the subsistence wage is not a function of the biological minimum to maintain survival but a function of habit, custom, and expectations.

Goals of Population Planning

From the discussion so far it is evident that the goals and objectives of population planning are almost always a reflection of the perceptions of the people involved in planning. Nevertheless, it is reasonable to assume that the goal of population planning is by and large to improve the well-being of a society in a well-defined spatial unit. The spatial unit in question could be a country, a region, or the world itself. The next step is to identify the objectives and indicators. As most of the economic indicators have materialistic overtones, a "context-independent" index such as the density of population can be used as an indicator for well-being. Unlike the traditional optimization problems, the purpose of population planning is not to maximize or minimize the density of population. Rather, the purpose is to adjust the level of population density to an optimum size so that the region in question can support the population at an adequate level of contentment. However, this contentment is a function of, among other things, the age distribution and the expectations of the population. To stretch this point somewhat, let us consider the following scenario.

Suppose that humanity survives! That is, humanity survives if, for instance, we abandon the idea that something—such as pollution, famine, depleted resources, poisoned environment, political disintegration, overcrowding, a thermonuclear war, pestilence, or a disease—is going to limit the human population. Then what should be the goal of planning? Improvement of the ubiquitious well-being still sounds like a decent goal. However, what one means by well-being now can take on a different meaning. If the socioeconomic and allied problems that plague us today are brought under control, on what terms could we manage a happy survival? If the present trend in development continues in the fields of science and technology, then our best chance to survive as civilized people on this planet requires a future with more emphasis on the aged and less on the young. This statement can be verified by performing simple calculations about the growth of human populations and the subsequent age distribution of that population. The world's population now stands at about four billion and is increasing at a rate of 2% a year. At this rate, unless disaster intervenes, the world population will surpass the seven billion mark by the year 2000. With energy shortages and worldwide inflation it is hard to see how the food supply can be increased in the coming decades. Even if we imagine new sources of energy, new sources of food, new methods of food distribution, and a new era of peace and order during the next generation, that will only give humanity a short breathing spell—unless we take steps to control the population before disaster does it for us.

Suppose we take charge of the situation! The only alternative left to us in controlling population is by controlling birth rates unless we decide to start killing people wholesale. This brings forth another disturbing consequence. During almost all the history of the human race, mankind has lived under conditions where the death rate was high. Life expectancy varied from 25 to 35, so that when the birth and death rates were equal, half the population was under 30. When the birth rate was considerably higher than the death rate, so that the population was growing rapidly in numbers at the young end of the scale, half the population might be under 15. Throughout most of the recorded history, the number of people over 40 was never more than perhaps 20% of the whole, while the number of people over 65 was never more than perhaps 1% of the total population. Thus, in almost all societies mankind has consisted largely of yound people, and the social structure over the millenia has evolved to adapt to this fact. This accent on the young has undoubtedly influenced our present mode of life—the nuclear family, the school systems, the tax structures, the employment and retirement patterns, and the social security system.

If birth rate was lowered as a means of population planning, the age distribution of the population would drastically change. For instance, in the United States, about 4% of the total population was over 65 years in 1900. By 1940, about 7% of the total population was over 65. In 1970 this percentage rose nearly to 10%. By 2000, there may be as many as 12–13% elderly people out of an estimated total population of 240 million. Thus, the "responsibility" of maintaining more and more elderly falls on the shoulders of fewer and fewer young workers. This "responsibility" becomes a burden if a society does not plan to adjust with the realities of progress. One reality of this situation is to recognize that "old" and "young" are relative terms. Second, in a society where most of the "work" will be handled by machines and computers, the "work" of the world will not be a matter of muscle and sinew.

Animal and Plant Populations

The population dynamics of plants is still in an embryonic stage, so most of the discussion in the sequel is primarily aimed at describing the dynamics of animal populations. For convenience, animal species can be divided into:

(1) the species we want to protect in order to prevent them from getting extinct;
(2) the species we want to harvest by hunting or fishing; and
(3) the species we want to maintain at a lower population level than their natural level because they could otherwise be harmful either to man, his domestic animals, and crops, or to the preservation of nature.

In all these cases, population dynamics will contribute in laying a foundation for effective management.

Protection of Wildlife

The intensive utilization of land for housing, industry, agriculture, fisheries, recreation, etc., and the widespread disposal of waste products is making demands on the adaptability of many organisms. Consequently, the areas over which the said organisms can spread and multiply is decreasing, resulting in many species becoming extinct or nearly so. However, it must be emphasized that humanity at large is increasingly willing to keep the animal world as rich and diverse as possible. This means that managing to preserve animal species should take place by means of far-reaching measures at a higher level of organization than managing for harvesting. Whenever a species is on the decline, one must determine whether regulation is possible at all under prevailing conditions. For example, protection of the Indian tiger in a territory in which only four old males are still alive is a senseless proposition. Creating the right conditions for the species to grow and assembling a sufficiently large number to mate and multiply thus becomes mandatory. Zoos can bring some respite to this problem. However, the concept of protecting an animal species only in the confined environment of a zoo ignores the fundamental fact that each species has a role to play in the complex ecology of a trophic web.

Cropping of Wild Animals

Wildlife can be an important source of food for human beings. The bison of the American West and several wild animals of the African continent are now being considered as useful sources of food. Fishing, whaling, and bird hunting are other examples of the utility of wildlife as a source of food. With overhunting and overcatching, however, population densities fall off rapidly, and this could lead to an eventual extinction. Also hunting done at the proper level at the proper time could have a regulatory effect. For instance, in temperate climatic zones, a natural shortage of greenery arises during late autumn through winter up to early spring. If the hunting of herbivorous animals took place before the shortage hit hard, it could possibly clear off any surplus before a shortage developed and help keep the population at a satisfactory level. Hunting should be distinguished from large-scale commercial fishing and other similar techniques. Hunting is selective in the sense that the best of the species have a better chance to escape the hunter.

Pests

Stated in a nutshell, pests are man-made calamities. It has been found that indiscriminate exploitation of natural resources results not so much in the destruction of nature directly but indirectly in the form of massive occurrences of animal species that are considered pests. Stated differently, insect pests are generally kept at low levels by natural regulatory mechanisms. The indiscriminate use of natural resources often tends to break the natural regulatory mechanisms. Why these regulatory mechanisms fail is not well understood as yet. To meaningfully control these incidents one must know, several months beforehand, when an outbreak of pests is likely to occur. Then, using biological control mechanisms or a combination of chemical and biological controls, one can take reasonable long-range actions.

Management of an Agricultural Ecosystem

Let us now address the problem of population control within a more specific context. For this purpose an agricultural ecosystem is considered. This ecosystem is the totality of interacting organisms and their abiotic (nonliving) environment, which includes weather conditions, solar radiation, water, oxygen, carbon dioxide, phosphates, and a number of organic substances that are the products of biotic activity. Thus, land used for an agricultural activity can be regarded as an ecosystem. The relevant attributes of this agricultural ecosystem are abiotic factors, producers, herbivores, carnivores, and decomposers. The abiotic factors include items such as pesticides, fertilizers, irrigation water, sunshine, rain, and wind. In this system, cultivated crops, weeds, and other uncultivated vegetation are called *producers* because they absorb solar energy and produce other forms of energy, The herbivores include man and a large number of animals, birds, and insects. Thus, man is in direct competition with other species and these other species, primarily insect species, are labeled "pests" by man. Predators of insects include other insects as well as birds and small animals. The pests usually feed on a number of crops, and the predators usually feed on a number of pests and other insects (nonpests). Thus, for a given crop we usually have one or two pest species to worry about. However, in an attempt to destroy these target pests, we generally use broad-spectrum insecticides and inadvertently kill a number of other species of insects as well as some of the predators. Released from natural enemies, a second pest species may now develop an ability to increase in numbers and cause considerable damage to the crop. Nevertheless, in order to survive, a society must manipulate and control its environment. Before implementing any control, one must understand the long-range impacts of certain policy decisions. Thus, a mathematical model of an ecosystem must be able to predict not only the monetary but also the environmental and social costs of various management policies and control strategies.

Chemical Control

The basic idea of chemical control is to use synthetic pesticides, such as DDT, to eradicate undesirable pest species. However, most of the synthetic pesticides are nonselective and they often destroy natural control agents such as predators, pathogens, and parasites. Second, chemical interference with natural control often leads to secondary outbreaks of pests. Part of the answer to this problem would be the development of more selective chemicals. However, the economics of research and development makes this approach unattractive. Furthermore, the well-known ability of insects to develop resistance to pesticides is another difficulty that arises from a strict reliance on chemical control.

Biological Control

The basic idea of biological control is to let natural enemies such as predators, parasites, and pathogens do the work for us. Biological control techniques have the advantage of being (a) density dependent, (b) ecologically selective, (c) flexible enough to retard the pest's development of resistance, and (d) long lasting. Despite these attractive features, biological controls alone have been found to be insufficient in a number of cases. The equilibrium levels established by biological controls may not be low enough to prevent economic loss caused by pests. Even if the equilibrium levels are low, the biological control agents may not be able to cope with the occasional outbreaks of pests.

Other Methods

In addition to chemical and biological controls, one can use other innovative techniques. One promising method is to inundate the area with millions of sterile males. The sterile males compete with normal males for females and thus reduce the reproductive rate. This technique is most effective if the female of the species mates only once per season.

Pheromones can also be used for pest control. Pheromones are chemical or physical stimuli that enable insects to communicate. These stimuli include vibrations, odors, and chemical solutions. Artificially created pheromones can be made to interfere and jam natural communications, thus reducing the chances of the pest species to survive.

There are a number of pest control techniques that depend upon the physical manipulation of the insect environment. Times of planting and harvesting, crop rotation, soil preparation, and water management are some of the widely used methods.

Integrated Pest Control

It should be evident by now that no one pest control technique is ideal. Integrated control is a pest management concept in which several pest control techniques are combined in such a way as to maximize control effectiveness. Many of the advances in integrated control have resulted from a combination of chemical and biological controls. Other combinations are also possible. Since the complexity of an ecological system seems to influence its stability, there are some integrated control techniques aimed at increasing the complexity. This approach requires one to evaluate the effect of marginal vegetation and adjacent fields on crop land ecology. Stability can also be increased by innovative harvesting techniques. For example, strip harvesting of alfalfa enables all insect life to find support in the unharvested strips through the summer months.

Thus, it is apparent that an integrated pest control and management of an agricultural system can be made effective if suitable mathematical models are available for various stages of management. For instance, one can develop a model to find the best policy to control a pest population. One can also simulate the spread of a pest and the damage caused by it under different control policies. Host–parasite and predator–prey models can be included in this simulation. Then it becomes necessary to determine the parameters in these models. This is an identification problem and merits study in its own right. One can also apply mathematical programming techniques to these problems. The concept of a queue in simulating the water stored in a reservoir was introduced in Chapter 5. The idea of inventory control in a production scheduling problem was also introduced. These ideas are applicable in the context of pest management. In inventory theory, one has a stock of goods that must be constantly replenished as the goods are sold. In pest control the situation is the opposite. The stock here represents the pest population, and the goal is to decrease this population by using a suitable control policy. Interested readers may pursue these topics with zeal by actually formulating a model, however crude it may be, and following it through the various stages of computation and solution.

EXERCISES

1. Formulate a model of a savings account system in which interest is paid at the rate of r compounded n times per year. Under this plan, a deposit made in a given period does not earn interest until the next conversion period. A withdrawal made in a given period reduces the principal, which is eligible for interest in that period. In this model, let $u_1(k)$ and $u_2(k)$ represent, respectively, the deposits and withdrawals in the kth period. Compare this model with the one discussed in the text. Which of these represent a typical savings account system?

2. Conduct a comparative study of the savings accounts currently available in your town. Discuss their merits and demerits.

3. If $b = b_1 N$, where b_1 is a constant, prove that the solution of Eq. (3.2) is now given by

$$N = \frac{m}{b_1 - [(b_1 N_0 - m)/N_0]e^{mt}}$$

Discuss the nature and behavior of the solution for small and large values of t. Plot the approximate shape of N versus t curve.

4. Define $r = (b - m) =$ the fertility rate. At present the fertility rate of the human population is approximately 1.9. If world population is growing according to

$$\frac{dN}{dt} = rN; \qquad N(t = 0) = N_0$$

how long will it take to double the present population N_0? If it is desired to delay this doubling period to one that is twice as long, what should the new fertility rate be?

5. A population is said to follow the *logistics growth law* if the growth is described by

$$\frac{dN}{dt} = rN\left(1 - \frac{N}{M}\right)$$

where r is the fertility rate and M is an upper bound on the population. How do you solve this equation? Compare this with the exponential growth law discussed in the test.

6. Let the total dollar value of an enterprise at time t be equal to $u(t)$. Suppose that the rate of increase of this value is directly proportional to the value itself.

(a) Write a differential equation to describe the growth of the enterprise; call this Eq. (A). Although such a growth is quite satisfactory, the principal aim of its management is to make some profit. Suppose the total profit made over an interval $[0, t]$ is $p(k)$. This profit has to come by reducing the dollar value of the business at a rate equal to a fixed fraction k of the value.

(b) Write a differential equation showing the rate of growth of profit. Call this Eq. (B). By writing Eq. (B), Eq. (A) is no longer valid; it has to be modified to account for the dollars taken for profit.

(c) Modify Eq. (A) to account for this. Display your initial condition on all equations clearly.

(d) Solve for $p(k)$.

(e) How does the enterprise maximize its profit over a fixed interval of time $[0, T]$? (It is clear that if $k = 0$, the firm grows rapidly but makes no profit. If $k = 1$, everything is being taken out of the business as profit, and so it cannot grow.)

7. Consider the problem of conflict between a host population H and its parasite population P. Let b_h and b_p be the birth rates of hosts and parasites, respectively. Let d_h and d_p be the death rates. d_h does not include deaths caused by parasite attacks. Let kHP be the death rate in hosts due to parasite invasion where k is assumed to be a constant. Let $b_p = k'kHP$ where k' indicates that only a fraction of the eggs hatch. Prove that the differential equations of this system are

$$\frac{dH}{dt} = (b_h - b_p)H - kHP \qquad \text{and} \qquad \frac{dP}{dt} = k'kHP - d_pP$$

If there is an additional host population E with birth and death rates b_e and d_e and death rates due to parasites given by hPE, then write a differential equation for this new system.

8. Suppose that there are two species of hunters that are in competition for the same food. Here these two species not only compete for a limited resource but kill *each other* whenever they encounter. This kind of competition occurs among predators wishing to establish territorial rights on their habitats and in business, where companies attempt to monopolize a product.

The equations describing the dynamics of the situation are very similar to those discussed under the host–parasite section with one exception. Write differential equations similar to those of Eq. (5.3) and point out the exception. Discuss the nature of the solution. Is the solution periodic?

9. A prey–predator model is given by

$$\frac{dN_1}{dt} = \alpha_1 N_1(k_1 - N_1 - M_1 N_2^2) \quad \text{and} \quad \frac{dN_2}{dt} = \alpha_2 N_2(k_2 - N_2 - M_2 N_1^2)$$

Give a physical explanation of the underlying mechanism that leads to the above formulation.

10. The motion of a simple pendulum with damping is given by

$$\ddot{\theta} + \epsilon\dot{\theta} + k \sin\theta = 0; \qquad k > 0, \quad \epsilon > 0.$$

Find the equilibrium points and linearize the equation about the equilibrium points.

11. Rapaport has formulated a model of the arms race between two nations as

$$\dot{x}_1 = -m_1 x_1 + a_{12} x_2 + b_{12} x_2^2, \qquad \dot{x}_2 = -m_2 x_2 + a_{21} x_1 + b_{21} x_1^2$$

where x_1 and x_2 are expenditures by rival nations. Discuss the nature of the equilibrium points.

12. Two persons A and B are in the business of manufacturing two different products. A produces an amount x of his product and B produces an amount y of his product. Each agrees to trade a fraction q, $0 < q < 1$ of his own product and a fraction p, $p = 1 - q$ of his competitor's product. Such a two-person economic system can be modeled by the following set of differential equations:

$$\frac{dx}{dt} = \frac{p}{px + qy + 1} - \frac{p}{2} \quad \text{and} \quad \frac{dy}{dt} = \frac{p}{qx + py + 1} - \frac{p}{2}$$

Develop this model along the lines of the host–parasite model discussed in the text. Show that the production either approaches an equilibrium or one of the two persons will drop out of business. Find the conditions for these two cases. In the second case, show that the question of which man drops out depends only on the original amounts produced by the two men.

13. Let $x(t)$ be the size of a population at time t. Let α stand for the birth rate (i.e., the number of births per unit population), σ for the life span, and τ for the gestation period. Then prove that the differential equation for the change in population is given by

$$\frac{dx(t)}{dt} = \alpha[x(t-\tau) - x(t-\tau-\sigma)]$$

Here $\alpha x(t-\tau)$ = the number of individuals born at time t, while $x(t-\tau-\sigma)$ = the number that expire at time t.

14. Show that the solution of the homogeneous equation

$$\frac{dx}{dt} = k(t)x(t)$$

is given by

$$x(t) = x(a) \exp \int_a^t k(x)d$$

(a) by differentiating the solution;
(b) without differentiating the solution.

15. (a) Solve $dy/dx = xy$; $y = 1$ if $x = 0$.
(b) Solve the above equation by assuming that $y = a_0 + a_1x + a_2x^2 + \cdots$. Substitute the series in the equation and equate the coefficients of various powers of x.

SUGGESTIONS FOR FURTHER READING

1. For an elementary introduction to differential equations, see
   R. Bellman, *Modern Elementary Differential Equations*. Addison-Wesley, Reading, Massachusetts, 1968.
2. An interesting discussion on the ecological aspects of growth with copious amounts of practical information can be found in
   K. E. F. Watt, *Ecology and Resource Management*. McGraw-Hill, New York, 1968.
3. For a mathematical presentation of growth processes, consult
   A. J. Lotka, *Elements of Physical Biology*, Williams & Wilkins, Baltimore, Maryland, 1925.
   A. J. Lotka, *Elements of Mathematical Biology*. Dover, New York, 1956.
   N. S. Goel, S. C. Maitra, and E. W. Montroll, *On the Volterra and Other Nonlinear Models of Interacting Populations*. Academic Press, New York, 1971.
4. For a good mathematical introduction to population dynamics, see
   N. Keyfitz, *Introduction to Mathematics of Population*. Addison-Wesley, Reading, Massachusetts, 1968.
5. For a recent survey of models dealing with population dynamics at an advanced level, refer to
   T. N. E. Grevelle (ed.), *Population Dynamics*. Academic Press, New York, 1972.
6. An in-depth discussion on various ecological aspects of forest management and pest control can be found in
   K. Stern and L. Roche, *Genetics of Forest Ecosystems*. Springer- Verlag, Berlin and New York, 1974.
   A. Woods, *Pest Control: A Survey*. Wiley, New York, 1974.

7. A discussion on the concept of optimum population can be found in

M. Gottlieb, Theory of optimum population for a closed society, in *Population Theory and Policy* (J. J. Spengler and O. Duncan, eds.). Free Press, New York, 1956.

8. For a discussion of the principle of competitive exclusion in population biology, refer to

G. F. Gause, *The Struggle for Existence*, Hafner Publ., New York, 1964.

9. For a presentation of issues relating to mathematical biology and social processes, consult

N. Rashevsky, *Mathematical Biology of Social Behavior*. University of Chicago Press, Chicago, Illinois, 1951.

N. Rashevsky, *Looking at History Through Mathematics*. M.I.T. Press, Cambridge, Massachusetts, 1968.

10. For mathematical aspects of some current problems of worldwide concern—such as overcrowding, food shortage, pollution, and inflation—see

E. W. Montroll and W. W. Badger, *Introduction to Quantitative Aspects of Social Phenomena*, Gordon & Breach, New York, 1974.

Appendixes

1

Sets and Relations

SETS AND ALGEBRA OF SETS

Any collection of individuals is a *set*, the individuals being termed its *elements*. The fact that a set A is formed by a collection of its elements $a_1, a_2, \ldots, a_n$ can be written as

$$A - \{a_1, a_2, \ldots, a_n\}$$

If the elements $a_1, a_2, \ldots, a_n$ possess a well defined property P, this fact can be denoted compactly as

$$A = \{a: P\}$$

Example 1. The collection of all rational numbers is a set. The collection of goods, called *commodity bundles* is a set. The collection of even numbers less than 7 is a set written as $\{2, 4, 6\}$. However, this procedure is impractical to describe, for example, the set of all real numbers. Then we can use the second notation and write $\{x: x \text{ is real}\}$. ■

The fact that a is an element of A can be denoted by $a \in A$. If a is not an element of A, it can be denoted by $a \overline{\in} A$ or $a \notin A$.

The union or *sum* of two sets A and B is the set containing all the elements that belong to A alone or to B alone or to both A and B. This is denoted by $A \cup B$ (read this as "A union B") or $A + B$ (read this as "A or B"). Thus,

$$A \cup B = A + B = \{x: \quad x \in A \quad \text{or} \quad x \in B\}$$

The intersection or *product* of two sets A and B is the set containing all the elements which belong to both A and B. This is denoted by $A \cap B$ (read this as "A intersection B") or $A \cdot B$ or AB (read this as "A and B"). Thus,

$$A \cap B = A \cdot B = AB = \{x \in A \quad \text{and} \quad x \in B\}$$

Two sets A and B are disjoint if they have no elements in common, that is, if

$$AB = \phi$$

In the future, it will be useful to consider not arbitrary sets, but only sets whose elements are taken from the largest set S, which is called the *universe set*.

The complement of a set A is the set of all elements that are not in A but are in S:

$$\bar{A} = \{x: \quad x \notin A, x \in S\}$$

The difference between two sets A and B is the set

$$A - B = \{x: \quad x \in A \quad \text{and} \quad x \notin B\}$$

The symmetric difference between two sets A and B is the set of all elements that are either in A or in B, but not in both:

$$A \oplus B = \{x: \quad x \in A \cup B \quad \text{and} \quad x \notin A \cap B\}$$

RELATIONS

Normally speaking any set $A = \{x, y\}$ which has only two elements is called a pair and is denoted by (x, y). The pair is an *ordered pair* if $x \neq y \Rightarrow (x, y) \neq (y, x)$.

Ordering is a very important concept in the study of large-scale systems. For instance, the concept of *technological order* is extensively used in planning activities. A set of activities $\{a_1, a_2, \ldots, a_n\}$ is said to be in technological order if activity a_i cannot be commenced unless all other activities a_j for which $j < i$ (i.e., all other activities *preceding* the ith activity) are completed.

Other examples of ordering abound. While storing and retrieving data in computers the concepts of a *file* and a *record* prevail. A *record* can be defined as an ordered set of data, while a *file* is a set of records. Thus, the concept of ordering is very general, and a given set can be ordered in different ways with different attributes of the elements as a *key*. If a set of students is arranged according to their heights, the key is height. The same set of students can be rearranged according to their weight in which case the key is weight.

If (x, y) is an ordered pair, x is termed the *predecessor* and y the *successor*. This predecessor–successor relationship can often be represented pictorially by a directed line from point x to point y. An ordered pair can also be represented by a vector in a two-dimensional plane. An ordered n-tuple $(x_1, x_2, \ldots, x_n)$ can be represented by a vector in an n-dimensional space.

The Cartesian product of two sets A and B is denoted by $A \times B$ and is defined as the set

$$C = A \times B = \{(x, y): \quad x \in A; y \in B\}$$

of all ordered pairs (x, y). The ordering is important here. That is,

$$A \times B \neq B \times A$$

If $A = \{a, b, c, d\}$ and $B = \{x, y, z\}$. Then, the set of pairs shown in the following array constitute the Cartesian product of sets A and B:

Set A ↓ \ Set B →	x	y	z
a	(a, x)	(a, y)	(a, z)
b	(b, x)	(b, y)	(b, z)
c	(c, x)	(c, y)	(c, z)
d	(d, x)	(d, y)	(d, z)

The Cartesian product can also be written as

$$C = A \times B = \{(a, x), (a, y), (a, z), \ldots, (d, y), (d, z)\}$$

An important special case of the Cartesian product is the product of a set by itself, AA, which is often represented as A^2.

Let A be a set and A^2 be the product of A by itself; let R be a subset of A^2. That is, $R \subset A^2$; then R is said to be a *relation* defined over A.

A relation is then a set of pairs. If $x \in A$ and $y \in A$, then either $(x, y) \in R$ or $(x, y) \notin R$; if $(x, y) \in R$, the pair (x, y) is said to verify the relation R. Instead of $(x, y) \in R$, other notations in vogue are $x \mathrm{R} y$ and $\mathrm{R}(x, y)$. The notation $x \mathrm{R} y$ is really not as strange and abstract as it presents itself at first sight. By replacing the letter R by familiar relational expressions like "greater than," "equal to," "heavier than," the strangeness quickly vanishes. Some conventional and unconventional modes of expressing relations are summarized in Table A1.1.

Typical relations are relations such as "not less than" (if A is the set of ordinary numbers), "every component of vector $\mathbf{x}$ is not less than the corresponding component of vector $\mathbf{y}$" (if A is a set of vectors), and "product A is not less desirable than product B" (if A is a set of products). Stated differently, R is a relation defined on a set A if, for every pair of elements

TABLE A1.1

Typical Relations Encountered in Practice

Representation	Explanation
1. $x \text{ R } y$	x is related to y
2. $x = y$	x is equal to y
3. $x < y$	x is less than y
4. $a \text{ H } b$	a is heavier than b
5. $s \text{ F } t$	s is the father of t
6. $w \text{ M } z$	w is a model of z
7. $u \text{ P } v$	u is the prey for v
8. $x \text{ P } y$	x preceds y
9. $w \bar{\text{M}} z$	w is *not* a model of z
10. $x \perp y$	x is orthogonal to y
11. $x \succ y$ or $x \text{ P } y$	x is preferred to y
12. $x \sim y$ or $x \text{ I } y$	x is equivalent (or indifferent) to y

$(x, y) \in A$, where the order is important, the statement $x \text{ R } y$ is either true or false. A relation $\text{R}(A, A)$ on A is called:

Reflexive if it contains pairs (a, a) for every $a \in A$.
Symmetric if whenever it contains a pair (a, b), it also contains the pair (b, a).
Transitive if whenever it contains both the pairs (a, b) and (b, c), it also contains (a, c).

For instance, Relation (5) in Table A1.1 is not reflexive because no one can be his own father. Similarly, $s \text{ F } t$ is not symmetric because if s is the father of t, then, obviously, t cannot be the father of s. By a similar argument it is easy to see that the relation $s \text{ F } t$ is also not transitive.

The relation $w \text{ M } z$, that is, w is a model of z can be reflexive because an object can be a model of itself. This relation is generally transitive and may or may not be symmetric. The fact that "w is a model of z" generally denotes that "z is *not* a model of w" except in some special cases where a one-to-one correspondence between the sets of elements w and z exists. Thus, in general, the modeling relation is antisymmetric, that is, $w \text{ M } z \rightarrow z \bar{\text{M}} w$. Finally, if w is a model of h and h is a model of y, then w is, in some sense, a model of y. That is, $w \text{ M } h$ and $h \text{ M } y \rightarrow w \text{ M } y$. Thus, the relation M is transitive. The power of modeling is due in large part to the transitive property which makes a hierarchical approach to large-scale systems possible.

A relation on A is called an *equivalence relation* if it is reflexive, symmetric, and transitive. An equivalence relation is often denoted by the symbol $\sim$. The relation between two straight lines, "to be parallel or coincident," is an equivalence relation. The relation of equality in arithmetic is also an equivalence relation.

Example 2. Let H be the set of all human beings alive at a given instance of time. Let R be a relation defined on H by x R y, provided x is a male and x and y have the same parents. In short, x R y, provided x is a male and x and y have the same parents. In short, x R y (i.e., x is related to y) if and only if x is the brother of y. This relation is transitive and reflexive but not symmetric. (Why?) ■

A relation R leads to a *quasi-ordering* (also called *preordering*) if R is only reflexive and transitive. For example, if A represents the set of all countries and R represents "entered the United Nations not later than," then (a) x R x is true because any country x entered the United Nations not later than itself and (b) x R y and y R z imply x R z because of obvious reasons.

A quasi-ordering becomes an ordering if x R y and y R x imply $x = y$. For instance, consider the set A of all real numbers. This set is ordered under the relation $\geq$ because $a \geq b$ and $b \geq a$ imply $a = b$. In contradistinction, consider a set A of commodity bundles. It is possible to pick two commodity bundles x and y and relate them so that "x is at least as desirable as y" and "y is as desirable as x" without the bundles being identical. Thus, if x R y and y R x and $x \neq y$, then one says that x and y are *equivalent* for the quasi-ordering defined by R.

An ordering or quasi-ordering over a set A is complete if for $x, y \in A$ we have x R y or y R x or both; otherwise, it is *partial*. As an illustration, let A stand for the set of all two dimensional vectors. Let the relation R stand for "both components of the vector **x** are not less than the corresponding components of the vector **y**." Then R is certainly both reflexive and transitive, and **x** R **y** and **y** R **x** together imply **x** = **y**. Thus, the relation R, defined as above, imparts an ordering on A. However, if R stands for "the first component of **x** is less than the first component of **y**, and the second component of **x** is greater than the second component of **y**," then we have neither **x** R **y** nor **y** R **x**. Therefore, this ordering is partial. Stated differently, a relation R is called a partial ordering of a set A, if and only if R is reflexive, antisymmetric, and transitive.

MAPPING AND FUNCTIONS

Relations can also be represented by a *mapping*. For example, consider two sets A and B. A mapping from A to B means a correspondence whereby to each point x in A exactly one point z in B is associated. The totality of all such ordered pairs (x, z) where $z \in B$ corresponds to $x \in A$ is called a *function* and is usually denoted by a single letter f. The z that corresponds in this way to a given x is called the *value* of the function and is denoted by $f(x)$. Thus, a function consists of three attributes: two nonempty sets like A and B and a rule f. This rule f is often called a *mapping*, or *transformation*, or *operator* to amplify the concept in different circumstances. The set A is called the *domain* of the function, and the set of all z's in B, which correspond to the x's in A, is called the *range* of the function.

The rule of correspondence may be described by a geometrical description, such as *$f(p)$ is the distance from p to the origin*; or by a physical relationship, such as *$f(p)$ is the pressure at the point p*; or by a varicty of formulas, such as $f(x) = 3x^2 - 4x; x \geq 0$.

Some of the functions that are encountered are so important that they are labeled with special names.

The signum function is defined by

$$f(x) = \text{sgn}(x) = \begin{cases} +1 & \text{if} \quad x > 0 \\ 0 & \text{if} \quad x = 0 \\ -1 & \text{if} \quad x < 0 \end{cases}$$

The domain of this function for example is the real line, and the range is the set of points $(-1, 0, +1)$. Suppose that a decision maker is faced with a ternary choice. Option A may lead to a profit, Option B may lead to a loss, and the status may remain unchanged if no action is taken. This situation can be modeled using a signum function.

The production function is a basic representation for the transformation of resources to products. Specifically, the production function is a mathematical description of the maximum output that can be obtained from a given set of resources $(x_1, x_2, \ldots, x_n)$. That is, a production function

$$z = f(x_1, x_2, \ldots, x_n)$$

where z is expressed in units of production and $x_1, x_2, \ldots, x_n$ are physical rather than monetary resources. A frequently used production function is of the type

$$z = Ax_1{}^{\alpha}x_2{}^{\beta}; \qquad \alpha, \beta < 1.0, \quad \alpha + \beta = 1.0$$

when the production or output z is expressed as a multiplicative function of input x_1 (usually labor), input x_2 (usually capital), and a constant A. Functions of this type are called Cobb–Douglas production functions and they exhibit decreasing marginal returns on investment.

Demand functions come in handy while building models to forecast the state of an economy. Unually demand functions are empirically derived relations. For instance, the demand function for beer in a community has been found to be

$$q = 1.058Q^{0.136}p^{-0.72}r^{0.914}s^{0.816}$$

where q is the quantity of beer consumed, Q is the aggregate real income of the community, p is the average retail price of beer, r is the average retail price level of other commodities, and s is an index of the strength of beer. The exponents of the variables carry significant information. For instance, the exponent of Q, namely, 0.136, is called the *income elasticity* of the consumption of beer. That is, an increase in 1% in the price of beer will bring about, other things being equal, an increase in beer consumption of approximately $\frac{1}{7}$ of 1%. *Price elasticity* and *preference elasticity* can be similarly defined. A large value for the preference elasticity signifies that consumers prefer a stronger beer over a lighter beer.

2

Elements of Probability

INTRODUCTION

Many students, brought up with a deterministic outlook of the physical world, find the subject of probability vague and difficult. The novice often mistakenly thinks that probability theory is used when our ignorance or inability prevents us from solving a problem exactly. In fact, the precise opposite is true. At a pragmatic level, there are no exact solutions. The so-called "exact solution," or, more accurately, the *deterministic* solution, is a special case of the more general solution involving random values. Indeed, indiscriminate deterministic idealization often leads either to a loss of information or to a misleading result.

To illustrate how probabilistic concepts are rooted in real life situations, a simple example is considered here. Consider a large "bucket" full of 5-ohm (5-Ω) electrical resistors. This bucket need not be a real bucket; it may stand for the set of all 5-Ω resistors manufactured by a certain manufacturer in a certain period of time. We are simply interested in finding the total resistance of two resistors, from the bucket, which are connected in series. In a deterministic procedure, the total resistance is simply 10 Ω. However, this answer is idealized for several reasons. Due to manufacturing process variations, the resistance of any given resistor is not likely to be 5 Ω. Even if the resistors were exactly 5 Ω, the instrument used to measure the resistance might be erroneous. However, common sense tells us that if we measure as accurately as possible a reasonably large number (a *sample*) of resistors with a nominal rating of 5 Ω, the arithmetic average of all the meter readings is likely to be 5 Ω even if none of the resistors in the sampled group has exactly this value. Thus, 5 Ω is the resistance of a fictitious idealized resistor that can be used as a representative value for a typical member from the total *population* of resistors manufactured. By using this 5-Ω value in any future calculation we are implicitly neglecting information we possibly could have in actually measuring a large number of resistors, namely, how the various true resistance values are distributed. For instance, actual measurement of a sample of 100 resistors might yield a distribution of resistor values such as the one shown in Fig. A2.1. In this figure the resistance of 25 resistors lies somewhere between 4.995 and 5.00 Ω, 14 resistors show a value lying between 4.99 and 4.995 Ω, and so forth. Thus, it is evident if the two resistors that were connected in series were taken from the lowest interval, namely, (4.97, 4.975), then the total resistance would be in the interval (9.94, 9.95). If, on the other hand, both the resistors were picked from the highest interval, namely, (5.025, 5.03), then the total resistance would lie in the interval (10.05, 10.06). Nevertheless, the result will always lie between 9.94 and 10.06 Ω. Thus, the idealized solution obtained by working with averages lies within this interval and so is acceptable even though the distribution of the actual value around this average is lost.

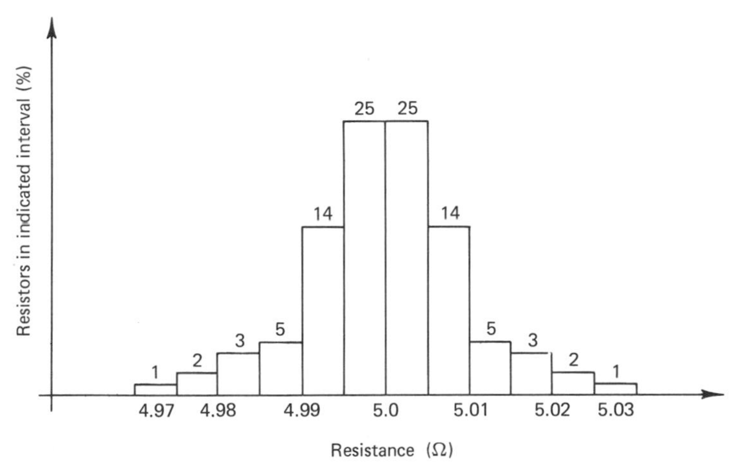

Fig. A2.1 Distribution of resistance.

Working with averages is not always this simple; at times it may lead to misleading results. For example the average life span of the people in the United States was 70.2 years in 1964. However, only about 7% of the people who died in 1964 were actually between 70 and 71 years of age. Thus, the average here does not convey the same meaning as the average in the resistor example discussed earlier. Thus, at an intuitive level it is evident that averages give approximately correct results only if the actual values assumed by a variable do not deviate very far from the average. In circumstances like this one has to invariably depend upon concepts that are deeply rooted in probability theory.

RANDOM EXPERIMENTS AND SAMPLE SPACES

Central to the theory of probability are the concepts of *random experiment* and *sample space*. The process of picking up a resistor from the bucket and measuring its value is a random experiment. The value of the resistance (say, 4.95 Ω) thus measured is the *outcome* of the experiment. Since one can pick any resistor and that resistor may show any value, the set of all possible outcomes is the set of all possible resistor values. Theoretically speaking, this sct may well be the set of all real numbers. Realistically, however, the kind of resistors under discussion cannot have negative values. This restriction makes it possible to identify the set of all possible outcomes with the set of all positive real numbers. However, no manufacturer can stay in business by producing 5-Ω rated resistors whose true values go all over the range of positive real numbers. Thus, a reasonable set of all possible outcomes is, say, the set of numbers in the interval (0, 10). This set S of all possible outcomes is called the *sample space*. Thus, the sample space is determined partly by the experiment and partly by the purpose for which the experiment is carried out. Subsets of the set S (i.e., the sample space) are called *events*. Three types of events can be identified at this time: the *sure event*, the *impossible event*, and the *random event*. If a coin is thrown into the air with an initial velocity v_0 at an initial angle θ_0, then one can possibly predict where the coin will land but one cannot predict whether the coin will land with heads or tails on the upper face. Thus, in this experiment, the event "heads" is a random event. So is the event "tails." However, the event "either heads or tails" is the sure event and the event "neither heads nor tails" is the impossible event. Formally, the sample space S for the coin tossing experiment is the set $S = \{\text{heads, tails}\}$. This has four subsets, namely the set $\{\text{heads}\}$, the set $\{\text{tails}\}$, the empty set $\varnothing$, and the set S itself. Here, the empty set has been identified with the impossible event, the set S with the sure event, and the sets $\{\text{heads}\}$ and $\{\text{tails}\}$ with the random events.

PROBABILITY

Consider a little cube (a die), each face of which has the number 1, 2, 3, 4, 5, or 6. In throwing this die, a number of events may occur. Some of these events can be described as "A_1 = number 1 appeared," "A_2 = number 2 appeared," . . . , "A_6 = number 6 appeared," "A_7 = an odd number," "A_8 = an even number," A_9 = "a number less than 4," "A_{10} = either 3 or 5," etc. Now select a set of those events such that (a) no two of them can happen simultaneously (i.e., mutually exclusive events) and (b) one of them must inevitably occur (i.e., exhaustive events). As one of the numbers on the faces must always appear and no two faces can appear simultaneously, an exhaustive and mutually exclusive set of events is readily seen to be $\{A_1, A_2, A_3, A_4, A_5, A_6\}$. Such a set is called the set of *elementary events*. If the die is not loaded (i.e., a fair die), common sense tells us that there is an equal chance for any of the elementary events to occur. In probability jargon, the elementary events are equiprobable. The probability of an event A_i, written as $P(A_i)$, is the ratio of the number (m) of "favorable" events to the number (n) of elementary events in the set. For example, if the event A_i is defined as "obtaining an even number," then the favorable elementary events are A_2, A_4, and A_6; hence, $m = 3$. The total number of elementary events $n = 6$. Therefore, $P(A_i) = \frac{3}{6} = \frac{1}{2}$. Notice that the probability of any event is always greater than or equal to zero and less than or equal to unity. Thus, probability can be viewed as a mapping between the set of all events in the sample space and the set of real numbers in [0, 1]. This basic idea can be treated more formally.

RANDOM VARIABLES AND PROBABILITY DISTRIBUTIONS

Thus far we have seen (a) how to associate a sample space with a random experiment, (b) how to assign probabilities to elementary events, and (c) how to calculate probabilities of some nonelementary events. That is, given an experiment such as the rolling of a die, we know how to calculate the chances of realizing an event such as "even numbers up." In the context of gambling, for instance, our interest generally does not stop at this stage. We would like to know whether there is a reward or a punishment for the event realized. Suppose that the rules of reward and punishment are as follows. If an even number i shows on the face of the die, then we win $5i$ dollars; if an odd number i shows, we win $-(3 + i)$ dollars (i.e., we lose). One can choose any other rule that appears attractive. Evidently, the amount of money that one wins or loses is a random quantity because it depends on the outcome of a random experiment—namely, the rolling of a die. That is, the money that is won or lost, say, X, depends upon the experimental outcome (as well as on the rules of gambling). In mathematical parlance, X is a function of the experimental outcome. This function X is called the *random variable*. Thus, the random variable is really a function.

Now consider once again the experiment of picking a resistor from the bucket of resistors. The outcome of the experiment is the resistor we have in hand. We measure the value of this resistor, and the meter reading is associated with this resistor. Thus, the resistance associated with the resistor is also a random variable. If we want to sample 100 resistors from the bucket and study how the resistance values are distributed, then, roughly speaking, we are talking about the distribution function F of the random variable X. Strictly speaking, the distribution function F really refers to how the resistor values in the population (i.e., the bucket) are distributed. However, this is difficult, or at times impossible, to get. For instance, if the bucket contains a million resistors, it would be impractical to measure all these resistors. Maybe the resistors are coming out of a production line, in which case all the resistors in the "bucket" are not even accessible for measurement. Therefore, in most of the statistical work, we are interested in taking a sample from the population, studying the properties of the sample, and inferring the properties of the population.

In order to develop a mathematical definition of a distribution function we proceed as follows. Consider the experiment of picking a resistor from the bucket. Let X be the random variable characterizing the resistance of this resistor. As one can pick any resistor from the bucket, X can assume any (reasonable) value. Now consider the event "the resistance of the resistor picked is less than or equal to 10 Ω." Mathematically, this event is written as $\{X \leq 10\}$. What is the probability of this event occurring? Allowing for a reasonably wide margin, if all the resistors are assumed to have a value in the interval $[0, 10]$, any resistor picked in the random experiment will certainly have a value less than or equal to 10 Ω. Thus, the event $\{X \leq 10\}$ is the sure event. That is,

$$P\{X \leq 10\} = 1$$

Similarly, one can argue that the event $\{X \leq 0\}$ is the impossible event. That is,

$$P\{X \leq 0\} = 0$$

Using an analogous argument, we can write $P\{X \leq x\}$ for *any* value of x. In fact, from an inspection of Fig. A2.1, it is evident for $x = 20$ the required probability is unity, for $x = -10$, for example, the required probability is zero, and for $x = 5.0$, the required probability is 0.5. Thus, $P\{X \leq x\}$ is a function of x and is denoted by $F_X(x)$. If we are talking about only one random variable X, then there is no ambiguity and the subscript X can be omitted. This function $F(x)$ is called the *distribution function* or the *cumulative distribution function*.

3

Elements of Matrix Methods

INTRODUCTION

A *vector* is a set of n numbers arranged in a definite order. It is customary to arrange these numbers as a column as shown

$$\mathbf{x} = \begin{bmatrix} x_1 \\ x_2 \\ \vdots \\ x_n \end{bmatrix}$$

The numbers so arranged are called *components* of the vector. The number n is the *dimension* of the vector.

To save space in writing, the above vector can be displayed as a row by defining the *transpose* of $\mathbf{x}$ as

$$\mathbf{x} = (x_1 \quad x_2 \quad \cdots \quad x_n)^{\mathrm{T}}$$

A vector $\mathbf{x}$ can be multiplied by a scalar a, and a vector $\mathbf{x}$ can be added on to another vector $\mathbf{y}$:

$$a\mathbf{x} = a \begin{bmatrix} x_1 \\ x_2 \\ \vdots \\ x_n \end{bmatrix} = \begin{bmatrix} ax_1 \\ ax_2 \\ \vdots \\ ax_n \end{bmatrix}, \mathbf{x} + \mathbf{y} = \begin{bmatrix} x_1 + y_1 \\ x_2 + y_2 \\ \vdots \\ x_n + y_n \end{bmatrix}$$

A *matrix* is a set of numbers (or elements, as they are often called) arranged in a square, or more generally rectangular, pattern. A typical element of a matrix is customarily denoted by a_{ij}, where the first subscript i refers to the *row* in which the element appears and the second subscript j refers to the *column*. For instance, the set of quantities

$$\begin{matrix} a_{11} & a_{12} & a_{13} \\ a_{21} & a_{22} & a_{23} \end{matrix}$$

is a matrix with two rows and three columns. A matrix with m rows and n columns is called an $m \times n$ matrix. The numbers m and n are called *dimensions* of the matrix. If $m = n$, it is a *square matrix*. The entire matrix is frequently denoted by upper-case letters such as **A** or by (a_{ij}). The elements $a_{11}, a_{22}, \ldots, a_{ii}, \ldots$ of **A** are called the *diagonal entries* or *elements of the principal diagonal*.

If all but k rows and s columns of an $m \times n$ matrix **A** are deleted, the resulting $k \times s$ matrix is called a *submatrix* of **A**. At times it is useful to *partition* a given matrix **A** into submatrices. For instance, a matrix **A** can be partitioned as

$$\mathbf{A} = \left|\begin{array}{cc|c} a_{11} & a_{12} & a_{13} \\ a_{21} & a_{22} & a_{23} \\ \hline a_{31} & a_{32} & a_{33} \\ a_{41} & a_{42} & a_{43} \end{array}\right| = \begin{bmatrix} \mathbf{A}_{11} & \mathbf{A}_{12} \\ \mathbf{A}_{21} & \mathbf{A}_{22} \end{bmatrix}$$

where

$$\mathbf{A}_{11} = \begin{bmatrix} a_{11} & a_{12} \\ a_{21} & a_{22} \end{bmatrix}, \mathbf{A}_{21} = \begin{bmatrix} a_{31} & a_{32} \\ a_{41} & a_{42} \end{bmatrix}$$

$$\mathbf{A}_{12} = \begin{bmatrix} a_{13} \\ a_{23} \end{bmatrix}, \qquad \mathbf{A}_{22} = \begin{bmatrix} a_{33} \\ a_{43} \end{bmatrix}$$

A notationally convenient form of partitioning is to treat each column of a matrix as a submatrix. That is, each submatrix of **A** is now a column vector. As it is customary to denote column vectors by lower-case bold face letters, a matrix **A** partitioned into columns can be represented as

$$\mathbf{A} = \begin{bmatrix} a_{11} & a_{12} & \cdots & a_{1n} \\ a_{21} & a_{22} & \cdots & a_{2n} \\ a_{41} & a_{42} & \cdots & a_{nn} \end{bmatrix} = (\mathbf{a}_1 \quad \mathbf{a}_2 \quad \cdots \quad \mathbf{a}_n)$$

Thus, a matrix can be viewed as a set of identically dimensioned column vectors or row vectors. Therefore, most of the rules that are applicable to matrices are also applicable to vectors.

MATRIX ADDITION AND MULTIPLICATION

Two matrices **A** and **B** are said to be equal if, and only if, they both are of the same dimension and if their corresponding elements are equal. That is,

$$\mathbf{A} = \mathbf{B} \qquad \text{if} \quad a_{ij} = b_{ij} \quad \text{for all} \quad i, j$$

The *sum* of two matrices is defined as

$$\mathbf{C} = \mathbf{A} + \mathbf{B} \qquad \text{if} \quad c_{ij} = a_{ij} + b_{ij} \quad \text{for all} \quad i, j$$

The product of a matrix **A** times a scalar α is the matrix $\alpha\mathbf{A} = (\alpha a_{ij})$.

The product of a matrix **A** times a vector **x** is another vector defined by $\mathbf{y} = \mathbf{A}\mathbf{x}$, where the components of **y** are given by

$$y_i = \sum_j a_{ij} x_j$$

In other words, the ith component of **y** is obtained by multiplying term by term the ith row of **A** by the components of **x** and adding. For example,

$$y_i = \begin{bmatrix} a_{i1} & a_{i2} & \cdots & a_{in} \end{bmatrix} \begin{bmatrix} x_1 \\ \vdots \\ x_i \\ \vdots \\ x_n \end{bmatrix}$$

or

$$y_i = a_{i1}x_1 + a_{i2}x_2 + \cdots + a_{in}x_n$$

The above definition has a meaning only if the number of columns of **A** is the same as the number of components of **x**. Therefore, an expression such as **xA** has no meaning.

The *product* **AB** of two matrices **A** and **B** is defined only if the number of columns of the first matrix **A** is equal to the number of rows of the second matrix **B**; otherwise, the product is undefined. If **A** is an $m \times n$ matrix and **B** is an $n \times p$ matrix, then the product $\mathbf{C} = \mathbf{AB}$ is an $m \times p$ matrix. In $\mathbf{C} = \mathbf{AB}$, **A** is said to be *postmultiplied* by **B** or **B** is said to be *premultiplied* by **A**. The elements of $\mathbf{C} = \mathbf{AB}$ are defined by

$$c_{ij} = a_{i1}b_{1j} + a_{i2}b_{2j} + \cdots + a_{in}b_{nj} = \sum_{k=1}^{n} a_{ik}b_{kj}$$

for

$$i = 1, 2, \ldots, m, \qquad j = 1, 2, \ldots, p$$

With the above definition of multiplication, it can be seen that the product $\mathbf{AB}$ is not necessarily equal to $\mathbf{BA}$. That is, in general, matrix multiplication is not *commutative*.

Matrix multiplication is *associative*, that is,

$$\mathbf{A}(\mathbf{BC}) = (\mathbf{AB})\mathbf{C}$$

and *distributive*, that is,

$$\mathbf{A}(\mathbf{B} + \mathbf{C}) = \mathbf{AB} + \mathbf{AC} \qquad (\mathbf{A} + \mathbf{B})\mathbf{C} = \mathbf{AC} + \mathbf{BC}$$

But the *commutative law*, namely, $\mathbf{AB} = \mathbf{BA}$, is not always valid. If either $\mathbf{A} = \mathbf{I}$ or $\mathbf{B} = \mathbf{I}$, then matrix multiplication is always commutative. That is,

$$\mathbf{AI} = \mathbf{IA}$$

If λ is a scalar, then $\lambda\mathbf{A}$ means that each element of $\mathbf{A}$ is multiplied by λ. The powers of a matrix $\mathbf{A}$ are defined as

$$\mathbf{A}^0 = \mathbf{I}, \qquad \mathbf{A}^2 = \mathbf{AA}, \qquad \mathbf{A}^3 = \mathbf{AAA}$$

and so forth.

A square matrix whose elements of the principal diagonal are unity and whose other elements are all zero is called the *identity matrix* and is usually denoted by $\mathbf{I}$. A matrix whose elements are all zero is called the *zero matrix* or *null matrix* and is denoted by $\mathbf{0}$. Both $\mathbf{I}$ and $\mathbf{0}$ satisfy

$$\mathbf{AI} = \mathbf{IA} = \mathbf{A}, \qquad \mathbf{0A} = \mathbf{A0} = \mathbf{0}$$

Matrix multiplication has another property for which there is no counterpart in the algebra of ordinary numbers—that is, the product of two non-null matrices may be the null matrix. For example,

$$\mathbf{AB} = \begin{bmatrix} 1 & 0 \\ 0 & 0 \end{bmatrix}\begin{bmatrix} 0 & 0 \\ 1 & 0 \end{bmatrix} = \begin{bmatrix} 0 & 0 \\ 0 & 0 \end{bmatrix}$$

$$\mathbf{A}^2 = \mathbf{AA} = \begin{bmatrix} 1 & i \\ i & -1 \end{bmatrix}\begin{bmatrix} 1 & i \\ i & -1 \end{bmatrix} = \begin{bmatrix} 0 & 0 \\ 0 & 0 \end{bmatrix}$$

If a matrix $\mathbf{A}$ possesses the latter property then $\mathbf{A}$ is said to be *nilpotent*. If $\mathbf{A}^2 = \mathbf{A}$, then $\mathbf{A}$ is said to be *idempotent*.

In several applications it becomes necessary to perform a multiplication called *elementwise multiplication*, which is defined as

$$\mathbf{A} \times \mathbf{B} = \mathbf{C}$$

where

$$c_{ij} = a_{ij}b_{ij}$$

ALGEBRA WITH BINARY MATRICES

In many applications, it is useful to define matrices whose entries are binary. While manipulating these matrices, it often becomes necessary to use Boolean rules of arithmetic which are defined in terms of *unions* and *intersections*:

$$\begin{array}{ll} 0 \cup 0 = 0 & 0 \cap 0 = 0 \\ 0 \cup 1 = 1 & 0 \cap 1 = 0 \\ 1 \cup 1 = 1 & 1 \cap 1 = 1 \end{array}$$

Using these rules, multiplication of a matrix $\mathbf{P}$ by a vector $\mathbf{s}$ can be done as follows:

$$\begin{bmatrix} 0 & 0 & 0 & 0 & 0 & 0 \\ 1 & 0 & 0 & 0 & 0 & 0 \\ 1 & 0 & 0 & 0 & 0 & 0 \\ 1 & 0 & 0 & 0 & 0 & 0 \\ 0 & 1 & 1 & 0 & 0 & 0 \\ 0 & 0 & 1 & 1 & 0 & 0 \end{bmatrix} (\cdot) \begin{bmatrix} 0 \\ 0 \\ 1 \\ 1 \\ 0 \\ 0 \end{bmatrix} = \begin{bmatrix} 0 \\ 0 \\ 0 \\ 0 \\ 1 \\ 1 \end{bmatrix} \cup \begin{bmatrix} 0 \\ 0 \\ 0 \\ 0 \\ 0 \\ 1 \end{bmatrix} = \begin{bmatrix} 0 \\ 0 \\ 0 \\ 0 \\ 1 \\ 1 \end{bmatrix}$$

$$\mathbf{P}\,(\cdot)\,\mathbf{s} = \mathbf{c} \cup \mathbf{d} = \mathbf{s}_p$$

The rule of this multiplication is to choose the ith and jth columns of $\mathbf{P}$ if the vector $\mathbf{s}$ has units in its ith and jth rows. As $\mathbf{s}$ has units in its third and fourth rows, the third and fourth columns of $\mathbf{P}$ are selected (namely, $\mathbf{c}$ and $\mathbf{d}$) and their union gives the answer, the vector $\mathbf{s}_p$. For this reason, the vector $\mathbf{s}$ is often called the *selection vector*. The above operation, denoted by the symbol $(\cdot)$ is consistent with regular matrix multiplication as can be readily verified.

The idea of Boolean multiplication can be generalized to multiplication of two matrices. For instance, $\mathbf{P}^2 = \mathbf{P}\,(\cdot)\,\mathbf{P} = \mathbf{PP}$ can be obtained as follows:

$$\begin{bmatrix} 0 & 0 & 0 & 0 & 0 & 0 \\ 1 & 0 & 0 & 0 & 0 & 0 \\ 1 & 0 & 0 & 0 & 0 & 0 \\ 1 & 0 & 0 & 0 & 0 & 0 \\ 0 & 1 & 1 & 0 & 0 & 0 \\ 0 & 0 & 1 & 1 & 0 & 0 \end{bmatrix} \begin{bmatrix} 0 & 0 & 0 & 0 & 0 & 0 \\ 1 & 0 & 0 & 0 & 0 & 0 \\ 1 & 0 & 0 & 0 & 0 & 0 \\ 1 & 0 & 0 & 0 & 0 & 0 \\ 0 & 1 & 1 & 0 & 0 & 0 \\ 0 & 0 & 1 & 1 & 0 & 0 \end{bmatrix} = \begin{bmatrix} 0 & 0 & 0 & 0 & 0 & 0 \\ 0 & 0 & 0 & 0 & 0 & 0 \\ 0 & 0 & 0 & 0 & 0 & 0 \\ 0 & 0 & 0 & 0 & 0 & 0 \\ 1 & 0 & 0 & 0 & 0 & 0 \\ 1 & 0 & 0 & 0 & 0 & 0 \end{bmatrix}$$

In the above multiplication, each column of the second matrix is regarded as a selection vector, which performs the appropriate selection operation on the first matrix.

The operation performed above can also be interpreted in a different manner. For instance,

$$p_{51}^2 = \bigcup_{k=1}^{6} p_{5k}p_{k1}$$

$$= \max[\min(p_{51}, p_{11}); \min(p_{52}, p_{21}); \min(p_{53}, p_{31}); \min(p_{54}, p_{41}); \min(p_{55}, p_{51}); \min(p_{56}, p_{61})]$$

$$= \max[\min(0, 1); \min(1, 1); \min(1, 1); \min(0, 1); \min(0,0); \min(0, 0)$$

$$= \max[0; 1; 1; 0; 0; 0] = 1$$

OTHER USEFUL DEFINITIONS

The *transpose* of a matrix **A** is obtained by interchanging rows with its columns and is denoted either by $\mathbf{A}^T$ or $\mathbf{A}'$. A matrix is said to be *symmetrical* if $\mathbf{A}^T = \mathbf{A}$. A symmetric matrix is necessarily a square matrix. A matrix is said to be *skew symmetric* if $\mathbf{A}^T = -\mathbf{A}$. A skew symmetric matrix is also a square matrix. The diagonal entries of a skew symmetric matrix are necessarily zero because $a_{ij} = -a_{ji}$ should be satisfied.

A matrix is said to be a *permutation matrix* if it has a single unity entry in each row and column and zero everywhere else and is usually denoted by the letter **P**.

Premultiplying a matrix **A** with **P** has the effect of interchanging the rows of **A**. Post multiplying **A** with **P** has the effect of interchanging the columns of **A**.

Some of the properties of transpose and permutation matrices are

$$(\mathbf{A}^T)^T = \mathbf{A}$$

$$(\mathbf{AB})^T = \mathbf{B}^T\mathbf{A}^T \qquad \mathbf{P}^T = \mathbf{P}^{-1}$$

$$(\mathbf{A} + \mathbf{B})^T = \mathbf{A}^T + \mathbf{B}^T$$

A matrix **A** is *positive* if all $a_{ij} > 0$. The matrix is *semipositive* or *nonnegative* if all $a_{ij} \geq 0$.

INVERSION OF A MATRIX

Division is not defined in matrix operations. However, something analogous to division can be accomplished by using the *inverse* of a matrix. Two matrices **A** and **B** are said to be inverses of each other if

$$\mathbf{AB} = \mathbf{BA} = \mathbf{I}$$

Whenever the above relation holds, one can write $\mathbf{A} = \mathbf{B}^{-1}$ or $\mathbf{B} = \mathbf{A}^{-1}$. This observation leads to the relation

$$\mathbf{B}^{-1}\mathbf{B} = \mathbf{B}\mathbf{B}^{-1} = \mathbf{I}$$

To understand the computational process of finding $\mathbf{A}^{-1}$, it is necessary to define a determinant.

The *determinant* of **A** is a single number that is defined only if **A** is square, and is denoted by det **A** or $|\mathbf{A}|$. It is difficult to "define" a determinant, so the concept is explained as follows.

For a 2×2 matrix **A**,

$$\det \mathbf{A} = \begin{vmatrix} a_{11} & a_{12} \\ a_{21} & a_{22} \end{vmatrix} = a_{11}a_{22} - a_{21}a_{12}$$

For a 3×3 matrix **A**

$$\begin{aligned} \det \mathbf{A} &= \begin{vmatrix} a_{11} & a_{12} & a_{13} \\ a_{21} & a_{22} & a_{23} \\ a_{31} & a_{32} & a_{33} \end{vmatrix} \\ &= a_{11}(a_{22}a_{33} - a_{32}a_{23}) - a_{12}(a_{21}a_{33} - a_{31}a_{23}) + a_{13}(a_{21}a_{32} - a_{31}a_{22}) \\ &= a_{11}\alpha_{11} - a_{12}\alpha_{12} + a_{13}\alpha_{13} \end{aligned}$$

Written this way, α_{ij} is called the *cofactor* of the element a_{ij}. Indeed, α_{ij} is nothing but the value of the determinant of what is left of the matrix **A** after deleting the ith row and jth column of **A**.

Some useful properties of determinants of an $n \times n$ matrix **A** are summarized here. An interchange of two columns of the matrix **A** changes the sign of det **A**. An interchange of two rows of the matrix **A** changes the sign of det **A**. If every element of the ith row or ith column of **A** is zero, then det $\mathbf{A} = 0$. Addition of a multiple of row k to row i, $i \neq k$ (or of a multiple of column k to column i, $i \neq k$ of **A**) does not change the value of det **A**. Operations such as these are called *row operations* (or columns operations). The determinant of **A** vanishes if any two rows or two columns of **A** are identical. If **A** and **B** are nth-order matrices, then det(**AB**) = (det **A**)(det **B**). A square matrix **A** is said to be *singular* if det $\mathbf{A} = 0$; it is *nonsingular* if det $\mathbf{A} \neq 0$.

Once a method for calculating the determinant of a matrix and its cofactors is available, the inverse of a matrix can be calculated using Crammer's rule, given by the formula

$$\mathbf{A}^{-1} = \text{adjoint of } A/\det \mathbf{A}$$

where the adjoint of **A** is a matrix equal to

$$\begin{bmatrix} \alpha_{11} & \alpha_{21} & \cdots & \alpha_{n1} \\ \alpha_{12} & \alpha_{22} & \cdots & \alpha_{n2} \\ \vdots & \vdots & & \vdots \\ \alpha_{1n} & \alpha_{2n} & \cdots & \alpha_{nn} \end{bmatrix}$$

in which α_{ij} are the cofactors of a_{ij} of **A**.

Some of the more important properties of the matrix inverse are

(1) The inverse of a nonsingular matrix is unique.
(2) $(\mathbf{AB})^{-1} = \mathbf{B}^{-1}\mathbf{A}^{-1}$.
(3) $(\mathbf{A}^{-1})^{-1} = \mathbf{A}$.
(4) If **A** is nonsingular and $\mathbf{AB} = \mathbf{0}$, then $\mathbf{B} = \mathbf{0}$.

Though the properties of a matrix inverse are well understood, actual computation of the inverse of large matrices is not always an easy problem. Even with high-speed digital computers, inverting an $n \times n$ matrix for large n is an expensive and time-consuming process. In many problems of practical interest it becomes usually necessary to invert large matrices of the order of 1000×1000, not once but many, many times. As it requires approximately n^3 arithmetic operations, such as multiplications, to invert an $n \times n$ matrix, 10^6 multiplications would be required to invert a 1000×1000 matrix once. If a multiplication can be done in 1 μsec, then it would take 1 sec to invert a matrix once. In the course of solving a problem, if a matrix inversion is called for 1000 times, the solution time would be a little over 16 min. In simulation studies, one generally solves the same problem again and again with minor changes, and the computation time for the simulation could easily become prohibitive.

To overcome these difficulties several specialized and sophisticated techniques for inverting a matrix have been developed over the past several years. None of these methods is good for any general situation, but some are good for some specialized situations. For instance, the method that is efficient if inverting a matrix is the final goal is not necessarily efficient if an inverse is sought in the course of solving a system of algebraic equations. Similarly, some methods are good for *symmetric matrices*, some for *positive definite matrices*, some for *sparse matrices*, and some others for *band matrices*.

Product Form of the Inverse

In many applications, such as in linear programming, one needs to compute the inverse of a matrix for which only one column is different from that of a matrix whose inverse is known. Suppose that **A** is an $(n \times n)$ nonsingular matrix and also suppose that $\mathbf{A}^{-1}$ is known. To facilitate presentation, let

$$\mathbf{A} = [\mathbf{a}_1 \quad \mathbf{a}_2 \quad \cdots \quad \mathbf{a}_n]$$

where $\mathbf{a}_i$ represents the ith column of $\mathbf{A}$. Now it is desired to find the inverse of $\mathbf{A}_b$ which is obtained by replacing the rth column of $\mathbf{A}$ by the column vector $\mathbf{b}$. That is,

$$\mathbf{A}_b = [\mathbf{a}_i \quad \mathbf{a}_2 \quad \cdots \quad \mathbf{a}_{r-1} \quad \mathbf{b} \quad \mathbf{a}_{r+1} \quad \cdots \quad \mathbf{a}_n]$$

To compute $\mathbf{A}_b^{-1}$, the following steps are followed.

(1) Compute $\mathbf{y} = \mathbf{A}^{-1}\mathbf{b}$.
(2) Form the vector

$$\boldsymbol{\eta}^{\mathrm{T}} = \left[-\frac{y_1}{y_r} \quad -\frac{y_2}{y_r} \quad \cdots \quad -\frac{y_{r-1}}{y_r} \quad \frac{1}{y_r} \quad -\frac{y_{r+1}}{y_r} \quad \cdots \quad -\frac{y_n}{y_r} \right]$$

(3) Construct a matrix $\mathbf{E}$, which is obtained by replacing the rth column of an $(n \times n)$ identity matrix with $\boldsymbol{\eta}$.

(4) Now $\mathbf{A}_b^{-1}$ is obtained from

$$\mathbf{A}_b^{-1} = \mathbf{E}\mathbf{A}^{-1}$$

Example 1. Consider the matrix $\mathbf{A}$ whose inverse is known to be

$$\mathbf{A}^{-1} = \begin{bmatrix} 2 & 0 & 4 \\ 5 & 1 & 6 \\ 7 & 9 & 3 \end{bmatrix}$$

The second column of $\mathbf{A}$ is replaced by $\mathbf{b} = [1 \quad 8 \quad 6]^{\mathrm{T}}$ to yield $\mathbf{A}_b = [\mathbf{a}_1 \quad \mathbf{b} \quad \mathbf{a}_3]$.

To compute $\mathbf{A}_b^{-1}$, first compute

$$\mathbf{y} = \mathbf{A}^{-1}\mathbf{b} = \begin{bmatrix} 2 & 0 & 4 \\ 5 & 1 & 6 \\ 7 & 9 & 3 \end{bmatrix} \begin{bmatrix} 1 \\ 8 \\ 6 \end{bmatrix} = \begin{bmatrix} 26 \\ 49 \\ 97 \end{bmatrix}$$

Next, $\boldsymbol{\eta}$ vector is constructed as

$$\boldsymbol{\eta} = \begin{bmatrix} -y_1/y_2 \\ 1/y_2 \\ -y_3/y_2 \end{bmatrix} = \begin{bmatrix} -26/49 \\ 1/49 \\ -97/49 \end{bmatrix}$$

The $\mathbf{E}$ matrix is

$$\mathbf{E} = \begin{bmatrix} 1 & -26/49 & 0 \\ 0 & 1/49 & 0 \\ 0 & -97/49 & 1 \end{bmatrix}$$

Finally

$$\mathbf{A}_b^{-1} = \mathbf{E}\mathbf{A}^{-1} = \begin{bmatrix} 1 & -26/49 & 0 \\ 0 & 1/49 & 0 \\ 0 & -97/49 & 1 \end{bmatrix} \begin{bmatrix} 2 & 0 & 4 \\ 5 & 1 & 6 \\ 7 & 9 & 3 \end{bmatrix}$$

$$= (1/49) \begin{bmatrix} -32 & -26 & 40 \\ 5 & 1 & 6 \\ -142 & 344 & -435 \end{bmatrix}$$

Some of the specific advantages of the product form of inverse can be seen while solving systems of simultaneous equations. A very popular method of solving simultaneous equations is by *Gaussian elimination* or *Gaussian Reduction.* The Gaussian elimination method of solving $\mathbf{Ax} = \mathbf{b}$ merely gives the answer $\mathbf{x} = (\mathbf{A}^{-1}\mathbf{b})$ but does not automatically give $\mathbf{A}^{-1}$ explicitly. The product form of the inverse gives not only the solution vector $\mathbf{x}$ but also the inverse. The difference between these two methods is illustrated by the following example.

Example 2. Consider the problem of solving using Gaussian elimination and by using the product form of the inverse:

$$\begin{aligned} x_1 + x_2 - x_3 &= 2 \\ -2x_1 + x_2 + x_3 &= 3 \\ x_1 + x_2 + x_3 &= 6 \end{aligned}$$

or, in matrix notation,

$$\begin{bmatrix} 1 & 1 & -1 \\ -2 & 1 & 1 \\ 1 & 1 & 1 \end{bmatrix} \begin{bmatrix} x_1 \\ x_2 \\ x_3 \end{bmatrix} = \begin{bmatrix} 2 \\ 3 \\ 6 \end{bmatrix}$$

The first step of Gaussian elimination is to make the (1, 1) entry equal to unity. This is already the case in the given example. The next step is to make every other entry in the first column equal to zero. This is accomplished by first adding 2 times the first row to the second row and then subtracting the first from the last row. The result is

$$\begin{bmatrix} 1 & 1 & -1 \\ 0 & 3 & -1 \\ 0 & 0 & 2 \end{bmatrix} \begin{bmatrix} x_1 \\ x_2 \\ x_3 \end{bmatrix} = \begin{bmatrix} 2 \\ 7 \\ 4 \end{bmatrix}$$

As the resulting matrix is upper triangular, the row operations are stopped and the resulting equations are solved backwards; thus,

$$2x_3 = 4 \qquad \text{or} \qquad x_3 = 2$$

$$3x_2 - x_3 = 7 \qquad \text{or} \qquad x_2 = (7 + x_3)/3 = (7 + 2)/3 = 3$$

and

$$x_1 + x_2 - x_3 = 2 \qquad \text{or} \qquad x_1 = 2 - x_2 + x_3 = 2 - 3 = +1$$

Thus,

$$x_1 = 1, \qquad x_2 = 3, \qquad \text{and} \qquad x_3 = 2$$

Thus, Gaussian elimination solved for x_1, x_2, and x_3, but $\mathbf{A}^{-1}$ is not available explicitly.

To solve the same system of equations using the product form of inverse, the first step of the procedure would be to arrange the coefficient matrix **A**, an identity matrix **I**, and the right side vector **b** as a rectangular array as

$$\mathbf{A} \,\vdots\, \mathbf{I} \,\vdots\, \mathbf{b}$$

If this array is premultiplied by $\mathbf{A}^{-1}$, we get

$$\mathbf{A}^{-1}\mathbf{A} \,\vdots\, \mathbf{A}^{-1}\mathbf{I} \,\vdots\, \mathbf{A}^{-1}\mathbf{b} = \mathbf{I} \,\vdots\, \mathbf{A}^{-1} \,\vdots\, \mathbf{x}$$

Thus, the matrix $\mathbf{A}^{-1}$ and the solution vector $\mathbf{x}$ are explicitly obtained if the original **A** is successively reduced to an identity matrix using row operations. For instance, in the present example, the starting array is

$$\left[\begin{array}{rrr|rrr|r} 1 & 1 & -1 & 1 & 0 & 0 & 2 \\ -2 & 1 & 1 & 0 & 1 & 0 & 3 \\ 1 & 1 & 1 & 0 & 0 & 1 & 6 \end{array}\right]$$

Using the same row operations as in the Gaussian reduction, namely, first adding 2 times the first row to the second row and then subtracting the first row from the last,

$$\left[\begin{array}{rrr|rrr|r} 1 & 1 & -1 & 1 & 0 & 0 & 2 \\ 0 & 3 & -1 & 2 & 1 & 0 & 7 \\ 0 & 0 & 2 & -1 & 0 & 1 & 4 \end{array}\right]$$

The process is not yet complete as the original position of **A** is not yet replaced by **I**. Subtracting $\frac{1}{3}$ of the second row from the first and dividing the second row by 3,

$$\left[\begin{array}{rrr|rrr|r} 1 & 0 & -\frac{2}{3} & \frac{1}{3} & -\frac{1}{3} & 0 & -\frac{1}{3} \\ 0 & 1 & -\frac{1}{3} & \frac{2}{3} & \frac{1}{3} & 0 & \frac{7}{3} \\ 0 & 0 & +2 & -1 & 0 & 1 & 4 \end{array}\right]$$

Adding $\frac{1}{3}$ of the third row to the first, then adding $\frac{1}{6}$ of the third row to the second and dividing the third row by -2,

$$\underbrace{\begin{bmatrix} 1 & 0 & 0 \\ 0 & 1 & 0 \\ 0 & 0 & 1 \end{bmatrix}}_{\mathbf{I}} \; \underbrace{\begin{matrix} 0 & -\frac{1}{3} & +\frac{1}{3} \\ \frac{1}{2} & \frac{1}{3} & \frac{1}{6} \\ -\frac{1}{2} & 0 & \frac{1}{2} \end{matrix}}_{\mathbf{A}^{-1}} \; \underbrace{\begin{matrix} 1 \\ 3 \\ 2 \end{matrix}}_{\mathbf{x}}$$

The process is terminated because **I** replaced the original position of **A**. The inverse of **A** and the solution vector **x** are clearly displayed in the final array.

The student is advised to compare this method with the simplex method of solving a linear programming problem. ■

EIGENVALUES AND EIGENVECTORS

Consider the equation

$$\mathbf{y} = \mathbf{A}\mathbf{x}$$

where **x** and **y** are n vectors and **A** is a $n \times n$ matrix. In a n-dimensional space, the vector **x** can be represented by a point whose coordinates are the components of **x**. Similarly, **y** can be represented by another vector. Thus, multiplication of vector **x** by matrix **A** results in a new vector **y**. The relative magnitude and orientation of **y** with respect to **x** depends on the nature of **A**. However, interesting things can happen if the new vector **y** has the same orientation as **x** but is merely elongated or compressed, that is, if

$$\mathbf{A}\mathbf{x} = \lambda\mathbf{x}$$

where λ is a scalar quantity. A vector **x** that satisfies the above equation is called an *eigenvector* and λ is called the *eigenvalue* of **A**.

What is happening here can be illustrated by means of an example. Consider a block of two-dimensional material under a pure compression along the y axis. Then the block elongates along the x axis as shown in Fig. A3.1.

Evidently the original vectors OA and OB, after the transformation represented by compression, changed into OA′ and OB′. That is, OA is shortened and OB is elongated with no change in their orientation. On the other hand, the vector OC not only changed its orientation but also its magnitude. Thus, OA and OB can represent eigenvectors, but not OC.

The magnitude and direction of change induced by the transformation can be determined by solving $\mathbf{A}\mathbf{x} = \lambda\mathbf{x}$ or

$$(\mathbf{A} - \lambda\mathbf{I})\mathbf{x} = \mathbf{0}$$

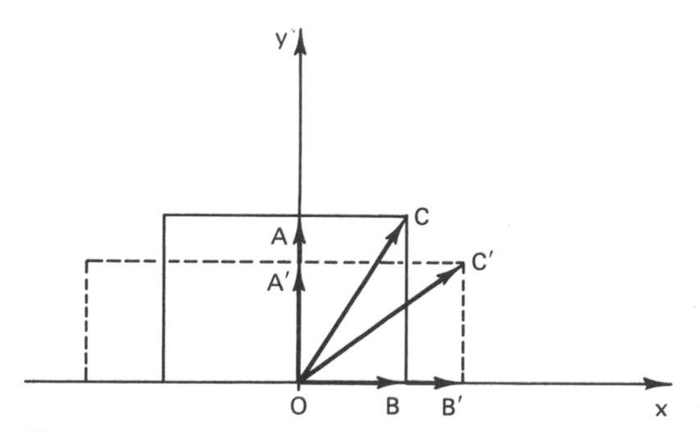

Fig. A3.1 Illustration of the concept of eigenvalues.

This is a system of n equations in n unknowns. Being homogeneous, the above system has a nontrivial solution if and only if $\det(\mathbf{A} - \lambda\mathbf{I}) = 0$. The quantity $\det(\mathbf{A} - \lambda\mathbf{I})$ expands into an nth order polynomial in λ. Therefore, the equation $\det(\mathbf{A} - \lambda\mathbf{I}) = 0$ is an algebraic equation called the *characteristic equation.* The n roots of the characteristic equation are variously called *characteristic roots* or *eigenvalues.* If λ_k is one of the eigenvalues, then

$$\mathbf{A}\mathbf{x} = \lambda_k \mathbf{x} \qquad \text{for} \quad k = 1, 2, \ldots, n$$

The values $\mathbf{x}_k$ of $\mathbf{x}$ that satisfy the above equation are called *characteristic vectors* or *eigenvectors.*

Since the system of equations under study is homogeneous, the eigenvectors can be determined only up to a scalar multiple. However, in many applications, it is inconvenient to have indefinite scalar multipliers. For this reason, the eigenvectors are usually *normalized* in the following way.

Let $\mathbf{x}_k$ be an eigenvector associated with the eigenvalue λ_k. Then

$$(\mathbf{x}_k)^{\mathrm{T}}(\mathbf{x}_k) = \sum_{i=1}^{n} (x_{ki})^2 = \mu^2$$

Now, define a new vector

$$\mathbf{v}_k = (1/\mu)\mathbf{x}_k$$

where μ is the positive square root and $\mathbf{v}_k$ is the normalized eigenvector.

Eigenvalues and eigenvectors play a fundamental role in matrix theory and applications.

For instance, if a matrix $\mathbf{A}$ has n *distinct* eigenvalues λ_k, $k = 1, 2, \ldots, n$, then it is always possible to find a transformation matrix $\mathbf{T}$ such that $\mathbf{T}^{-1}\mathbf{A}\mathbf{T} = \mathbf{\Lambda}$, where $\mathbf{\Lambda}$ is a diagonal matrix whose n diagonal entries are the eigenvalues of $\mathbf{A}$. The process of transforming $\mathbf{A}$ into $\mathbf{\Lambda}$ is called *diagonalization.* This diagonalization cannot be accomplished if any two (or more) eigenvalues are equal to each other. Diagonalization is so useful that the problems raised by nondistinct eigenvalues, which prevent diagonalization, are serious.

Example 3. Consider the problem of finding the eigenvalues and eigenvectors of the matrix

$$\mathbf{A} = \begin{bmatrix} 3 & 2 & 2 \\ 1 & 4 & 1 \\ -2 & -4 & -1 \end{bmatrix}$$

The characteristic equation of this matrix is

$$\det(\mathbf{A} - \lambda\mathbf{I}) = \begin{vmatrix} 3-\lambda & 2 & 2 \\ 1 & 4-\lambda & 1 \\ -2 & -4 & -1-\lambda \end{vmatrix} = 0$$

That is,

$$\lambda^3 - 6\lambda^2 + 11\lambda - 6 = 0$$

The roots of this equation are 1, 2, 3. Therefore, the eigenvalues of the matrix $\mathbf{A}$ are 1, 2, 3.

The eigenvector corresponding to the eigenvalue 1 must satisfy the equation

$$\mathbf{Ax} = \mathbf{x}$$

If the vector $\mathbf{x}$ is written as $\mathbf{x} = (x_1, x_2, x_3)^T$, then the above equation becomes

$$\begin{aligned} x_1 + x_2 + x_3 &= 0 \\ x_1 + 3x_2 + x_3 &= 0 \\ x_1 + 2x_2 + x_3 &= 0 \end{aligned}$$

Solving this we see that $x_2 = 0$ and $x_1 = -x_3$. Thus an eigenvector corresponding to the eigenvalue 1 is $(-1, 0, 1)^T$. Notice that the eigenvector is determined here only up to a scalar multiple. Another eigenvector associated with the eigenvalue 1 is $(1, 0, -1)^T$. Other vectors are also possible. It can be verified that the eigenvector corresponding to the eigenvalue 2 is $(-2, 1, 0)^T$ and the eigenvector corresponding to the eigenvalue 3 is $(0, 1, -1)^T$.

Now the matrix $\mathbf{A}$ can be diagonalized by defining the transformation matrix $\mathbf{T}$ such that each column of $\mathbf{T}$ is one of the above eigenvectors. That is,

$$\mathbf{T} = \begin{bmatrix} -1 & 2 & 0 \\ 0 & 1 & 1 \\ 1 & 0 & 1 \end{bmatrix}$$

The inverse of **T** can be computed using one of the standard methods to yield

$$\mathbf{T}^{-1} = \begin{bmatrix} 1 & 2 & 2 \\ -1 & -1 & -1 \\ 1 & 2 & 1 \end{bmatrix}$$

Therefore,

$$\mathbf{\Lambda} = \mathbf{T}^{-1}\mathbf{A}\mathbf{T} = \begin{bmatrix} 1 & 0 & 0 \\ 0 & 2 & 0 \\ 0 & 0 & 3 \end{bmatrix}$$

4

Differential Equations

INTRODUCTION

It may be safely stated that differential equations enjoy the widest field of applications and play a fundamental role in mathematical model building. Derivatives represent rates of change and as such they enter naturally in translating natural phenomena into mathematical equations.

It should be recognized that not all systems problems involve differential equations, and the description of a physical process is not necessarily done via a differential equation. Several problems in the socioeconomic sciences, switching circuits, etc. lead to equations that are by no means differential in nature. Similarly, it is not always necessary to formulate a problem as a differential equation. For example, the vibration of a beam can be described by a biharmonic partial differential equation or by an integral equation; the radiative transfer of heat may be characterized by a differential difference equation or by an integral equation. Some processes in biology, for example, lead to higher forms of functional equations.

Similarly, the deterministic causal approach, which is often used in the characterization of physical systems, is not the only method of obtaining satisfactory answers. One important alternative, based on the probabilistic approach, leads to the so-called Monte Carlo or random-walk models, queuing models, etc.

There is no standard way, or cookbook recipe, that tells us which kind of formulation is "best." However, if the given equation is not a differential equation, it is worthwhile to explore the possibility of transforming the given equation into a differential equation. The reasons are twofold: On the one hand, there is a well-developed theory of differential equations and on the other hand, automatic computers are ideally suited to solve ordinary differential equations. For instance, the electronic differential analyzer, the so-called analog computer, is essentially a differential equation solver. Even the routine task of function generation is usually done by treating the function as a solution of a differential equation. These are precisely the reasons for the importance of differential equations in systems engineering.

WHAT ARE DIFFERENTIAL EQUATIONS?

A differential equation is an equation connecting a function y and its derivatives. Since derivatives represent rates of change, we can construct an especially simple differential equation by asking the following question: What is the function y whose rate of change at a point x is equal to the value of the function itself evaluated at that point? The rate of change of y with respect to x is dy/dx. Therefore, the function y is obtained by equating this rate of change to y, that is, by solving the differential equation:

$$\frac{dy}{dx} = y \tag{1}$$

Solution of Eq. (1) can, of course, be obtained by inspection as

$$y = e^x \tag{2}$$

Many of the differential equations encountered in practice are more difficult to solve and consequently more challenging to study. Certain differential equations occur in so many different fields that they are labeled with distinct names such as the wave equation, Laplace's equation, Matheu's equation, and Bessel's equation.

Solving a differential equation is similar to solving an algebraic equation in a very broad sense. While solving an algebraic equation, one finds *numbers*, real or complex, with the property that, when substituted back in the given equation, an identity is obtained. Similarly, while solving a differential equation such as

$$\frac{d^2u}{dt^2} = u \tag{3}$$

the goal is to find out all *functions* u that, when substituted back in Eq. (3), yield an identity. If an algebraic equation is a polynomial of degree n, it has exactly n solutions (or roots), whereas a statement of an analogous nature is not possible in the case of differential equations. For example, $u = \cos t$ is a solution of Eq. (3); so also are $u = \sin t$ and $u = \alpha \cos t + \beta \sin t$, where α and β are arbitrary constants.

At first sight, this abundance of solutions may appear to be a stumbling block. If the differential equation were to represent a physical system, one would expect the solution of the differential equation to describe the behavior of the system. If there are so many mathematically acceptable solutions, how do we choose the one solution that best describes the actual behavior of a given physical system? It is precisely at instances like this that the freedom in the choice of one out of a multitude of possible solutions helps to bridge the gap between physical reality and mathematical precision.

Most differential equations are not solvable in "closed form," that is, in a form containing a finite number of "known" or familiar functions of analysis. Some equations become solvable in a "closed form" if the number of "known" functions is extended to include thus far unfamiliar functions. For example, solution of (the dot indicates differentiation with respect to time)

$$\ddot{u} = -(g/l) \sin u \tag{4}$$

arising in the study of the motion of a pendulum may be expressed in elliptic integrals—a new type of function generally not found in books on elementary calculus. Similarly, the $Si(x)$ function and $Ei(x)$ function occurring in the study of diffraction phenomena are further examples of the new functions. It is always advisable to try new kinds of transformations of the dependent variables that may reduce a given equation to some other one whose solution is already known.

Nomenclature

As stated earlier, differential equations occur in many fields of study and it would be useful to classify the equations and study the properties of a typical member in each class:

(1) The derivatives of a function of one independent variable are called *ordinary* derivatives or simply *derivatives* of the function. The derivatives of functions of at least two independent variables are called *partial derivatives* of the function. An equation containing only ordinary derivatives is called an *ordinary differential equation* (or o.d.e.) whereas an equation containing at least some partial derivatives is called a *partial differential equation* (or p.d.e.)

An *o.d.e. of order n* is a relation of the form

$$f(x, y, y^{(1)}, \ldots, y^{(n)}) = 0 \tag{5}$$

which in general involves a function of $(n + 1)$ variables, the independent variable x, the unknown function $y(x)$, and its first n derivatives, namely, $y^{(1)}, y^{(2)}, \ldots, y^{(n)}$. The highest derivative $y^{(n)}$ must necessarily be included in Eq. (5).

The *degree of a term* in a differential equation is the sum of the exponents of the unknown function and its derivatives in the term. The degree of a differential equation is the maximum degree of the terms in the equation. A differential equation is *linear* if its degree is *unity*. A differential equation is *nonlinear* if its degree is at least two. Notice that the nonlinearity of a differential equation is related to the degree of the unknown variable and its derivatives. There are other ways of defining linearity and they will be discussed later. A linear equation is said to be *homogeneous* if every term contains either the dependent variable or one of its derivatives; otherwise, it is *nonhomogeneous*. The additional terms of a nonhomogeneous equation are often referred to as the *forcing function*. *General systems* theorists tend to call systems characterized by homogeneous differential equations by the term *closed systems* and those characterized by nonhomogeneous equations by the term *open systems*. Open and closed systems are not to be confused with *open-loop* and *closed-loop* systems of *control theory*.

Example 1. The equation

$$a\frac{dy}{dx} + xy^2 = c$$

is an ordinary differential equation because it contains only ordinary derivatives (i.e., derivatives of a function of one independent variable). The above equation is *not* a homogeneous equation because the right side term contains c, which does *not* contain y or its derivatives. In this case, c is the forcing term. The above equation is nonlinear because the dependent variable y appears with a second degree.

However, the degree of

$$\{1 + (y')^2\}^{3/2} = y''$$

is not evident at the outset. As the degree, by definition, is required to be an integer, the above equation can be rationalized as follows:

$$(y')^6 + 3(y')^4 - 9(y'')^2 + 1 = 0$$

Now this equation may be regarded as being of degree *six* in y' or of degree *two* in y''.

By this logic, the degree of

$$\ddot{y} + (g/l)\sin y = 0$$

is infinite because sin y can be expanded as

$$\sin y = y - y^3/3 + y^5/5 - y^7/7 + \cdots$$

The equations

$$a\,\frac{\partial u}{\partial x} + u = c + \frac{\partial u}{\partial t}$$

$$\frac{\partial^2 u}{\partial x^2} = \frac{\partial^2 u}{\partial t^2}$$

are typical examples of partial differential equations. They contain two independent variables. The equations are linear and therefore are necessarily of first degree.

(2) A function $y = \phi(x)$ is said to be a solution of Eq. (1) if this function and its first n derivatives, when substituted in Eq. (1), yield an identity. The process of finding a solution to a differential equation is sometimes called *integration of the equation.* Many kinds of solutions exist.

The function

$$y = \pm\sqrt{c - x^2}$$

satisfies the equation

$$y'y + x = 0$$

The solution is called a *general solution* because it contains an arbitrary constant c. A general solution of this form where the variables are separated is called an *explicit form.* If the general solution is written as

$$x^2 + y^2 = c$$

then, it is an *implicit form.* If the value of the arbitrary constant c is specified, then the solution is a *particular* solution.

The function

$$y = 1 - (x - c)^2/4$$

satisfies

$$y' = \sqrt{1 - y}$$

where c is once again an arbitrary constant. The general solution is a family of parabolas as shown in Fig. A4.1. By assigning a value to c, a member of this family can be picked. The straight line $y = 1$ is the *envelope* of these parabolas of which the vertices are all the points $(c, 1)$. The line $y = 1$ is also

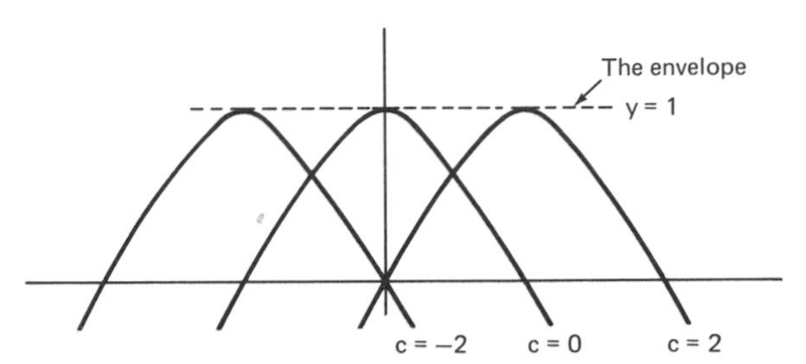

Fig. A4.1 A family of solutions.

a solution of the above differential equation. Since this is not a member of the parabola family, it is called a *singular solution.* The graph of the singular solution is an envelope of the curves defined by the general solution. A *singular point*, on the other hand, is the location where either a coefficient or the solution itself become infinitely large or zero.

(3) If the independent variable in a differential equation is time, then systems described by such equations are called *dynamical systems.* Solutions of dynamical systems are sometimes called *trajectories.*

(4) The symbols used to represent the dependent and independent variables may vary from problem to problem. Sometimes, the notation adopted may have a physical significance, such as t may stand for time, x for distance, and ϕ for potential; at other times, it is merely a matter of choosing some convenient letters to avoid ambiguity. The fact that t need not necessarily represent time is very important to remember when programming analog computers to solve differential equations.

(5) Most of the equations that are encountered in this book are either of first or second order. Surprisingly enough, a large number of problems of practical interest are of the first or second order. Even if one encounters a higher-order equation, it is always possible to rewrite that as a set of first-order equations.

Example 2. Convert the equation

$$\frac{d^3y}{dt^3} + 2\frac{d^2y}{dt^2} + 3t\left(\frac{dy}{dt}\right)^2 + \sin y = e^{-t^2}$$

into a set of three first-order equations. The procedure would be to define three new variables, y_1, y_2, and y_3 as follows:

$$y \triangleq y_1$$
$$\dot{y}_1 \triangleq y_2$$
$$\dot{y}_2 \triangleq y_3$$

The next step is to compute $\dot{y}_3$ by expressing it in terms of y_1 and y_2. As $\dot{y}_3$ is nothing but $\dddot{y}$, it can be obtained by rewriting the given equation as

$$\dot{y}_3 = \dddot{y} = e^{-t^2} - 2\ddot{y} - 3t(\dot{y})^2 - \sin y$$

Now it is only necessary to identify those terms on the right side with the newly defined variables. This yields

$$\dot{y}_3 = e^{-t^2} - 2y_3 - 3t(y_2)^2 - \sin y_1$$

Thus, the three first order equations that are equivalent to the given third order equation are

$$\dot{y}_1 = y_2$$
$$\dot{y}_2 = y_3$$
$$\dot{y}_3 = e^{-t^2} - 2y_3 - 3ty_2{}^2 - \sin y_1.$$

■

This representation is often called the *normal form* or *state space representation*. If the latter terminology is used, then y_1, y_2, and y_3 are called *state variables*. The function e^{-t^2} is the *forcing function*, *input* or *control variable*.

(6) In between the concepts of linearity and nonlinearity lies what is referred to as a *bilinear equation*. A bilinear equation is one in which products between state variables and input variables appear. Consider a simple first-order equation such as

$$\frac{dy(t)}{dt} = -ky(t) \tag{6}$$

If k is assumed to be a constant, then the above equation is obviously linear. If k is a function of time, then Eq. (6) becomes

$$\frac{dy(t)}{dt} = -k(t)y(t) \tag{7}$$

In the above equation $k(t)$ changes with time or *can be* changed with time. In the latter case $k(t)$ can be regarded as a control variable. From this viewpoint, the right side of Eq. (7) is a product of the state variable $y(t)$ and the control variable $k(t)$. Equation (7) is a simple example of a bilinear equation. Bilinear equations are important in many application areas such as physics, biology, and economics.

Conditions for a Solution

In practice, it is possible to come across a differential equation that has no solution at all or it may have more than one solution for the same set of prevailing conditions. If, for a given set of conditions, there is *one and only one* solution, the solution is said to *exist* and to be *unique*. A computer can solve only those problems for which one and only one solution exists, It is also true that, due to limitations on the capabilities of the computers involved, even equations with unique solutions cannot always be solved.

The set of prevailing conditions imposed on a differential equation can be variously called *auxiliary conditions, initial conditions*, or *boundary conditions*. For an nth-order system, a *necessary condition* for a unique solution is that it have exactly n independent conditions specified. If all these extra conditions are specified for a single value of the independent variable, then it is called an *initial value problem*; otherwise it is a *boundary value problem.*

Consider Eq. (3) once again. There it was stated that a very general solution to

$$\frac{d^2u}{dt^2} = u$$

is

$$u = \alpha \cos t + \beta \sin t$$

where α and β are arbitrary constants. That is, unless these constants are precisely determined, the equation in question generates, not one, but a family of solutions. If one wants to use the above differential equation as a model to a physical phenomenon, it is necessary to precisely state what the values of α and β are. As there are two unknown quantities to be determined, namely, α and β, two additional conditions are required. The same conclusion can be arrived at by observing that the given differential equation is of second order, as two such additional conditions are required to get a unique solution.

Where is one to impose these conditions? The general rules to be followed for applying *initial conditions* are:

(1) *In an nth order differential equation initial conditions are imposed on y and its first $(n - 1)$ derivatives.*

(2) *In a set of n first-order equations initial conditions are imposed on the set of normal variables $y_1, y_2, \ldots, y_n$.*

A differential equation with a complete set of initial conditions specified is the easiest to solve. However, differential equations can also be solved when the condition of the dependent variable and its derivatives are specified at several values of the independent variable. Such a problem is called a boundary value problem. The most common of this variety, called the two-point boundary value problem, is one that has conditions imposed at two values of the independent variable.

The boundary value problem is much harder to solve than the initial value problem. First, it is hard to establish whether a solution exists or is unique. There is only a necessary condition which states that the number of independent boundary conditions must be equal to the order of the differential equation.

Example 3. The problem of solving

$$\frac{d^2y}{dt^2} + y = 0$$

requires two conditions as it is a second order equation. If both these conditions are stated as, say,

$$y(t = 1) = 5$$

$$\frac{dy}{dt}(t = 1) = 0.3$$

then it is an initial value problem because both y and y are specified at the same value of t, the initial time.

If, on the other hand, the conditions are stated as one of the following sets, then it is a two-point boundary value problem.

Set 1	Set 2	Set 3
$y(t = 1) = 5$ $y(t = 10) = 2$	$y(t = 1) = 5$ $\dot{y}(t = 10) = 1$	$\dot{y}(t = 1) = 0$ $y(t = 10) = 2.5$

Linear versus Nonlinear Equations

There are important differences between linear and nonlinear differential equations. A linear differential equation, such as Eq. (3), satisfies the *principle of superposition.* That is, as stated earlier, if $u_1(t)$ and $u_2(t)$ are solutions of a linear differential equation, then $\alpha u_1(t) + \beta u_2(t)$ is also a solution to that equation for an arbitrary choice of α and β. A nonlinear equation, such as Eq. (4), does not satisfy the superposition property. Therefore, there is considerable merit in attempting to approximate a nonlinear model by a linear model. However, in a large number of cases linearization, such as the approximation of $\sin u$ by u for small u, is not sufficient. Further, it happens frequently that new phenomena occur in nonlinear models that cannot occur in linear models. For example, the period of oscillation of a simple pendulum, modeled by Eq. (4), depends upon the initial conditions, whereas the period of oscillation of a simple harmonic oscillator modeled by Eq. (3)

is independent of the initial conditions. Similarly, the location of singular points of a linear differential equation is independent of the initial conditions; furthermore, the points are exhibited by the coefficients. The singularities of a nonlinear equation are not necessarily exhibited by the equation itself. Thus, one can make costly mistakes by indiscriminately linearizing a nonlinear differential equation.

Index

D

E

F

G

H

I

J

K

L

M

N

O

P

Q

R

S

A
B
C
D
E
F
G
H
I
J